VITAL DUST

VITAL DUST

LIFE AS A COSMIC IMPERATIVE

Christian de Duve

BasicBooks
A Division of HarperCollins*Publishers*

Published by BasicBooks, A Division of HarperCollins Publishers, Inc.

Designed by Ellen Levine

LIBRARY OF CONGRESS CATALOGING-IN-PUBLICATION DATA

De Duve, Christian.
Vital dust : life as a cosmic imperative / [Christian René de Duve].
p. cm.
Includes bibliographical references and index.
ISBN 0–465–09044–3
1. Life—Origin. 2. Life (Biology). 3. Evolution (Biology).
I. Title.
QH325.D42 1994
577—dc20 94–12964
CIP

95 96 97 98 ◆/RRD 9 8 7 6 5 4 3 2 1

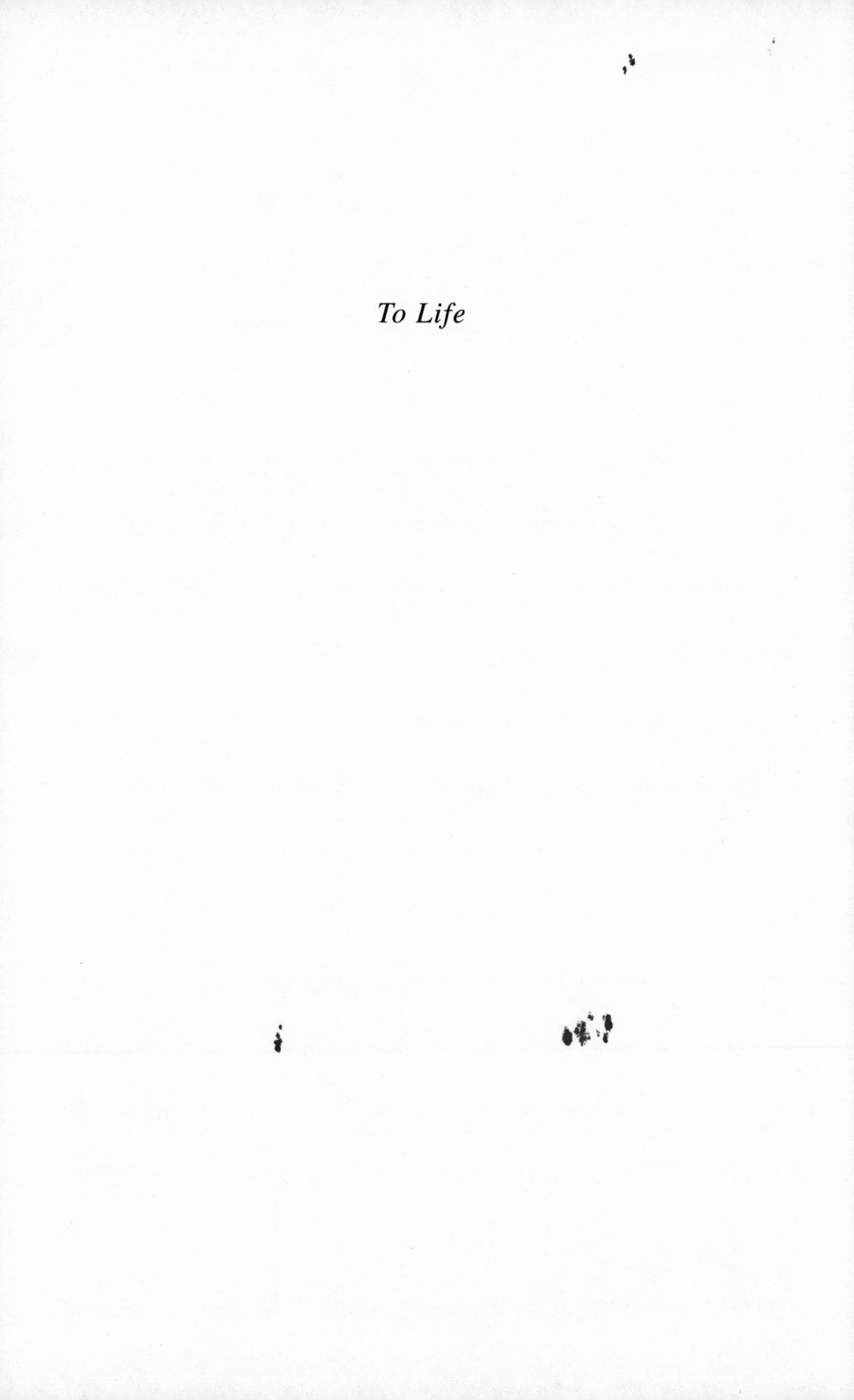

To Life

Contents

PART II
THE AGE OF INFORMATION

PART III
THE AGE OF THE PROTOCELL

PART IV
THE AGE OF THE SINGLE CELL

PART V
THE AGE OF MULTICELLULAR ORGANISMS

PART VI
THE AGE OF THE MIND

Tables and Figures

Preface

It is enough for me to contemplate the mystery of conscious life perpetuating itself through all eternity, to reflect upon the marvelous structure of the universe which we can dimly perceive, and to try humbly to comprehend even an infinitesimal part of the intelligence manifested in nature.

—Albert Einstein

PLENTY OF BOOKS have been written on the origin of life, genes, cells, evolution, biodiversity, the advent of humankind, brain, consciousness, society, the environment, the future of life, its meaning or lack of it. No one has had the temerity to handle all these subjects at the same time, for the simple reason that no one can master more than one or two of them, let alone all. Though no exception to such limitation, I have ventured beyond the boundaries of my competence because I feel that the attempt must be made if we are to understand the universe and our place in it. Life is the most complex phenomenon known to us, and we are the most complex beings so far produced by life.

This book represents my attempt to look at the "bigger picture." It goes back to a naive dream, conceived almost sixty years ago when, as a young medical student at the Catholic University of Louvain, Belgium, I first entered the field of science. What lured me into the laboratory, besides the fun of tackling problems, was the urge to understand. It seemed to me that science, by its insistence on rationality and objectivity, offered the best way to approach truth. The study of life looked particularly promising. It was going to be my pathway to the truth: *per vivum ad verum.*

The dream soon receded. The demands of schooling and training—first in medicine, then in chemistry, finally in biochemistry—the struggle to establish a research group in postwar Belgium, the excitement of discoveries that led me to join the small band of investigators exploring living cells with modern methods, an appointment in 1962 that led me to share my time between my Belgian alma mater and the Rockefeller Institute (now the Rockefeller University) in New York, the duties and obligations of academic life, the additional burden of founding a biomedical research institute in Brussels, and, in the middle of it all, a disruptive trip to Stockholm in 1974, all conspired to keep me busy with day-to-day problems, leaving little time for wider issues. Active science narrows the mind more often than it broadens it, the reason being the increased specialization of facts, concepts, and techniques. As we dig deeper, our scope shrinks.

An invitation to deliver the 1976 Alfred E. Mirsky Christmas Lectures at the Rockefeller University started to draw me out of my hole. The lectures were addressed to an audience of some 550 selected high school students from the New York area. I chose to "shrink" my young auditors a millionfold, equip them with appropriate "cytonaut" gear, and take them on a visit of the main sites to be found in a cell. By a combination of circumstances, a four-hour excursion became a four-year expedition, which finally got into print in 1984 under the title of *A Guided Tour of the Living Cell.* To write and illustrate that book, I had first to turn into a cytonaut myself, move outside my immediate research area, and explore parts of the cell with which I had only a passing acquaintance. It was an enjoyable experience, the first step in a voyage of discovery that was to keep me busy for the following ten years.

The next step came when I began to reflect on the origin of the cells I had just toured. First their formation from primitive bacteria, a topic on which a number of revealing clues had been uncovered, leading back further to the origin of the first bacteria themselves. On this second question, I had always, like most biologists, uncritically accepted the standard version of a thickening soup of prebiotic chemicals somehow self-assembling into cells. I started looking more closely and soon found myself engrossed in the subject. It became my new research interest, resulting in *Blueprint for a Cell,* published in 1991, which took a fresh look at the origin of life. This book ended with an affirmation—life is an obligatory manifestation of the combinatorial properties of matter—and with some questions: How about the further evolution of life? How about us?

These questions defined the rest of my itinerary. The resulting trip, more hurried and sketchy than I would have liked—but time has become short—represents the nearest I shall ever get to accomplishing the dream of my youth. I offer this account with misgiving, aware of its inadequacies but hoping that it will stimulate others to further reflection. Even showing where I went wrong will be helpful.

A warning: All through this book, I have tried to conform to the overriding rule that life be treated as a natural process, its origin, evolution, and manifestations, up to and including the human species, as governed by the same laws as nonliving processes. I exclude three "isms": vitalism, which views living beings as made of matter animated by some vital spirit; finalism, or teleology, which assumes goal-directed causes in biological processes; and creationism, which invokes a literal acceptance of the biblical account. My approach demands that every step in the origin and development of life on Earth be explained in terms of its antecedent and immediate physical-chemical causes, not of any outcome known to us today but hidden in the future at the time the events took place.

Within this context, *Vital Dust* seeks to retrace the four-billion-year history of life on Earth, from the first biomolecules to the human mind and beyond. It takes the reader through seven successive "ages," corresponding to seven levels of complexity: the Age of Chemistry, the Age of Information, the Age of the Protocell, the Age of the Single Cell, the Age of Multicellular Organisms, the Age of the Mind, and, challenging our insight, the Age of the Unknown, which includes the future and the timeless.

The Age of Chemistry brings us straight to the essence of life, its universal aspect. Life, a chemical process, is to be understood in terms of chemistry. It started through the spontaneous formation and interaction of small organic molecules widely distributed in the universe. Given the physical-chemical conditions that prevailed on prebiotic Earth, these molecules were caught in a reaction spiral of growing intricacy, eventually giving rise to the nucleic acids (RNA and DNA), proteins, and other complex molecules that dominate life today. This network of chemical reactions, formed almost four billion years ago, continues to provide the underpinning of all present-day manifestations of life.

Although chemistry pervades this book, the reader will find no formula more complicated than H_2O or CO_2. I focus on principles common to all forms of life on Earth. An important conclusion emerging from this consideration is that there must be congruence between protometabolism—the set of chemical reactions that first put life on track—and metabolism—the set of chemical reactions that support life today. Thus, our knowledge of present-day metabolism yields insights into life's beginnings.

Another lesson of the Age of Chemistry is that life is the product of deterministic forces. Life was bound to arise under the prevailing conditions, and it will arise similarly wherever and whenever the same conditions obtain. There is hardly any room for "lucky accidents" in the gradual, multistep process whereby life originated. This conclusion is compellingly enforced when one considers the development of life as a chemical process.

The Age of Information introduces molecular complementarity—the lock-and-key relationship—as a universal mechanism of biological recognition, which rules such diverse phenomena as enzyme specificity, self-assembly, communication among cells, immunity, hormonal effects, drug actions, and many other biological events. Its most fundamental manifestation is base pairing, the two-by-two joining of the main constituents of nucleic acids, first uncovered by Watson and Crick as the key to the double-helical structure of DNA, now known to govern all forms of genetic information transfer.

In reviewing this crucial stage in the development of life, I emphasize underlying mechanisms. Base pairing arose from chemical events that had nothing to do with information transfer. Molecular replication, the offshoot of base pairing, was a fringe benefit of prebiotic chemistry. Once it emerged, however, replication opened the way to hereditary continuity—based on accurate copying of genetic messages—and to evolution—by way of mutations of the messages and screening by natural selection. But for this to happen, a machinery had to be put together for expressing the messages in a form suitable for natural selection to act on. Every step in the construction of this machinery was the product of deterministic chemical processes, modulated by selection.

A key factor that first came into play with the Age of Information is contingency. Mutations are chance events, which fact, it is often claimed, implies a view of evolution as being ruled by chance. While not denying the role of contingency in evolution, I point out that chance operates within constraints—physical, chemical, biological, environmental—that limit its free play. This notion of constrained contingency runs as a leitmotiv throughout my reconstruction of the history of life on Earth.

The Age of the Protocell was a long period during which the main attributes of cellular organization were progressively assembled. Its outcome was an organism ancestral to all forms of life present on Earth today. The contention that all living organisms are derived from a common ancestor rests on overwhelming evidence. This organism emerged around 3.8 to 3.7 billion years ago.

The Age of the Single Cell was dominated by two events. One was the evolution and diversification of bacteria, or prokaryotes, which now occupy almost every available niche on our planet. A fateful occurrence in this evolution was the appearance of organisms capable of using the energy of sunlight to extract from water the hydrogen needed for self-construction, thereby releasing molecular oxygen. This event is responsible for the rise in atmospheric oxygen that occurred between 2.0 and 1.5 billion years ago. It posed a major threat to the anaerobic forms of life that occupied Earth at that time. Exposed to increasing amounts of the, for them, toxic oxygen molecule, organisms had to adapt or perish. Many bacterial species succumbed to the "oxygen holocaust"; those that survived did so with innovations that played a crucial role in further evolution.

The second key event during the Age of the Single Cell was the prokaryote-eukaryote transition, the transformation of an ancestral bacterial cell into the much larger, more complex cells that make up algae, amoebae, yeasts, and many other unicellular organisms, as well as all plants, fungi, and animals, including humans. This epochal transformation, which may have taken as long as one billion years, led to the development of a primitive phagocyte (eating cell), a large, highly organized cell able to engulf and digest bacteria and other bulky objects. Cells of this type occasionally established mutually advantageous relationships with engulfed bacteria, which were retained as permanent guests, or endosymbionts, and evolved into functional cell parts, including mitochondria and chloroplasts. The need to adapt to oxygen may have precipitated this evolution.

With the Age of Multicellular Organisms, life entered the phase most familiar to us. The Earth, which for some three billion years had harbored only invisible microorganisms, became progressively occupied, first in the waters and later on land, by a gamut of increasingly complex plants and animals. This evolution was marked by successive improvements in reproductive strategy adapted to changing environments. A major step was the development of sexual reproduction. In the plant world, the progress of this development was from spores to seeds to flowers and fruits. In the animal world, haphazard aqueous fertilization gave way to copulation; fertilized egg cells were first laid and allowed to develop in water, then on land, in the sheltered confines of an amniotic egg, and finally inside a womb—for a short developmental stage in marsupials and, later, for a longer one in placentals.

This evolution seems dominated by biodiversity, a profusion of species, products of chance mutations that happened to confer an advantage in a particular environment. Within this variability, however, there is a trend toward complexification. The two features explain the structure of the "tree of life." First, there is the trunk, shaped by a series of "fork organisms," each affected by a mutation that significantly changed the body plan in the direction of greater complexity. Then there is the system of increasingly ramified branches, expressing increasingly trivial alterations of established body plans, the main source of diversity within each major group. This distinction reconciles two views of life that have often been opposed to one another in the past; it puts chance and necessity in correct perspective. Also important in the development of this tree was the expanding web of interrelationships linking living organisms with each other and with the environment in increasingly complex ecosystems.

A concomitant of animal evolution was development of a brain. Once neurons appeared—which they did very early—they joined into increasingly elaborate networks, driven at each step by the resulting evolutionary advantages. Out of the brain arose consciousness, inaugurating, in a manner that defies comprehension, the Age of the Mind. The latest stages in this evolution have been amazingly rapid, leading in only a few million years to the conversion of primates into humans.

This event has dramatically modified the history of life on Earth, largely substituting the fast, human-directed process of cultural evolution for the slow process of Darwinian evolution by natural selection. Art, science, philosophy, ethics, religion are products of this new age. So are medicine and technology, which have transformed the face of the Earth in the space of a few centuries, creating immense problems that urgently challenge human ingenuity and wisdom. If we do not, in the near future, deal satisfactorily with these problems, especially the demographic explosion, which is at the root of most of them, natural selection will do it for us, but with consequences that may be tragic for humankind and for much of the living world. Such is the message we receive when we use our understanding of the history of life to peer into the Age of the Unknown.

Whatever happens, life will recover, as it has so many times in the past after major planetary catastrophes. Most likely, it will continue to evolve toward greater complexity. There is no reason why we should view ourselves as the pinnacle of a process that still has another five billion years to go. What form the next step will take, where and how it will happen, even what extant species will be involved, are unanswerable questions. What will be recognized tomorrow as a fork organism is a mere terminal twig on the tree of life today.

In the last chapter, I try to put it all together. From the perspective of determinism and constrained contingency that pervades the history of life as I have reconstructed it, life and mind emerge not as the results of freakish accidents, but as natural manifestations of matter, written into the fabric of the universe. I view this universe not as a "cosmic joke," but as a meaningful entity—made in such a way as to generate life and mind, bound to give birth to thinking beings able to discern truth, apprehend beauty, feel love, yearn after goodness, define evil, experience mystery. I make no explicit mention of God because this term is loaded with multiple interpretations linked to a variety of creeds. As a scientist, I have chosen to provide a summary of available evidence and to share my personal interpretation of this evidence, leaving it to readers to draw their own conclusions. Lest I be misunderstood, let me stress once more that the key word is *chemistry,* not some preconceived notion of how things ought to be.

To whom is this book addressed? To everyone. The topic, which includes our nature, origin, history, and place in the universe, is of interest to all of us. Faced with a number of burning issues affecting the future of life on Earth, perhaps even the survival of humankind, it is imperative that we contemplate these problems within their natural context. We must learn to "think biologically" and act accordingly.

Like most history books, *Vital Dust* includes parts that may interest some readers more than others. Although a thread of continuity runs through all seven parts, each has been written in such a way as to encourage browsing.

This book is very much the product of personal reading and reflection. I owe an immense debt of gratitude to the many authors who have helped me by their thoughtful, well-documented, and enlightening expositions of their fields. I have done my best to give them due credit in the notes and references cited at the end of the book.

I have also benefited greatly from conversations and discussions with a number of colleagues and friends. Thankfully mentioning their names in no way implies that they are ready to vouch for my presentation of scientific facts, even less that they approve my interpretations or share my ideas. I have in mind my long-time associate, friend, and present "boss," Miklós Müller, who has greatly helped me with his encyclopedic knowledge of microorganisms; my newly made friends in the origin-of-life field, including Gustaf Arrhenius, Manfred Eigen, Albert Eschenmoser, Stanley Miller, Leslie Orgel, William Schopf, Arthur Weber, and many others; Stuart Kauffman, who has introduced me to the intricacies of "artificial life"; and Francis Crick and Gerald Edelman, who have tried their best—largely in vain, I regret to say—to convert me to thinking correctly about the brain. My son Thierry deserves my thanks for guiding me through the intricacies of Kant.

My greatest debt is to my former publisher and editor, my faithful friend Neil Patterson, who has sacrificed endless hours of his valuable time to put this book in acceptable shape. He has not only trimmed flowery adjectives, chatty asides, irrelevant remarks, ponderous constructions, and other infelicities. He has drawn my attention to a number of mistakes and obscurities, clipped some of my more exuberant or incautious statements. My thanks extend to Ippy Patterson for so beautifully drawing the tree of life, something of a symbol of this book.

My gratitude goes also to my publishers at Basic Books, especially Susan Rabiner, who has made a number of valuable suggestions concerning the organization of the book, and to Suzanne Wagner, the copyeditor, and Michael Mueller, managing editor, who have been most helpful in putting the book in its final shape.

Finally, I wish to thank my children, Thierry, Anne, Françoise, and Alain, for the word processor they gave me on the occasion of my seventieth birthday. This frightening gift—I had never even used a typewriter in my life—has become a trusted and valued assistant. Its use has not, however, diminished my constant calls on the services of two highly competent and devoted "flesh and blood" assistants, Anna Polowetzky (Karrie), in New York, and Monique Van de Maele, in Brussels. Only my wife, Janine, can tell how much she has endured while I was struggling with my unwieldy project. I thank her with my love.

Nethen and New York
January 31, 1994

VITAL DUST

Introduction

THIS BOOK IS ABOUT the history of life on Earth—from its birth, shrouded in the depths of the past, to the variegated pageantry of living beings that cover our planet today. It is the most extraordinary adventure in the known universe, an adventure that has produced a species capable of influencing in decisive fashion the future unfolding of the natural process by which it was born.

The history of life is marked by a series of innovations, each introducing a new level of complexity, each to be accounted for in terms of the natural laws of physics and chemistry. Before we embark on this voyage of discovery, I shall define a few general notions that will be with us all along the way.

THE UNITY OF LIFE

Life is one. This fact, implicitly recognized by the use of a single word to encompass objects as different as trees, mushrooms, fish, and humans, has now been established beyond doubt. Each advance in the resolving power of our tools, from the hesitant beginnings of microscopy little more than three centuries ago to the incisive techniques of molecular biology, has further strengthened the view that all extant living organisms are constructed of the same materials, function according to the same principles, and, indeed, are actually related. All are descendants of a single ancestral form of life.

This fact is now established thanks to the comparative sequencing of proteins and nucleic acids. These two groups of substances, which are the most important constituents of all forms of life, are entirely different chemically but are both long chains made by the stringing together of a large number of molecular units—up to several hundred for the proteins, often considerably more for the nucleic acids. Think of strings of beads of different colors, of trains made of different kinds of cars hooked end to end, or, more appropriately, of very long words assembled with different letters. The beads, cars, or letters that make up proteins are called amino acids; those that form nucleic acids are called nucleotides. Protein "words" are made with an alphabet of twenty kinds of amino-acid letters; nucleic-acid "words" with an alphabet of four kinds of nucleotide letters.

High-performance methods now exist for establishing the exact order in which the building blocks of these natural macromolecules follow each other in a given chain. These techniques allow scientists to decipher with great accuracy the sequences of amino acids in proteins and of nucleotides in nucleic acids, the "spelling" of molecular words. The fine print in the book of life is now legible.

One fact of immense importance has emerged from this newly gained molecular literacy. Organisms as different as a microbe, a corn plant, a butterfly, and a human being contain similar proteins and nucleic acids. These similarities are much closer than could possibly be accounted for on the basis of chance. They enforce the inescapable conclusion that these molecules and, therefore, all organisms throughout the living world are related to each other, derived from a common ancestor. As a simple analogy, compare the English word *assembly* with the French word *assemblée* used to convey the same notion. The words obviously did not arise independently in the two languages; they are related through a common ancestral word from which both are derived. The words are not identical because they have changed differently since they diverged from their common ancestor. The same is true for related macromolecules. Their sequences differ because they have suffered changes transmissible from generation to generation—that is, mutations—as the various organisms that contain them have evolved after diverging from a common ancestor.

This unity within diversity simplifies our task. We are trying to trace the history of life, not of lives. The common ancestor divides our itinerary into two parts. First, we must reconstruct the manner in which the common ancestor emerged from whatever materials were available before life appeared. Next, we must find out how all extant living organisms evolved from the common ancestor.

THE TREE OF LIFE

It is common knowledge that life has left fossil traces of its history. Patient decrypting of these vestiges has allowed paleontologists to conjure up from the remote past the ghosts of ancient plants and animals and to piece together a rough evolutionary history of the organisms that inhabit our planet today. However, the fossil record is very incomplete. Quite often, a single bone or tooth, the imprint of a leaf, or the hollow cast of a worm is all that is available to reconstruct an entire organism. In addition, the fossil record hardly goes back further than 600 million years. The earlier record is extremely sparse. Countless organisms must have existed that have left no trace or whose traces have not yet been unearthed. Were only fossil documents available, however numerous and well preserved, a complete history of life could not be written or even contemplated. Our information comes not so much from dead remains as from living beings. The whole history of life is written into present-day organisms. All we need in order to reconstruct this history is to be able to read that text.

This we can now do using the comparative sequencing of related macromolecules from different kinds of organisms. Such analysis can serve to evaluate the evolutionary distance between a pair of organisms—sisters, first cousins, ten-times-removed cousins, and so on—taking as a yardstick the number of differences between the compared sequences. The more numerous these differences—so the assumption goes, subject to a considerable number of caveats and qualifications—the longer the time the molecules have been evolving separately, that is, the longer the time since the species possessing the molecules diverged from their last common ancestor. With enough information of this kind, it is possible, in principle, to reconstruct the entire tree of life on the basis of the properties of organisms living around us today.

In the linguistic analogy, such an approach amounts to molecular etymology. Imagine a linguist confronted with only contemporary texts in French, Italian, Spanish, and Romanian. Even without knowledge of the past, such an expert would be able to conclude from the many similarities among words of the same meaning that these four languages are related. By careful comparative studies based on the assumption that words change only gradually with time, he might even succeed in reconstructing the ancestral Latin, as well as the manner in which the four languages evolved from it. This reconstruction would be shaky at first, easily led astray by coincidental similarities, appropriations by one language from another, and other red herrings. But it would become increasingly secure as more words were examined, analyzed, and compared.

One of the earliest examples of molecular etymology—now a classic—concerns a protein called cytochrome *c*.[1] This small protein, about one hundred amino acids long, participates in the utilization of oxygen by many living beings. The human version of cytochrome *c* differs from that of the rhesus monkey by a single amino acid and from those of the dog, rattlesnake, bullfrog, tuna fish, silkworm, wheat, and yeast by 11, 14, 18, 21, 31, 43, and 45 amino acids, respectively. Such figures provide estimates of the increasingly remote times when each of these species branched from the last ancestor they have in common with us. These estimates, which fit with the fossil record for the animals, go further back in time, to kinships for which no fossil evidence exists. Note that even the wheat and yeast cytochrome *c* molecules share more than fifty amino acids with each other and with the human molecule, indisputable proof that these three widely different species have a common ancestry.

Comparative sequencing of cytochrome *c* was done more than twenty years ago. Since then, many proteins, as well as nucleic acids, have been similarly compared, and more are being analyzed every day. Interpretation of the data is not simple. Nevertheless, even though many uncertainties and controversies are left to be resolved, we are beginning to know in some detail and with some degree of confidence the manner in which extant living forms originated from their common ancestor by a progressive branching of the tree of life. In its upper part, the molecular tree agrees with that drawn by paleontologists on the basis of fossil evidence, except for a number of details that have been added or corrected with the help of the new data. The lower part of the tree is new to us. It has proved surprising.

THE ANTIQUITY OF LIFE

The shape of the tree is gradually becoming clear. But what of its time scale? In paleontological work, the time coordinate is provided by the extensive geological and geochemical investigations that allow the age of a given rock formation to be estimated. If a fossil is found in a terrain believed by geologists to be 200 million years old, we know that the organism that left the remains lived 200 million years ago, give or take a few million years. For molecular trees, the unit of measure is not time but number of mutations: the changes, transmissible from generation to generation, that the molecules have undergone in the course of evolution. Or, more accurately, the number of mutations compatible with survival and proliferation (the "tolerable" mutations), since other changes are eliminated by natural selection[2] and leave no trace in extant molecules. In order to convert this unit into units of time, we must know the tolerable mutation rate. The time scale of a given tree will be very different if tolerable mutations are taken to occur, on average, every one million, two million, or ten million years. This is one of the major uncertainties of the molecular method. The best way to resolve the problem is to compare the molecular trees with the paleontological trees. This works for the upper part of the tree, for which paleontological data are available. But what about the lower part? The answer has come, just in recent decades, from bacterial fossils.

Bacteria are small entities, usually no more than a few hundred-thousandths of an inch in size, with shapes ranging from globular to threadlike, visible only with a good microscope. There are plenty of bacteria in the world today. For most of us, the name "bacteria" raises specters of plague, cholera, tuberculosis, leprosy, diphtheria, and other dreaded ills. However, disease-causing microbes are only a small minority among a wide diversity of harmless or useful forms, which occupy almost every possible kind of habitat, from the balmy shelter of the human gut to the brine of drying seas and the boiling waters of volcanic springs. The richest source of bacteria is the soil, where these invisible organisms accomplish the all-important decomposition of dead plants and animals, recycling the constituents of life.

Bacteria are the simplest forms of life and, as we long suspected and now know, the earliest ones. Fossil traces of these organisms would therefore be invaluable in reconstructing and timing the lower part of the tree of life. Such traces have been found in the last decades.[3] They are of two very different dimensions. At the visible level, the evidence comes from special layered rocks called stromatolites. Such formations are derived by fossilization from huge bacterial colonies composed of superimposed mats, each consisting of a different kind of bacterial species. The top layers of the colony are made of organisms, termed "phototrophic," that use the energy of sunlight to make their own constituents; later, after they die, their substance provides food for the underlying layers. Colonies of this sort cover large areas in certain coastal regions, for example, in Baja California in northwest Mexico. With time, the colonies fossilize into stromatolites by a process of which every stage is known from certain representative rocks. Stromatolites have been found in diverse terrains, in many different parts of the world, spanning all geological ages. Some date back to as much as 3.5 billion years ago, which, for all practical purposes, represents the limit of the useful geological record. It is possible that stromatolite-generating colonies existed even before that time, but their traces could not have survived geological transformations.

The second type of evidence of early bacterial life is microscopic. Most bacteria are enclosed within a solid shell, or wall. This has allowed ancient bacteria to leave their imprints in mud, which later solidified into rocks, just as long-extinct ferns have left their delicate tracings, except that elaborate techniques and a healthy dose of critical discrimination are needed to identify a genuine bacterial microfossil and distinguish it from spurious traces and recent contaminants. A number of authentic imprints are now known. Interestingly, these traces are often found in stromatolites, providing additional proof, if any were needed, of the bacterial origin of these rocks. A few microfossils also date as far back as 3.5 billion years.

So life is at least 3.5 billion years old. Such is the startling message delivered by stromatolites and microfossils. Compare this age with the limit of about 600 million years beyond which virtually no trace of plant or animal has been found and you can appreciate the enormous size of the hidden lower part of the tree of life: four to five times the size of the upper part, which encompasses the entire evolutionary history of plants and animals. During the immensely long time, almost three billion years, that preceded the emergence of the first plants and animals known to us by their fossils, life seems to have remained almost at a standstill. Stromatolites and microfossils do not look very different whether one billion or three billion years old. This appearance of stagnation is misleading, though. Events of cardinal importance took place in the shadow of the stromatolites, preparing the great explosion of life forms that burst forth around 600 million years ago.

According to their fossil traces, the bacteria that lived 3.5 billion years ago were diverse and advanced. They may even have included representatives of the most elaborate forms of phototrophic organisms known today. No doubt, these early life forms were preceded by more rudimentary ones, themselves preceded by the common ancestor of all life. When did this ancestral organism arise? Perhaps as early as 3.8 billion years ago, as suggested by physical analyses of fossil carbon deposits (kerogen) dating back to that time. These deposits show an enrichment in carbon atoms with atomic mass 12 (that is, with a mass equal to 12 times the mass of the hydrogen atom) over carbon atoms with atomic mass 13. Such enrichment of the lighter over the heavier carbon isotope[4] is a characteristic feature of biological carbon assimilation. An upper limit of four billion years ago for the earliest life form is set by the conditions that probably obtained on Earth in the beginning of its history. Experts tell us that the Earth first condensed from a cloud of gas and dust some 4.5 billion years ago. For the next 500 million years, the young planet, battered by falling asteroids and racked by violent volcanic eruptions, remained unfit for life.[5]

The common ancestor of life probably appeared on Earth some time between 4.0 and 3.8 billion years ago. Recognizing that these dates are uncertain, we shall adopt them as the most reasonable estimates available from the present state of our knowledge.

THE CRADLE OF LIFE

Where did life originate? The obvious answer to this question—that life originated on Earth—is not accepted by everyone, partly for reasons of time. It appears from the evidence just discussed that a maximum of 200 million years may have been available for the emergence of the common ancestor of life from the materials our lifeless planet could offer. Although short with respect to the whole history of life on Earth, this still amounts to a very long time span in absolute terms. If we represent the whole Christian era—two thousand years—by one inch, the time available for the emergence of life could measure as much as 1.5 miles. Yet there are some who see this time as too short for the development of something as complex as a bacterial cell. This attitude goes back to an early belief, no longer shared by most scientists, that life originated through an exceedingly slow and long process, perhaps too slow and too long for our planet to harbor. This belief is one reason why the suggestion has been made that life came to Earth from outer space.

The possibility that life may have an extraterrestrial origin has been repeatedly considered.[6] The theory was proposed at the turn of the century and defended with almost mystical fervor by Svante Arrhenius, a Swedish Nobel Prize–winning chemist who coined the term "panspermia" to express his belief that seeds of life exist everywhere in space and are showered continually on the Earth. More recently, a modified version of this theory has been advocated with equal forcefulness by Fred Hoyle, a celebrated British astronomer, and his colleague Chandra Wickramasinghe, an astronomer from Sri Lanka, who have claimed that viruses and bacteria continually arise on the tails of comets and fall on the Earth with particles of cometary dust.[7] Some of these germs could be pathogenic and start epidemics, which, according to these two scientists, may have played important roles in shaping human history. The human nose, they even speculated, may have developed as an evolutionary protection against diseases caused by inhaling raindrops contaminated by extraterrestrial germs. Another proposal has been made, under the name "directed panspermia," by Francis Crick, of double-helix fame, and Leslie

Orgel, a pioneer in prebiotic chemistry. The two British-born American scientists, now at the Salk Institute for Biological Studies in La Jolla, California, have suggested that the first germs of life reached the Earth by a spaceship sent by some distant civilization.[8]

With such distinguished proponents, panspermia can hardly be dismissed without a hearing. Critics of the theory have objected that living organisms could not withstand the intense radiation to which they would be exposed in outer space. But this claim has been disputed. Supporters of the theory have stated that life could not have originated on Earth for lack of time. On what basis they estimate 200 million years to be insufficient for the development of life is not clear. The real question is whether there is solid evidence on which to base a surmise. There is none for a spaceship or for its senders. Things are different for comets and other celestial objects, such as meteorites. These bodies do contain organic molecules of the kind found in living organisms. In the opinion of most investigators, however, these substances are produced by simple chemical reactions that take place "out there." They are not made by living organisms. Of the existence of such organisms, there is as yet no convincing, or even suggestive, sign.

Fairness demands that the matter be left open until the controversy is settled. But common sense and economy counsel that it be ignored in our further discussions. The best reason for doing so is that, even if we accept that life came to Earth from outer space, we are still left with the problem of how it originated. I shall, therefore, assume that life was born right where it actually is: here on Earth.

THE PROBABILITY OF LIFE

How did life originate? Would it emerge if we could move back in time and let events unfold in the same setting, or if the setting were duplicated on some other planet? If so, would it be the kind of life we know or something different? These are questions science has so far failed to answer. What we have instead is a profusion of theories, slanted by the scientific specializations, philosophical attitudes, and ideological biases of their authors. Two schools even go so far as to claim that the origin of life is not a valid problem for science to investigate. They do so for very different reasons, although both their reasons are rooted in the belief that life is an extremely improbable phenomenon. So improbable, according to the creationists, that nothing short of direct divine intervention can explain the emergence of even the simplest of living organisms. The more rationalist believers in the improbability of life reject this claim, pointing out that chance produces extremely improbable events all the time. However, for the very reason they are improbable products of chance, such events are unique and nonreproducible, and therefore inaccessible to scientific investigation. To explain this point of view, I shall take an example from the game of bridge.

Bridge is a card game played by four players with a deck of fifty-two cards, including thirteen each of spades, hearts, diamonds, and clubs. The cards are shuffled and dealt around the table one by one. Suppose you are one of the players and you pick up all thirteen spades. Without doubt, you would speak of a fantastic stroke of luck. You would be right. The odds of being dealt all thirteen spades are one in 635 billion. Let armies of bridge players play day and night for centuries and the thirteen spades may never turn up even once. Indeed, to my knowledge, such an event has never been recorded in the annals of bridge.[9] As the first recipient of this astonishing gift of chance, you would achieve instant world fame. Your name would appear in every bridge column and book. All very true and understandable, except that any other bridge hand has exactly the same probability of being dealt—one in 635 billion. Most hands, however, are not sufficiently spectacular to make history.

Note that I have not taken into account in my estimate the cards the other players are getting. If the complete distribution of the cards is to be specified, the probability is of the order of one in fifty billion billion billion (5×10^{28}). If all the human beings that ever existed had done nothing but play bridge day and night during their whole lives, the odds that a distribution dealt this evening at your bridge club had ever occurred before would still be very low. Yet in no bridge club do players exclaim at being witness to an extraordinarily improbable event every time the cards are dealt.

This example illustrates the simple fact, not always recognized, that single events of very low probability take place all the time without anybody paying any notice unless there is something special about the event. The emergence of life, it has been said, could have been such an event, a fantastic stroke of luck, like getting thirteen spades at bridge, but no transgression of the laws of probability.

If this were so, we would be wasting our time trying to explain the origin of life in scientific terms. A number of eminent scholars have made this claim. Some have even pushed it to its logical conclusion, that if life is a highly improbable product of chance, it has no place in any sort of cosmological view we may entertain. Let billions of planets go through the same history as that of the Earth. Let even billions of big bangs give rise to billions of universes similar to ours. Nowhere would there be life. Its emergence was a *lusus naturae,* a cosmic joke. In the words of the late Jacques Monod, one of the greatest French biologists: "The universe was not pregnant with life."[10]

This statement has profound philosophical implications. These I shall address later. For the time being, I wish merely to examine the scientific validity of the probability argument. Its logic is impeccable, provided we are dealing with a *single event.* But the emergence of life cannot possibly have happened as a single event. To illustrate this impossibility, Hoyle has used the analogy of a Boeing 747 arising ready to fly from a tornado-swept junkyard.[11] The possibility of a living cell coming together in one shot is immeasurably less plausible than the spontaneous assembly of a Boeing 747—if degrees of impossibility are to be envisaged. Only instant creation—a miracle—could accomplish such feats, and miracles, by definition, fall outside the boundaries of scientific investigation. They are a last recourse, when all attempts at a rational explanation have failed—a point, incidentally, that is difficult to identify, since the explanation may have to await new knowledge, as has so often been the case in times past. But we are far from having reached such a point with respect to the origin of life. The field is burgeoning with an abundance—almost a surfeit—of informative facts and stimulating ideas.

A Boeing 747 is built piecemeal in a very large number of steps. Raw materials are first refined or synthesized and worked into a multitude of separate parts. These are then joined, in modular fashion, to make the engines, the body and wings, the flaps, the landing gear, the electronic circuits, and all the other parts of the aircraft. These various parts are then brought together for final assembly. The steps in the construction of a living cell are different, but the principle is the same. Because of the high complexity of the final product, there must, by necessity, be a very large number of steps, often modular in nature.

This consideration completely alters the probability assessment. We are being dealt thirteen spades not once but thousands of times in succession! This is utterly impossible, unless the deck is doctored. What this doctoring implies with respect to the assembly of the first cell is that most of the steps involved must have had a *very high likelihood of taking place under the prevailing conditions.* Make them even moderately improbable and the process must abort, however many times it is initiated, because of the very number of successive steps involved. In other words, contrary to Monod's affirmation, the universe was—and presumably still is—pregnant with life.

To me, this conclusion is inescapable. It is based on logic, not on an a priori philosophical tenet. It does not, however, imply that the emergence of life followed a rigid, preordained course. Even less does it mean that only one kind of life was or is possible. There is room in a deterministic pathway for bifurcations, alternative routes, accidents, even chaos, just as there are many ways for rainwater to run down a mountain. What counts are the constraints of the terrain. A smooth top can lead in many directions. Even a pebble can alter the course of a rivulet. On the other hand, a crater leading into a gorge will force water to flow in a single direction.

FORESIGHT EXCLUDED

In the making of a Boeing 747, all steps are intentional, designed and organized according to a detailed blueprint of the final objective. Things cannot have been the same in the making of the first living cell. Every step had to stand on its own and cannot be viewed as preparation for things to come. This kind of objectivity is difficult to sustain because we know the outcome and also because our whole thinking about life is permeated by intentionality. Cells are so obviously programmed to develop according to certain lines, organs adapted to perform certain functions, organisms suited to certain environments, that the word *design* almost unavoidably comes to mind. A whole school of thought has been inspired by these appearances of design, maintaining that living organisms are actuated by final causes, in the Aristotelian sense of the term. Called finalism, this doctrine is close to vitalism, the belief that living organisms are animated by a vital principle. Both views are now largely discredited. Design has given place to natural selection. The vital principle has joined ether and phlogiston in the cemetery of discarded concepts. Life is increasingly explained strictly *in terms of the laws of physics and chemistry.* Its origin must be accounted for in similar terms.

THE AGES OF LIFE

History is a continuous process that we divide, in retrospect, into ages—the Stone Age, the Bronze Age, the Iron Age—each characterized by a major innovation added to previous accomplishments. This is true also of the history of life, which, so far, has gone through six major ascending planes of complexity (see table INT.1).

First, there is the Age of Chemistry. It covers the formation of a number of major constituents of life, up to the first nucleic acids, and is ruled entirely by the universal principles that govern the behavior of atoms and molecules.

Then comes the Age of Information, thanks to the development of special information-bearing molecules that inaugurated the new processes of Darwinian evolution and natural selection particular to the living world.

TABLE INT.1
The Seven Ages of Life on Earth

Age	Millions of Years
The Birth of Earth	4,550 before present
Chemistry Information The Protocell	4,000–3,800 before present
The Single Cell	3,800–3,700 before present
Multicellular Organisms	700–600 before present
The Mind	6 before present
The Unknown	present
The End of Earth	5,000 after present

The third stage in the history of life is the Age of the Protocell, the first living unit surrounded by a membrane and capable of acquiring a number of key properties linked to this feature. This age ends with the emergence of the common ancestor of all life on Earth.

Next comes the Age of the Single Cell, spanning more than two billion years divided into two major phases, the prokaryotic phase, leading to today's bacteria, and the eukaryotic phase, characterized by a much higher degree of organization and represented in the present world by a variety of microorganisms called protists.

The eukaryotic cell spawned the Age of Multicellular Organisms, with its new principles of cellular association, differentiation, patterning, communication, and collaboration. To this age belong all plants, fungi, and animals, including humans, with each group itself organized hierarchically along an ascending scale of complexity, exemplified at each level by extant organisms.

Finally, there is the Age of the Mind, with all its social and cultural implications and its attending moral responsibilities.

In the following chapters, I shall take the reader through each of these successive ages. I shall conclude with the Age of the Unknown, which encompasses the future of life and its timeless aspects.

PART I

THE AGE OF CHEMISTRY

Chapter 1

The Search for Origins

VIRTUALLY ALL the organic matter in the living world can be summarized symbolically, if not euphonically, by the formula CHNOPS, which stands for carbon (C), hydrogen (H), nitrogen (N), oxygen (O), phosphorus (P), and sulfur (S). These six elements, in myriad molecular combinations, make up the bulk of living matter. They were the main actors in the chemical birth of life as well.

In order to reconstruct this momentous event, we must find out in what form the six biogenic elements were present on the primitive Earth and how, driven by the special physical-chemical circumstances that prevailed, they were first caught in a spiral of increasing complexity out of which life was born. First, what do we know of the setting in which life arose?

THE SETTING

The Earth, four billion years ago, was beginning to recover from the battering by celestial bodies that accompanied its violent birth.[1] It had cooled sufficiently for water to condense on its surface. Islands were rising in the primeval oceans and starting to merge into continents. The lands were barren and the waters lifeless, but the scene was far from calm. Still in the throes of intense volcanic activity, the young Earth was pitted by red-hot craters spewing thick clouds of dust and fumes. It was ravined by deep cracks through which water seeped down to the molten core, later to erupt back up again, pressurized, super-heated, and laden with vapors extracted from the seething lava. Think of Yellowstone National Park, or of the solfataras of Sicily, the Hekla region in Iceland, the flanks of Mount Fuji in Japan, or the hot springs of Rotorua in New Zealand. One memory invariably comes to mind: the smell! The all-pervasive stench of rotten eggs, the characteristic odor of hydrogen sulfide. Indeed, there is every likelihood that the cradle of life reeked of hydrogen sulfide. This fact has rarely been taken into account in origin-of-life scenarios. It deserves to be.

There was no oxygen in the atmosphere around the Earth four billion years ago. Free oxygen is a product of life. This is as close to certain as anything can be in science. In consequence, the state of many minerals was very different from what it is today. This is particularly true of iron. Leave an iron object outside for some time and it turns to rust, which results from the interaction of iron with wet oxygen. There was no rust, that is, no iron oxide, on the prebiotic Earth. Instead, iron was abundantly present in the oceans in a form, called ferrous, that is not found today because it would react immediately with atmospheric oxygen.

The composition of the primitive atmosphere is still the object of debate. The prevalent view for a long time, popularized by the celebrated Urey-Miller experiment (see pp. 18–19), was that the atmosphere consisted of hydrogen (H_2), methane (CH_4), ammonia (NH_3), and water vapor (H_2O), and thus was very rich in hydrogen. This is now seriously doubted. According to many experts, carbon was probably not present in combination with hydrogen (methane) but with oxygen (mostly carbon dioxide, or CO_2). Nitrogen most likely existed as molecular nitrogen (N_2) or in one or more of its associations with oxygen, not as ammonia. Traces, at most, of molecular hydrogen were present. If these new estimates are correct, the source of the hydrogen needed for the formation of the first biomolecules raises a serious problem (see chapter 3, "The Case of the Missing Hydrogen").

Another problem we encounter when looking at the prebiotic scene concerns phosphorus. This element, in the form of phosphate, is a conspicuous constituent of many important biomolecules, in particular nucleic acids. The fact is surprising, since phosphate is hard to find in the present world, at least in solution. The Earth contains plenty of phosphate, but locked up in water-insoluble calcium phosphate, the constituent of the mineral apatite. In marine and fresh waters, the concentration of phosphate is very low; indeed its availability is often a limiting factor in sustaining life in these environments. This was made clear when phosphates were added to cleaning powders. Contamination of lakes by phosphate-containing waste water led to eutrophication, an excessive proliferation of algae nurtured by the newly available phosphate, which alters the food chain in such a way that oxygen becomes scarce and animal life severely hampered.

How the rare phosphate molecule turned out to play its central biological role is an intriguing question. A possible answer to the problem could be acidity, a physical property that is associated with sourness when mild—think of vinegar or lemon juice—and with bitingly corrosive powers when strong—think of aqua fortis (nitric acid), used in etching, or of vitriol (sulfuric acid), the favorite weapon of betrayed Victorian ladies. Apatite readily releases phosphate when exposed to even a mild acidic medium. Perhaps the primeval waters in which life originated had this property.[2]

What was the temperature of the prebiotic world? Little solid evidence is available. This is unfortunate, because temperature is a critical parameter that severely limits the life span of relatively fragile biomolecules such as proteins, nucleic acids, and many of their building blocks. With this fact in mind, many chemists concerned with the origin of life have opted in favor of a cold environment, even below freezing point.[3]

Geochemists, on the other hand, do not favor a cold prebiotic world. Their estimates are in the upper range, near the temperature of boiling water or even higher, compensated by a sufficiently high atmospheric pressure to keep the oceans from boiling. High temperature and high pressure are typical characteristics of underwater hydrothermal vents of the kind that have been discovered in several deep-lying areas of present-day oceans and that were no doubt more abundant on our young, volcanically convulsed planet.[4] Also favoring a hot cradle for life is the finding that the most ancient organisms, according to comparative sequencing, are bacteria living in such vents or in volcanic springs at temperatures of up to 110°C (230°F).

How about sunlight? The sun was cooler four billion years ago. It sent out about 25 percent less light energy to the Earth than today. However, this was probably offset by the greenhouse effect of atmospheric carbon dioxide, which may have been as much as one hundred times more abundant in prebiotic times than today. Ultraviolet radiation most likely was strong despite a cooler sun because, there being no oxygen, there was no ozone shield (ozone is made of three oxygen atoms).

One more point deserves to be mentioned about the prebiotic environment: It was devoid of life. This sounds like a tautology but it holds an important implication, already noted by Charles Darwin more than a century ago. In an oft-quoted letter to a friend, Darwin wrote: "It is often said that all the conditions for the production of a living organism are now present, which could ever have been present. But if (and oh what a big if) we could conceive in some warm little pond, with all sorts of ammonia and phosphoric salts, light, heat, electricity, etc., present, that a protein compound was chemically formed ready to undergo still more complex changes, at the present day such matter would be instantly devoured or absorbed, which would not have been the case before living creatures were formed."[5] As a rule, this passage is cited for its reference to a "warm little pond." But Darwin makes a very pertinent statement. There was nothing in the prebiotic world that might "biodegrade" organic molecules. These could survive and accumulate for a very long time, subject only to much slower physical and chemical degradation.

Summing up, we may state first that there was plenty of water on the Earth when life originated. It could hardly have been otherwise, since water is the vital element *par excellence*. Wake up in the desert after a night sprinkle and you witness the miracle of water. Everywhere, seeds left to dry in the barren soil spring to life in an enchantment of colors. This does not mean that the prebiotic waters would have been our preferred spot for a South Sea vacation. They were probably scaldingly hot, perhaps bitingly acidic, and loaded with ferrous iron, phosphate, and other minerals torn from the inner depths of the Earth. The atmosphere over the waters was heavy with carbon dioxide, nitrogen, hydrogen sulfide, and water vapor, but most likely poor in hydrogen. A pale but unshielded sun filtered through to bathe the surface waters in ultraviolet radiation, light, and warming infrared rays trapped by the carbon dioxide cover.

On such a planet, two possible settings existed for life to unfold: shallow surface waters where the "soup" could thicken and "cook" under sunlight; or dark, deep-seated hydrothermal vents harboring strange chemistries. Or, perhaps, currents between the two settings allowed some steps of biogenesis to occur in one environment, and others in the other.

CHEMISTS ON THE TRAIL

Once information started to become available concerning the physical-chemical setting in which life arose, the next obvious step was to re-create conditions in the laboratory under which the first stages in the origin of life could be reproduced. The Soviet biochemist Alexander Oparin is generally credited with originating this new field of inquiry. He first published a booklet on the origin of life in 1924 and later expanded it to a full-size book that went through a number of revised editions, some of which were translated into English.[6] Largely inspired by the cellular theory of life and by what was known in his days as colloid chemistry, Oparin's concept of the origin of life looks naive today. But he had the great merit of actually testing his ideas in the laboratory, where he prepared and studied a number of molecular aggregates that he saw as possible precursors of the first cells.

For a long time, Oparin did not attract much of a following. The feeling was—it certainly was my opinion at the time—that it did not make much sense searching for the origin of something that was so poorly understood. Things changed in the early 1950s. On April 23, 1953, the British magazine *Nature* published a brief note entitled "A Structure for Deoxyribose Nucleic Acid," by the American James D. Watson and the Englishman Francis Crick.[7] This epoch-making paper, which earned its authors a Nobel Prize nine years later, introduced the now famous double helix, which has become a symbol of the recent revolutionary advances in our understanding of life. Three weeks later, an equally brief and momentous note, titled "A Production of Amino Acids under Possible Primitive Earth Conditions," appeared in the May 15, 1953, issue of the magazine *Science*, the American counterpart of *Nature*. Written by a young graduate student, Stanley L. Miller, this paper inaugurated modern research on the origin of life.[8]

Miller worked in the Chicago laboratory of Harold Urey, a physicist who was awarded the 1934 Nobel Prize in chemistry for the discovery of heavy hydrogen, or deuterium. In later years, Urey became interested in the formation of the planets.[9] It was he who defended the view that the early Earth's atmosphere was a hydrogen-rich mixture of molecular hydrogen, methane, ammonia, and water vapor.

Miller decided to find out how lightning might have affected such an atmosphere. With the reluctant consent of his mentor, who considered this project too iffy for a doctoral thesis, he simulated primitive thunderstorms by producing repeated electric discharges inside a sealed glass enclosure containing a gaseous mixture of methane, ammonia, and hydrogen, through which water was continually recycled by evaporation and condensation, as would have happened over a primeval ocean. Imagine young Miller's surprise when he saw the "ocean" assume a pinkish glow in a matter of a few days; his eagerness when he opened the container and removed the water for chemical analysis; his delight when the analysis showed the presence of several amino acids and other organic molecules typically found in living organisms. The results exceeded his wildest expectations and propelled him instantly into the firmament of celebrities.

This historic experiment alerted organic chemists to the origin of life as a chemical problem. It sparked the birth of a new discipline, termed abiotic (without life) or prebiotic (before life) chemistry, concerned with the spontaneous formation of biological substances under conditions that might have prevailed on our planet some four billion years ago. Many important molecules have indeed been obtained in this fashion, though frequently under conditions somewhat more contrived than one would like for a truly abiotic process.[10] In this rich crop, Miller's original experiment remains a paradigm, virtually the only one conceived exclusively with the aim of reproducing plausible prebiotic conditions, with no particular end product in mind.

Ironically, the relevance of the conditions of this experiment is now seriously questioned. The prebiotic atmosphere may well have been much less rich in hydrogen than Urey thought. As Miller himself found out, if, in agreement with current views, methane is replaced by carbon dioxide, and ammonia by molecular nitrogen, in the gas mixture he used in his celebrated experiments, and if molecular hydrogen is left out, the yield of organic substances falls practically to zero. The verdict is not in yet. Estimates of the composition of the early atmosphere are very uncertain and may be revised again in the future. In the meantime, unexpected support for the validity of Miller's findings, if not his experimental conditions, has come from outer space.

SEARCHING THE SKIES

One of the most powerful techniques used to probe the cosmos is spectroscopy. Put simply, this technique allows analysis of incoming light after it has been separated into its component wavelengths by passage through a prism—in the same way that sunlight is broken into a rainbow of colors (wavelengths) by droplets of water. With appropriate decomposers and amplifiers, the same technique can be extended to nonvisible kinds of electromagnetic waves, such as ultraviolet, infrared, or radio waves, even of very low intensity. Substances present in outer space act as filters that absorb radiations of certain characteristic wavelengths (colors or their equivalent). In consequence, the absorbed radiations are found to be missing or attenuated in the recorded spectra (dark bands in the rainbow). Alternatively, certain wavelengths may be enhanced by emission from energetically excited substances. In many cases, the substances causing the absorption or emission can be identified from the spectral patterns, which serve as fingerprints of the substances involved. The analysis of microwave radiation—the kind that, at much higher intensities, heats ovens—has turned out to be particularly fruitful in this respect.

Spectroscopic probing has revealed that the cosmic spaces are permeated by an extremely tenuous cloud of microscopic particles (interstellar dust) containing a number of potentially biogenic molecules, mostly highly reactive combinations of carbon, hydrogen, nitrogen, oxygen, and, sometimes, sulfur or silicon that would hardly remain intact under Earth conditions but could give rise to biologically significant compounds.[11] This presumably happens in the formation of comets. Long seen as fiery objects hurtling through space while scattering a stream of sparks behind them, comets really consist mostly of dust and ice loaded with a variety of organic compounds. This has been learned by spectral analysis and, thanks to the recent passage in the Earth's neighborhood of the famous comet discovered in 1681 by the English astronomer Edmund Halley, by direct chemical tests with the help of instruments carried by a spacecraft.

Even more solid is the evidence brought to us by meteorites. For example, the Murchison meteorite, which fell in 1969 in Murchison, Australia, was found to contain a number of amino acids remarkably similar in nature and in relative quantity to those obtained by Miller in his experiments. This kind of evidence, while providing a considerable boost to the significance of Miller's results, tells us further that organic compounds can survive the scorching crash of celestial bodies through the atmosphere.

There is thus ample evidence that a number of biogenic compounds can form spontaneously under primitive Earth conditions, in interstellar space, and on comets and meteorites. Most likely, such compounds provided the first seeds of life. How much was made locally, how much was brought in from outer space, is still widely debated. The Belgian-born American astrophysicist Armand Delsemme, from the University of Toledo, believes that virtually all the building blocks of life, as well as all the terrestrial water, were carried to the Earth by comets that contributed to the final accretion of our planet.[12] According to Miller, on the other hand, the chemical precursors of life were formed mostly on Earth itself.

THE FIRST STEPS

Given the evidence from simulation experiments on Earth and the analysis of extraterrestrial objects, complemented by plausible conjectures, the following scenario may reasonably be proposed for the birth of life on Earth some four billion years ago. The seeds of life arose in space and in the atmosphere, in the form of various combinations of carbon, nitrogen, hydrogen, oxygen, and, as we shall see later, sulfur. Under the influence of electric discharges, radiation, and other sources of energy, the atoms in these combinations were reshuffled to produce amino acids and other basic biological building blocks.

Brought down by rainfall and by comets and meteorites, the products of these chemical reshufflings progressively formed an organic blanket around the lifeless surface of our newly condensed planet. Everything became coated with a carbon-rich film, openly exposed to the impacts of falling celestial bodies, the shocks of earthquakes, the fumes and fires of volcanic eruptions, the vagaries of climate, and daily baths of strong ultraviolet radiation. Rivers and streams carried these materials down to the seas, where the materials accumulated until "the primitive oceans reached the consistency of hot dilute soup," to quote a famous line from the British geneticist J. B. S. Haldane.[13] In rapidly evaporating inland lakes and lagoons, the soup thickened to a rich purée. In some areas, it seeped into the inner depths of the Earth, violently gushing back as steamy geysers and boiling underwater jets. All these exposures and churnings induced many chemical modifications and interactions among the original components showered from the skies.

The major outcome of all this geological cookery most likely consisted of some sticky, brownish, water-insoluble goo of indefinite composition, only too familiar to organic chemists who invariably see it staining the walls of their flasks when something goes wrong in their concoctions. It coated Miller's vessel as well but hardly seemed to him worth mentioning since life could not possibly arise out of such material. Somewhere on the primitive Earth, however, the seeds of life were saved from turning into goo and were channeled in the direction of productive chemical complexification. What was this direction? The most widely accepted answer to this question is not what one would expect.

Consider the following three statements, which all happen to be true: (1) Amino acids are among the most conspicuous products of abiotic chemistry, both on Earth and in space. (2) Amino acids are the building blocks of proteins. (3) Proteins are centrally important biological constituents. Most important for our purposes, the vast majority of enzymes, the catalytic agents responsible for the chemical reactions that take place in living organisms, are made of proteins. What conclusion do you draw? I can hear the whole class answering in triumphant unison: The next step on the way to life was the formation of proteins, which, in turn, provided the first enzymes whereby the biogenic process further unfolded. Right?

Wrong! At least according to current majority opinion. Proteins, it is argued, must have been preceded by ribonucleic acid (RNA). The main reason for this belief is that, in the present living world, RNA molecules provide both the catalytic machinery and the information—itself derived from deoxyribonucleic acid (DNA), as we shall see later—for the assembly of amino acids into proteins. No doubt, those in the class with a little biochemical savvy will immediately note the flaw in this argument. In the present living world, no RNA molecule could arise without the help of protein enzymes. In other words, protein makes RNA, which makes protein, which makes RNA . . . , and so on. What came first: protein or RNA? It is the old "chicken or egg" problem, which, it is said, kept a Chinese mandarin pondering fruitlessly all his life.

Molecular biologists have escaped this sorry fate by calling on Crick's "Central Dogma"—not a real dogma, of course, but a logically derived postulate that the codiscoverer of the double helix enunciated as early as 1957, when little of the empirical evidence that now overwhelmingly supports it was available. This postulate states that information flows only from nucleic acids to proteins, never in the reverse direction.[14] Hence, RNA came before protein. This affirmation gained a great boost in the early 1980s when two American investigators, Thomas Cech from the University of Colorado at Boulder and Sidney Altman from Yale University, who shared a Nobel Prize in 1991 for their discovery, found independently that certain RNA molecules were endowed with catalytic activity.[15] This fact suggested that RNA enzymes—called ribozymes by Cech—could have done the catalytic work of proteins in what is now known as the "RNA world," an expression coined in 1986 by the Harvard chemist Walter Gilbert,[16] whose method of DNA sequencing gained him a Nobel Prize in 1980. Gilbert defines the RNA world as a stage in the early development of life in which "RNA molecules and cofactors [were] a sufficient set of enzymes to carry out all the chemical reactions necessary for the first cellular structures."[17]

I shall have more to say on this topic. In the meantime, we may take it as most likely that, whatever came before RNA, the molecules ancestral to present-day proteins came after. The evidence supporting a primary role of RNA in the birth of these molecules leaves little room for doubt, as will become clear in part II. If we accept this premise, we must now face the chemical problems raised by the abiotic synthesis of an RNA molecule. These problems are far from trivial.

THE WAY TO RNA

RNA molecules are long, chainlike assemblages made of a large number—up to many thousands—of units called nucleotides. Each nucleotide consists of three parts: phosphate, ribose, which is a 5-carbon sugar, and a base, of which there are four different kinds—adenine, guanine, cytosine, and uracil. All four are flat, ring-shaped molecules of some complexity, made of carbon, nitrogen, hydrogen, and (except for adenine) oxygen atoms. Adenine and guanine belong to the group of purines, built from two fused molecular rings. Cytosine and uracil are pyrimidines, with a simpler, one-ringed structure.

In RNA, the nucleotides are linked by bonds between the ribose of one and the phosphate of another. As a result, all RNA molecules share the same backbone (except for its length) of alternating phosphate and ribose molecules. The bases are attached to each ribose unit of this backbone, as shown in the following diagram, in which each boxed unit is a nucleotide:

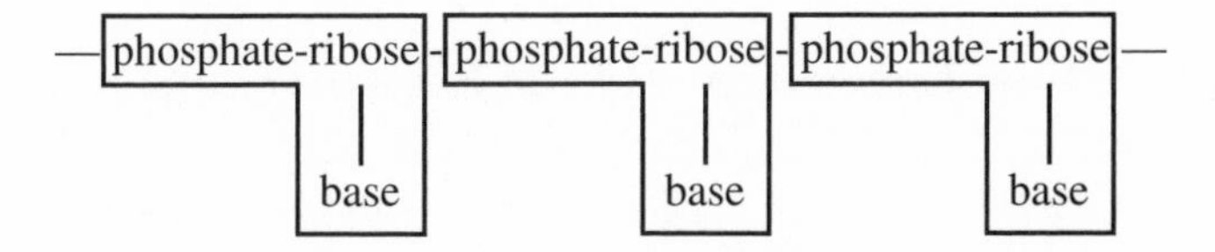

Chemists have achieved some success in making each of the five organic components of RNA, but with poor yields and under conditions far removed from a plausible prebiotic setting and different for each substance. Hitching the components together in the right manner raises additional problems of such magnitude that no one has yet attempted to do so in a prebiotic context.

In any RNA molecule, the bases provide the informational part. They are the four letters with which RNA words are constructed. The phosphate-ribose backbone, on the other hand, plays a purely structural role. Accordingly, one line of research has been to look for simpler backbones to carry the same bases and to support the same kind of information. Orgel and his coworkers at the Salk Institute have pursued this line particularly diligently and ingeniously. They have produced a number of interesting molecules but have not yet solved the problem to their own satisfaction.[18] Recently, a young Danish investigator, Peter Nielsen, has attracted considerable interest with a molecule he calls peptide nucleic acid (PNA), in which the backbone consists of amino-acid derivatives strung together as in proteins.[19] No evidence is yet available to evaluate this intriguing way of combining the chicken with the egg.

None of these molecules, if they ever existed, have left traces in extant organisms. Furthermore, it is far from clear how they could have arisen and how they could subsequently have been replaced by authentic RNA molecules. It is fair to state that no mechanism has yet been found that could account satisfactorily for the prebiotic synthesis of RNA, despite considerable efforts by some of the best chemists in the world. Even the staunchest defenders of the RNA world have expressed despondent views on the future prospects of this line of research.[20]

Could chance be the answer? All thirteen spades in the same hand? A single, highly improbable combination of circumstances that led to the spontaneous formation of a few RNA molecules somewhere in the prebiotic world? This possibility has been evoked on the grounds that self-replication could then have ensured the propagation of the RNA. The RNA world would thus have been born from a single molecular seed, itself the product of a chance event. This explanation does not hold water. In replication, existing RNA provides only information. The actual making of new RNA molecules requires the same chemical complexity as that of the first. We need thirteen spades many times in succession.

This is all the more true because the RNA world was not a fleeting, transient moment in the history of life. It was a long, drawn-out period that lasted all the time that was needed for a protein-synthesizing machinery to arise and turn out the various protein enzymes that eventually took over the job of sustaining emerging life catalytically. We don't know whether this took millennia or years, but it certainly took a long enough time to be possible only with a sturdy chemical underpinning.

THE LESSON OF METABOLISM

The conclusion is clear. We need a pathway, a succession of chemical steps leading from the first building blocks of life to the RNA world. Chemistry, however, has so far failed to elucidate this pathway. At first sight, the kind of chemistry needed seems so unlikely to take place spontaneously that one might be tempted to invoke, as many have done and some still do, the intervention of some supranatural agency. Scientists, however, are condemned by their calling to look for natural explanations of even the most unnatural-looking events. They must even, in the present case, eschew the facile recourse to chance, as I hope to have made clear.

The pathway to life must have been *downhill* all the way, with at most a few rare humps that could be negotiated with the help of the acquired momentum. One would expect such a roadway to be readily visible. Yet, so far, like some artfully hidden jungle trail, it has eluded every search, despite extensive experimentation and much imaginative theorization and speculation. Many devotees of abiotic chemistry, encouraged by the numerous positive results that have already been achieved, continue to believe that further attempts at reproducing early synthetic reactions in the laboratory will eventually uncover the trail. Others, however, impressed with the complexity of the many molecular assemblages needed for the continuing operation of even a skeletal RNA world, tend to be less sanguine.

A pathway does exist, of course, for everyone to see. It is being followed in every corner of the Earth by trillions of trillions of living cells. The green cells of plants and trees and many bacteria do so even without the celestial manna that seeded life in the beginning. These cells construct all their constituents from such simple materials as carbon dioxide, water, nitrate, sulfate, and traces of a few other mineral salts. These pathways make up metabolism. They are known in great detail. Why look elsewhere when nature points the way?

No reason, except that nature's way appears so strange and tortuous to the chemically trained mind that one cannot avoid the feeling that a simpler and more straightforward way must have existed before. Impressions can be deceptive, however. If life did start along pathways that have nothing to do with those of present-day metabolism, why were the early pathways replaced by new ones? And, especially, how? Biologists have a ready answer to the first question. They simply assume that the new pathways were better than the old ones, and they call on natural selection, the universal motor of biological evolution and progress, to promote the change.

The second question, however, is not as easily answered. It simply won't do to visualize a brand-new network of pathways as developing independently of the old ones and taking over only after being completed. That is what we do when we superimpose a new railroad or superhighway network over a primitive road system. But we do it with the kind of prescience and design that we have agreed the biogenic process did not enjoy. The naturally occurring replacement of prebiotic pathways by present-day pathways must necessarily have taken place gradually, one step at a time. It demands some sort of *congruence* between the early and the later pathways.[21]

To understand this point, imagine an old road map dating back to horse and buggy days. It consists of a network of roads linking the various cities and villages of the country. Suppose now that a benevolent contractor builds a better road leading, say, from site A to site B. Clearly, his gesture will be wasted if sites A and B are not parts of the old network. What would be the use of a road linking two points that each are in the middle of nowhere? On a (proto)metabolic map, the cities and villages are chemical intermediates, and the roads between them, drawn as arrows, represent chemical transformations of one or more intermediates into others. Most often, the connecting arrows indicate the presence of a catalyst, or enzyme, responsible for the transformation. On such a map, the new road built by a benevolent contractor would correspond to a new enzyme—catalyzing, say, the conversion of intermediate A to intermediate B—arising by chance through the operation of the protein-synthesizing machinery of the RNA world. This enzyme, part of the new metabolic network that is replacing the old protometabolic network, will be useless, and, therefore, will not be retained by natural selection, if it does not fit within the old network, that is, if intermediates A and B are not parts of the protometabolic map. This is the essence of my argument in favor of congruence. It will become clearer after we have considered the mechanism of selection in part II.

Chapter 2

The First Catalysts of Life

FROM THE FIRST building blocks of life well into the RNA world. Such is the hidden trail we must try to uncover. We know the outcome: complex organic substances made of carbon, nitrogen, oxygen, hydrogen, and phosphorus atoms linked with each other in molecular structures that are known with great accuracy. Our task is to find out how such arrangements arose naturally from simpler arrangements of the same atoms present in the prebiotic environment. Our major clue comes from the requirement for congruence. The metabolic maps reproduced in all biochemistry textbooks are modernized versions of the ancient networks and should help us in our task.

Enzymes are the signposts on metabolic maps. Virtually every one of the thousands of chemical reactions that take place in any living cell is catalyzed by an enzyme. Most of these reactions would not occur at all without enzymes. Hundreds of fatal or severely disabling genetic diseases characterized by a single enzyme deficiency attest to this fact. It is very unlikely that protometabolism could have done without catalysts. If these were not proteins, what were they?

CATALYSIS WITHOUT PROTEINS

Just as metabolism cannot operate without enzymes, protometabolism, whatever pathways it may have followed, could not have functioned without catalysts. Textbooks define catalysts as substances that specifically help reacting molecules to get together and interact, but are not themselves consumed in the reaction and, therefore, can serve an indefinite number of times in succession. Why did emerging life need catalysts? Chemical reactions take place all the time in the physical world without the help of catalysts.

There are two reasons for the requirement for catalysts: rapidity and yield. Uncatalyzed reactions most often are very slow. This means, in the prebiotic setting, that important reaction products might well have been destroyed almost as rapidly as they were made, never reaching a sufficient level to participate in a subsequent step. Without catalysts to speed up reactions, protometabolism would have been in the sorry predicament of the fifty Danaid sisters condemned to pour water eternally into a bottomless barrel.

Low yield raises an equally thorny problem. Take the following case. In 1960, the Catalonian-born American chemist Juan Oró discovered that adenine, one of the constituents of RNA, could form in a single step from ammonium cyanide, which is considered a compound likely to have been present on the prebiotic Earth.[1] This remarkable finding of a central biomolecule arising in such simple fashion was hailed as almost equivalent to Miller's historic experiment and has become a textbook paradigm of the power of abiotic chemistry. Yet the highest yield of adenine in this reaction was 0.5 percent, which means that 99.5 percent of the reaction mixture consisted of other stuff. Another example is the synthesis of ribose, another RNA component, from formaldehyde—also a classic of abiotic chemistry. In this case, the yield is less than 0.1 percent, and at least forty other sugars are present in the mixture.[2]

Such low yield of any one product is typical of many organic-chemical reactions when they are not subjected to the kind of strict constraints imposed by chemists in the laboratory. There are always side reactions, and these are all the more numerous the looser the constraints. Problems of this sort are compounded when a process requires several successive steps. Imagine a short sequence of only three steps—from *A* to *B,* from *B* to *C,* and from *C* to *D*—each with a yield (high for a prebiotic reaction) of 1 percent. In terms of *A*, the yield of *B* will be 0.01, that of *C,* 0.0001, and that of *D,* 0.000001, or one in one million. Even under the best conditions, chemists have to fight against this kind of vanishing act. They often purify a significant intermediate between two steps and change conditions at each step to maximize yield. Prebiotic chemists are acutely aware of these difficulties and have come up with a number of more or less plausible mechanisms whereby relevant intermediates could have been selectively concentrated from the prebiotic soup. Let us apply the rule of congruence and ask how nature copes with the problem.

Nature's solution lies in enzyme specificity. Biological catalysts are truly remarkable in this respect. They are by far superior to the best catalysts devised by human ingenuity; as a consequence, the large-scale production and engineering of enzymes for industrial purposes has become an important branch of modern biotechnology. Many enzymes catalyze a single reaction or a set of closely similar reactions that would hardly occur without a catalyst.

Enzymes are proteins or, exceptionally, RNA molecules (ribozymes); they could not have been present in a pre-RNA world that had not yet developed the RNA-based machinery of protein synthesis. We must look elsewhere for the catalysts of protometabolism. To most investigators, elsewhere has meant the inorganic mineral world, since we are talking of a time when the organic world was still in its infancy. In contemporary biochemistry, many enzymes act with the help of an auxiliary substance of inorganic nature, most often an atom of a metal such as iron, copper, calcium, magnesium, zinc, molybdenum, cobalt, or manganese.

A difficulty with metals is that they often need some supporting structure—usually a protein framework—to interact productively with the molecules that participate in the reaction they help catalyze. Thus, a great deal of attention has also been given to mineral surfaces likely to provide the necessary frameworks and also, perhaps, to act catalytically by themselves. Favorites are clay particles, already advocated more than fifty years ago by the British physical chemist John Desmond Bernal, one of the pioneers in the study of the origin of life.[3] Clays come in many different microcrystalline forms, and some, indeed, display catalytic activity. Montmorillonite, for example, which owes its name to the French town of Montmorillon, near which it is quarried, has been found to facilitate the assembly of short RNA-like chains from suitably prepared nucleotides.[4] Unlike metals, however, clays, which are aluminum silicates, have left no trace in present-day life to suggest that they played any role as catalysts in protometabolism.

As an alternative to clays, Gustaf Arrhenius, from the Scripps Institution of Oceanography in La Jolla, California, has advocated a catalytic role for what he calls "positively charged, double layer hydroxide minerals," especially in the synthesis of sugar phosphate molecules.[5]

The German chemist and patent attorney Günter Wächtershäuser has constructed an elaborate model—by far the most detailed such model on record—of the development of protometabolism on the surface of pyrite crystals.[6] Known as fool's gold because of its golden tinge, pyrite is a mineral composed of iron and sulfur. In Wächtershäuser's model, the mineral owes its catalytic role to the fact that electrically charged objects attract each other if they bear charges of opposite sign (and repel each other if they bear charges of the same sign). Pyrite, which is positively charged, offers a surface on which, according to the German author, negatively charged molecules bound by electrostatic attractions could be brought into close neighborhood with each other and caused to interact in various ways. Incidentally, Wächtershäuser accounts for the importance of phosphate in metabolism by his model: Phosphate is negatively charged and thus allows the molecules to which it is attached to bind to the pyrite surface.

Chemically, Wächtershäuser's model is largely inspired by present-day metabolism and obeys the congruence rule. But some of the mechanisms invoked are speculative and need experimental testing. Furthermore, his catalyst lacks specificity, except for the very broad property of binding, with variable strength, anything that is negatively charged. The model relies heavily on autocatalysis. A number of other workers have appealed to this concept as a solution to the problem of prebiotic catalysis.[7] Autocatalysis occurs when the product of a chemical reaction is catalytically helpful to the reaction: *B* catalyzes the conversion of *A* to *B*. A slow-starting reaction may thereby progressively accelerate, sometimes to the point of becoming explosive. There is no question of emerging life ever exploding, but the idea is that processes difficult to initiate can become self-sustaining by autocatalysis once they are primed. This is a way of catching chance events and turning them into "going concerns."

All the mechanisms mentioned so far may well have played a role in protometabolism. But could they have accomplished the whole job without the help of proteins? Several workers have expressed serious doubts about this and insisted on the early participation of protein catalysts, at the risk of contravening the Central Dogma.[8] There is much to say for this view, especially if the term "protein" is replaced by "peptide," defined as any chainlike assemblage of amino acids, not just the particular kind made from twenty specified varieties of amino acids by an RNA-dependent machinery.

THE CASE FOR PREBIOTIC PEPTIDES

Amino acids, the building blocks for making peptides, were probably among the earliest organic substances to be present in the prebiotic world. More than twelve kinds of amino acids formed in significant amounts in Miller's flask, and the same kinds have been extracted from meteorites. Some of these amino acids are found in proteins today, others are not. No matter, all possessed the essential characteristics that allow amino acids to join into peptides: the carboxyl group (-COOH) responsible for the acid nature of the substances, and the ammonia-derived amino group ($-NH_2$). In peptides and proteins, these two groups are joined to form a peptide bond (-CO-NH-), with the release of one water molecule.

Could the primeval amino acids have joined into peptides under prebiotic conditions? What looked like a simple positive answer to this question was found in 1958 by the American biochemist Sidney Fox, long of the University of Florida, now at the University of South Alabama.[9] His recipe: Just heat a dry mixture of amino acids for three hours at 170°C (338°F). Water comes out and you get a plastic-like solid that, when ground and mixed with water, yields up to 15 percent of its weight as a water-soluble product made, on average, of some fifty amino acids joined together. To this product Fox gave the name proteinoid, a cautious choice since proteinoids are far from having the regular chainlike structure of peptides.

For Fox, this discovery initiated a lifelong avocation. He found that proteinoids spontaneously form microscopic vesicles, or "microspheres," which he saw as the first cells, and spent his whole career pursuing these studies. Few origin-of-life experts are as sanguine as Fox concerning the significance of his results. It has been objected that the conditions required for the formation of proteinoids are not likely to have obtained on the prebiotic Earth, that the resulting material has more in common with primeval "goo" than with proteins, and that the microspheres are a far cry from anything that could be called a cell. I tend to share these misgivings but retain as possibly significant two of Fox's findings: proteinoids possess some weak, enzyme-like catalytic properties; and the amino-acid composition of proteinoids is specific and reproducible despite the disordered conditions of their formation. This means that the bonds between amino acids did not form purely at random, but that certain associations were advantaged and others excluded.

A more orthodox way of getting peptides was discovered in 1951, before Fox's findings, by a German chemist, Theodor Wieland.[10] At that time, biochemists had just discovered the thioester bond, which turned out to be of cardinal importance in all present-day living organisms, probably also in the origin of life. This exceptional situation warrants a brief excursion into biochemistry.

INTRODUCING THIOESTERS

An ester arises from the joining of a hydroxyl group (-OH), characteristic of alcohols, with a carboxyl group (-COOH), characteristic of organic acids. One molecule of water is removed in the process, and the two building blocks are linked by what is called an ester bond (-O-CO-). A thioester arises similarly from the joining, with the removal of water, of a thiol with an acid. Thiols (from the Greek *theion,* sulfur) are the equivalent of alcohols in which the oxygen atom is replaced by one of sulfur. They are characterized by the thiol group -SH. Thioester bonds have the structure -S-CO-.

Wieland became interested in thioesters as a student of Feodor Lynen, the discoverer of the first known natural thioester, a compound of acetic acid with a thiol designated coenzyme A in biochemical jargon.[11] Coenzyme A, a molecule of central importance, was discovered in 1947 by the German-born American biochemist Fritz Lipmann, the "father" of bioenergetics, who received the 1953 Nobel Prize in medicine for this discovery. Lynen, who was similarly rewarded in 1964, found that thioesters are the natural intermediates in the synthesis of esters from acids and alcohols.

The main problem in the making of an ester from an alcohol and an acid is that a molecule of water needs to be extracted. Such a reaction—the closure of a bond with the release of water—is called a condensation. Condensation reactions do not take place spontaneously in an aqueous medium because there is too much water around. The spontaneous direction of the reaction—that is, the direction that does not cost any energy but, on the contrary, releases energy—is the reverse of the condensation, the splitting of the bond with the help of water, or hydrolysis. For example, in the presence of a suitable enzyme, esters are hydrolyzed into alcohols and acids. For the reverse process of condensation of alcohols and acids into esters to occur, energy must be spent, the water molecule must be forcibly extracted. Chemists do this with special reagents called condensing agents. Nature uses a different device. It starts—spending energy in the process—by condensing the acid with a thiol (coenzyme A) into a thioester. This is the energy-requiring, water-extracting step. In a second step, the acid is transferred from coenzyme A to the alcohol, and coenzyme A is released, ready to participate in a new round. Group-transfer reactions of this kind play a primordial role in the innumerable condensation reactions that underlie the biosynthesis of all complex biological molecules, including not only proteins and nucleic acids, but also carbohydrates, fats, and many others.[12]

Back to Wieland. As a firsthand witness of the discovery of the biological process of ester formation by group transfer from a thioester, he decided to find out whether this would also work for peptides, which likewise are formed by condensation reactions, but between amino acids. So, Wieland synthesized amino-acid thioesters and simply threw them together in water. Amazingly, it worked! Peptides were formed, even though no catalyst was present.[13]

There is an amusing historical twist to this finding. When the mechanisms of protein synthesis were unraveled in the late 1950s and early 1960s, Wieland's results were found to be irrelevant. Proteins are indeed formed by group transfer—that much remains true—though not from thioesters, but from esters (of amino acids and RNA molecules). Wieland's vindication came a few years later, when Lipmann made the startling discovery that certain bacterial peptides—for example, the antibiotic gramicidin S—are synthesized in nature from thioesters.[14] The thiol involved in this process was found to be pantetheine, which is itself the business end of coenzyme A, the central thiol discovered by Lipmann twenty years before. So do the mysterious wheels of science turn.

In discussing his finding, Lipmann suggested that the thioester-dependent mechanism of peptide formation may have preceded the RNA-dependent mechanism of protein synthesis in the development of life. I have adopted this suggestion and transposed it to the earliest steps of the biogenic process. For reasons that I shall explain in greater detail later, I believe that thioesters played a key role in the development of life. This belief fits with two central requirements of the trail we are trying to uncover: (1) congruence—thioesters are immensely important in present-day metabolism; and (2) the physical-chemical setting of the cradle of life—the thiol group is derived from hydrogen sulfide (H_2S), the putrid but vital gas that pervaded the prebiotic world.

It is my suggestion that thiols were part of the early organic molecules that seeded the development of life on the prebiotic Earth. Given the primeval setting, this suggestion appears eminently plausible, but the means of testing it have long been lacking because abiotic chemists, for reasons of their own, tended to shy away from sulfur chemistry. The omission has been repaired. A recent contribution from Miller's laboratory describes a plausible procedure for the prebiotic synthesis of two natural thiols.[15] One is coenzyme M, a metabolic cofactor of particularly ancient, methane-producing bacteria known as methanogens. The other is cysteamine, a constituent of pantetheine, which we have seen is the key component of coenzyme A and the natural cofactor involved in the synthesis of bacterial peptides. The Miller group has, in fact, been able to obtain the entire pantetheine molecule under plausible prebiotic conditions.

I have made a further, more controversial suggestion, namely, that conditions on the prebiotic Earth favored the formation of thioesters from the primeval thiols and the amino acids and other acids that were presumably present also in large amounts. This possibility is more questionable because it involves the spontaneous occurrence of an energy-requiring condensation reaction. I shall discuss this in the next chapter when considering the sources of protometabolic energy. Let us take it as a working hypothesis for the time being.

CATALYTIC MULTIMERS TO THE RESCUE

Granting the presence of thioesters of amino acids, we know from Wieland's results that peptides are likely to assemble spontaneously from these materials, even without a catalyst. In addition to peptide bonds, such assemblages could have included ester bonds as well, since hydroxy acids (with an alcohol group) were probably present also in large amounts in the primeval soup, according to Miller's results. Rather than peptides, therefore, which are made exclusively from amino acids, I have chosen to call the resulting molecules multimers.[16]

Why this linguistic monstrosity—which combines the Latin *multus,* many, with the Greek *meros,* part—and not the more orthodox *polymer* (from the Greek *polys,* many) or *oligomer* (from the Greek *oligos,* few)? Because *polymer* sounds too long, at least to me, and *oligomer* too short, and because both terms evoke images of regularity and homogeneity that I wish to avoid. The multimers of my model are a mixed bunch, containing more than a few building blocks, but fewer than the average polymer.

My final suggestion, which many may find the most controversial, is that catalysts performing, albeit in crude form, the main activities carried out by enzymes in present-day metabolism were present in the multimer mixture and served as the main catalysts, or protoenzymes, of protometabolism. I have no proof of this, only some presumptions.

According to my hypothesis, the multimers arose from random interactions among whatever thioesters were present. This does not mean that the resulting mixture was random, in the sense of containing all sorts of associations in a completely disordered fashion, without rule or reproducibility. On the contrary, we may take it that, as long as conditions remained the same, the mixture would have had a constant and reproducible composition, corresponding to a tiny subset of all the possible associations that could be made from the available building blocks. A great many such associations would have been excluded either at the level of formation—they were made too slowly or not at all—or at the level of breakdown—they were destroyed too quickly. Solubility in water would have been an additional selective factor, although it is conceivable that some molecules were catalytically active in insoluble form. Finally, there is the possibility that catalytically active molecules were protected against breakdown by the molecules on which they acted, as many enzymes are by their substrates today. Only molecules that passed this multiple screening would have made a significant contribution to the resulting mixture. Because of the strictly physical-chemical nature of the factors involved in the screening, the composition of the mixture would have stayed the same as long as the conditions did not change. This point is important. It makes this part of the biogenic process reproducible and deterministic despite being dependent on random interactions.

Whether this reproducible mixture would have included the protoenzymes required for protometabolism is conjectural but not implausible. The reasons for this are the following. First, we know that some of Fox's proteinoids, in fact even single amino acids or mixtures of amino acids,[17] can display crude catalytic activities. The same should be true of the multimers I postulate. Next, the molecular configurations that would be expected to confer stability, such as a large enough molecular size and a compact or cyclic conformation, are also the configurations that a protein chemist would consider most likely to be required for catalytic activity. Third, as I shall explain in chapter 7, present-day enzymes must have started as relatively short peptides, probably no more than twenty to thirty amino acids long, perhaps much shorter. This fact makes the presence of catalytic molecules in the multimer mixture more likely. Finally, there is the rule of congruence. We are looking for activities that, in extant living organisms, are carried out by protein molecules, not by clays or other mineral surfaces. Barring authentic peptides of the protein type, whose formation and faithful reproduction under prebiotic conditions are most unlikely, the multimers of my hypothetical mixture appear as the next-best molecules for building the kind of three-dimensional structures that determine enzyme catalysis. This does not by any means preclude the participation of metals and other cofactors in protometabolism. On the contrary. It is intriguing in this regard that a molecule like pantetheine could conceivably have been part of the multimer mixture.

Chapter 3

The Fuel of Emerging Life

PROTOMETABOLISM could not have unfolded without a supply of energy, together with the means for productively exploiting this energy. The life-building complexification process was uphill all the way. To make it downhill and, therefore, able to occur spontaneously, a sufficient supply of energy was essential. There were plenty of energy sources on the prebiotic Earth, in the form of sunlight, ultraviolet radiation, electric discharges, shock waves, heat, and chemical upheavals of various sorts. Which among these various sources of energy did emerging life exploit? And, especially, how was the raw power in the prebiotic setting converted into productive life-creating events?

THE PROBLEM OF PRIMEVAL MEMBRANES

If, in accordance with the congruence rule, we ask present-day life for a hint with regard to these questions, we immediately run into a problem. The most important energy generators of living organisms today depend on the operation of highly complex substances organized within the fabric of an intricate, filmlike structure, or membrane. Could such arrangements have arisen early enough to satisfy the energy requirements of emerging life?

Several authors believe so. In a book devoted to bioenergetics, Franklin Harold, a biochemist from the University of Colorado, does not hesitate to head an important section with the declaration: "In the beginning was the membrane."[1] Clair Folsome, from the University of Hawaii, has proposed that primitive membranous vesicles might have formed from some oily "scum"—what I have called "goo" in chapter 1—which must have been abundant in the prebiotic world, and that these vesicles might have developed into photochemical "protobionts" by association with some light-catching molecule.[2] This and other similar proposals cannot be dismissed but seem to me incompatible with the necessarily rudimentary nature of the first energy-providing systems. Even if we assume the existence of some membrane-bound light-trapping system, we still have to account for the channeling of the trapped energy into productive chemical processes, rather than useless heat.

Could incipient life have done without a membrane? Such is the question I shall ask in this chapter. As we shall see, there are good reasons for believing that life could, indeed, have done so. Properly interrogated, it even tells us how. A good introduction to the topic is provided by what may have been the first energy obstacle to be overcome in the development of life.

THE CASE OF THE MISSING HYDROGEN

The problem is most easily defined in contemporary terms. As an example, take a bowl of spinach leaves and carefully dry them so as to remove all the water but no other volatile constituent. As every cook knows, not much will be left since spinach is "mostly water." What is left, however, is what gave Popeye his superstrength. Give it to a chemist for elementary analysis and he will tell you that the material consists largely of carbon, oxygen, nitrogen, and hydrogen. In numbers of atoms, the proportions are roughly: C, 60; O, 40; N, 2; and H, 100. Consider now the nature of the "foodstuffs" with which the spinach plant makes its constituents. Carbon comes from atmospheric carbon dioxide (CO_2), nitrogen from soil nitrate (NO_3^-), and hydrogen from water (H_2O). Now, try to make dry spinach from these building blocks and you find that you end up with a large surplus of oxygen: 120 atoms (60×2) brought in with carbon dioxide, 6 with nitrate (2×3), and 50 with water; a total of 176 oxygen atoms, or an excess of 136 atoms over the 40 that are needed. An alternative way of expressing this excess is in terms of the hydrogen that should be provided to convert the oxygen to water, that is, in the example considered, 272 atoms of hydrogen, two for each excess atom of oxygen. In conclusion, contemporary autotrophic (self-constructing) life needs a source of hydrogen. How about emerging life?

If the atmosphere had been as imagined by Urey, there would have been no problem. Take the carbon from methane (CH_4), the nitrogen from ammonia (NH_3), and the oxygen from water, and you already have a large surplus of hydrogen (326 atoms against the 100 atoms needed), not counting the molecular hydrogen that Urey added for good measure. But with carbon dioxide as the source of carbon, you are in great trouble, as Miller found out experimentally. Even if Urey should finally turn out to be right, the evil moment would only be postponed. Sooner or later—probably sooner than later—hydrogen would be missing, as it is today. Where was it found?

A naive sleuth, entrusted with the case of the missing hydrogen, might well answer: "What's your problem? There was more hydrogen in the oceans than you could ever use." Only too true, except that it wasn't there for the taking. Our sleuth forgot the golden rule that "you can't have your cake and eat it," which, in the chemical world, in the whole universe, in fact, translates to "you can't have it both ways." You can't take hydrogen from water and then use it to convert oxygen back to water. If you could, you would have invented perpetual motion, a dream pursued by generations of cranks, always in vain, because it is incompatible with one of the most profound laws of nature: For every natural event, there is an allowed direction and a forbidden direction. Apples fall; they don't jump up to their branches. Sugar lumps dissolve in your coffee; they don't form by themselves in a cup of sweetened coffee. Hydrogen combines with oxygen to form water; water does not dissociate spontaneously into a mixture of hydrogen and oxygen. All nature's streets are *one way.* You can take them in the wrong direction, of course, but you have to *work* for it: lift the apple, extract the sugar, wrench the hydrogen off the water molecules, for example, with electricity.

This is the absolutely fundamental way of nature, expressed by what scientists call the second law of thermodynamics, often shortened simply to the Second Law: *If you have to work for it, you are going in the forbidden direction.* In the allowed direction, on the other hand, the phenomenon, properly harnessed, may work for you, though it will never give you as much work as you would have to carry out yourself to reverse it. With a rope and pulley, you can use a falling apple to lift another apple, but only if the other apple is lighter. A convenient image to describe the two directions is downhill for what is allowed and uphill for what is forbidden. Perfectly horizontal means no work in either direction. It is the state of equilibrium, or perfect balance.

Let us now return to the problem of the missing hydrogen and ask contemporary organisms how they solve it. Where does spinach get the extra hydrogen it needs for growth? The answer is: right where our sleuth pointed, from water. But, as demanded by the Second Law, the spinach has to work for it. Or, rather, it makes the sun work for it. Chlorophyll, the green stuff of plants, does exactly that. It uses the energy of sunlight to tear off hydrogen from water and lift it to a high enough level of energy so that the hydrogen can, in turn, tear off the oxygen from carbon dioxide and nitrate and replace it, doing everything under its own power, that is, going downhill. This concept of energy level is crucial. We can readily picture it by an image borrowed from gravity, although we are, of course, dealing with chemical energy. The higher a weight, the greater the amount of work that can be performed by its fall. In chemistry, height is replaced by other concepts, such as pressure, concentration, potential, and suchlike. We shall do without these complications and adopt the less rigorous but readily grasped notion of energy level.

Chlorophyll is a very complex molecule, which, in addition, has to be associated with other complex molecules embedded in a membrane to do its job. It is highly unlikely that such a system could have arisen spontaneously in early prebiotic days. We must search elsewhere for the prebiotic supply of hydrogen. To do this, we must first take a closer look at the hydrogen atom.

Hydrogen is the smallest of all atoms. It consists of a single, positively charged particle, or proton, which forms the nucleus of the atom and bears most of its mass, and of a single, peripheral electron, a negatively charged particle less than one-thousandth the mass of the proton. In the simplified picture of the atom known by the name of the great Danish physicist Niels Bohr, the electron is viewed as gravitating around the nucleus like a planet around the sun. Quantum mechanics provides a more accurate but less intuitive image. The Bohr model of the atom will suffice for our purpose.

As it happens, small quantities of free protons exist naturally in water as a result of the spontaneously occurring dissociation of the water molecule into positive hydrogen ions, H^+, which are no other than protons, and negative hydroxyl ions, OH^-. In pure water, only one water molecule in ten million is thus dissociated (which still amounts to more than one million billion protons and hydroxyl ions in a teaspoonful of water). The number of free protons in water increases, and that of hydroxyl ions decreases, with increasing acidity (decreasing alkalinity), and vice versa. By definition, acids are substances that release protons in aqueous solution. Alkalis, or bases, are substances that pick up protons.

This little excursion into physical chemistry was needed to clarify a point of fundamental importance: *It is possible to get hydrogen from water by supplying electrons.* These can combine with protons, produced by the dissociation of water molecules, to form hydrogen atoms. However, the Second Law should not be forgotten. If we wish the hydrogen to perform the job we have in mind—tear off oxygen from carbon dioxide and nitrate—we need the hydrogen to be at a sufficiently high level of energy so that it can go downhill henceforth. This implies that the electrons must themselves be supplied at a sufficiently high level of energy so that the hydrogen atoms they make by combining with protons be lifted to the energy level needed to tear off the oxygen from carbon dioxide and nitrate.

In conclusion, in the presence of water to either absorb or provide protons, free hydrogen atoms and electrons are interchangeable. In biochemical jargon, the participation of protons is often made implicit, and one speaks simply of electrons. By definition, the gaining of electrons (or hydrogen) by a substance is called reduction; the loss of electrons (or hydrogen), oxidation. The two types of reactions are obligatorily linked. For one substance to be reduced, another has to be oxidized so as to provide the necessary electrons (or hydrogen atoms). Thus, we are always dealing with oxidation-reduction reactions, or, as more commonly termed nowadays, electron-transfer reactions. The notion of energy level must be kept in mind. When electrons are transferred, the electron donor is, by definition, the substance in which transferable electrons occupy the higher energy level, and the electron acceptor is the substance that, when reduced, has electrons occupying a lower energy level. Electrons move from donor to acceptor in the downhill direction, like every other happening in the world.

Equipped with this information, we may now search the prebiotic setting for a suitable source of electrons to effect the required reactions, which will henceforth be referred to as biosynthetic reductions. Several solutions to this problem have been proposed. I shall mention only two, which happen both to involve iron. The first mechanism uses sunlight energy to remove hydrogen from water, as does the plant system, but has the immense advantage that it needs no intricate catalyst to operate. The reaction takes place in simple aqueous solution, its only requirement being the presence of iron atoms in the form of ferrous ions (Fe^{2+}), which carry two positive charges.[3] (Ions are electrically charged atoms or molecules.) We have seen that the prebiotic oceans contained considerable amounts of this material. The source of energy is ultraviolet (UV) light, rather than visible light, but this poses no problem since the prebiotic Earth was exposed to strong UV radiation. When a ferrous ion is excited by a photon of UV light, it relinquishes an electron and changes to the ferric form (Fe^{3+}), which carries three positive charges. The electron combines with a proton to give rise to a hydrogen atom. In this process, electrons are transferred from ferrous iron (the donor) to protons (the acceptor). In the reverse reaction, hydrogen would be the donor and ferric iron the acceptor. In the absence of UV light, the downhill direction of the transfer would actually be the latter. Thanks to the energy provided by the UV light, the spontaneous direction of the reaction is reversed and the electrons released by ferrous iron are lifted to a conveniently high energy level that allows them to serve in prebiotic reductions.

A possible sign that such a reaction occurred is found in deposits of the mineral magnetite, a mixed oxide of ferrous and ferric iron found in iron-rich geological strata called banded iron-formations[4] because of their striped appearance. The age of banded iron-formations ranges from 1.5 to over 3.5 billion years. It is usually accepted that these formations arose from the interaction of ferrous iron with the oxygen produced by light-utilizing bacteria, but the possibility that the UV-supported reaction just described contributed to their creation is not to be excluded.

Another possible source of prebiotic electrons is hydrogen sulfide, a typical part of the prebiotic setting. Wächtershäuser has proposed a reaction in which, in the presence of ferrous iron, two sulfide ions (SH^-)—which form from hydrogen sulfide in aqueous solution—would be converted into a disulfide ion (S_2^{2-}) with the release of molecular hydrogen. In this case, the iron does not abandon an electron. Instead, it drives the reaction by combining with the generated disulfide to form the highly insoluble ferrous disulfide (FeS_2), thereby removing the product of the reaction from solution and allowing more to be formed. The validity of this model has been proved experimentally.[5] Ferrous disulfide is the constituent of pyrite, the mineral assumed to provide a catalytic surface in Wächtershäuser's protometabolic model discussed in the preceding chapter.

We thus have two model systems that could theoretically account for the missing hydrogen. They are not mutually exclusive. The two reactions could have occurred side by side or in different environments. By definition, the UV-supported reaction could have taken place only in surface water layers, whereas the pyrite-generating process could have occurred in the dark depths of the ocean.

In conclusion, whatever the exact composition of the prebiotic atmosphere, we may consider it likely that our young planet, abundantly supplied with ferrous iron, enveloped in hydrogen sulfide fumes, and exposed to strong ultraviolet radiation, was exuding hydrogen through every pore, while iron combinations destined one day to become the minerals magnetite and pyrite were laid down at the bottom of the oceans. Electrons were indeed available at a high enough energy level to support early biosynthetic reductions.

Interestingly, iron and sulfur are both key constituents of catalysts that participate in electron-transfer reactions in present-day living organisms. The most ancient such catalysts may well be proteins called iron-sulfur proteins,[6] in which the catalytic center is an iron atom, which oscillates between the ferrous and the ferric state, surrounded by a sulfide cluster. The hint is unmistakable.

THE CASE OF THE EXCESS WATER

A good supply of electrons solves only half of the prebiotic energy problem. The other half concerns the joining of molecules in a watery medium. Examples of such reactions already encountered are the formation of esters from alcohols and acids, of thioesters from thiols and acids, of peptides from amino acids, of nucleotides from phosphate, ribose, and a base, and of RNA from nucleotides. Many more such condensation reactions occur in living organisms. All have in common that they take place with the removal of water. In aqueous solution, this is the forbidden direction, as was explained in chapter 2.

Nature's universal answer to the problem is ATP. Almost as famous as DNA, this acronym stands for *a*denosine *t*ri*p*hosphate. Adenosine is the combination of the purine base adenine with ribose. In association with one phosphate molecule, adenosine forms adenosine monophosphate (AMP), one of the four nucleotides that serve to make RNA. With a second phosphate attached to the phosphate of AMP, one gets adenosine diphosphate (ADP), which, with an additional phosphate attached to its terminal phosphate, gives ATP.

The two bonds linking the three phosphates of ATP are called pyrophosphate bonds, after inorganic pyrophosphate (PP_i), a combination of two phosphates that arises when inorganic phosphate (P_i) is heated at a high temperature (*pyr* means fire in Greek). The formation of pyrophosphate bonds is a typical condensation reaction. It takes place with the removal of a water molecule. In the reverse reaction, the bonds are split with the help of water, by hydrolysis. As for all such bonds in an aqueous milieu, hydrolysis goes downhill, condensation uphill.

ATP is the universal biological condensing agent, that is, chemical water extractor. Its hydrolysis, either to ADP and P_i or to AMP and PP_i, serves to extract the water that needs to be removed in order to seal the created bond. This takes place by special mechanisms known as sequential group transfers that act in such a way that the water molecule involved is transferred directly from the joining molecules that generate it to the molecule (ATP) that consumes it for hydrolytic splitting; it never appears in free form nor mixes with the surrounding water. In such a transfer, the water molecule takes the direction of least resistance (downhill). If less energy is required to seal a bond between X and Y than between ADP and P_i or between AMP and PP_i—that is, if ATP hydrolysis releases more energy than is needed for joining X and Y—X-Y will be formed and a pyrophosphate bond of ATP will be split. The reverse will occur if the opposite is true. If the formation of the two bonds requires equal amounts of energy—the reaction is freely reversible—the exchange will be partial according to the rules of chemical equilibrium.

As it happens, most of the bonds in biological substances require less energy for their formation than do the pyrophosphate bonds of ATP, which explains why ATP is an efficacious condensing agent. The biochemist Fritz Lipmann has named the pyrophosphate bonds of ATP high-energy bonds for this reason.[7] The bonds in proteins and other natural substances that are sealed at the expense of ATP hydrolysis are low-energy bonds.

ATP is not the ultimate source of energy for condensation reactions. We don't get ATP from our food, and our cells contain this vital substance in only small amounts. Life would quickly grind to a standstill were not ATP continually reassembled from its hydrolysis products. This problem will be examined later in this chapter. First, let us ask whether a substance such as ATP could have been available to emerging life.

Surely not ATP itself. It is too complex a molecule for this early phase of prebiotic events. With ATP, we are well into the RNA world, not at the onset of protometabolism. But what about the simpler inorganic pyrophosphate molecule?

The pyrophosphate bond of inorganic pyrophosphate is not as energetically powerful as the similar bonds of ATP, but it is sufficiently energy-rich to substitute for the ATP bonds in many processes. There is ample evidence in the present living world that inorganic pyrophosphate can carry out the same basic functions as ATP. Most researchers believe that pyrophosphate preceded ATP as the first bearer of productive high-energy bonds.[8] This role could also have been played by polyphosphates, associations of a larger number of phosphate groups linked by pyrophosphate bonds, which are also found in a number of organisms. Accordingly, many scientists have searched the geological record for the possible prebiotic occurrence of such substances.

The results of this search are not promising. I mentioned in chapter 1 the rarity of soluble inorganic phosphate and the problem this poses in relation to the overwhelming biological importance of phosphate. The problem is compounded in the case of pyrophosphates and polyphosphates, which are much less abundant than phosphates and also locked up in water-insoluble combinations. However, acid, which I suggested as a possible means of solubilizing phosphate, could have done the same for pyrophosphate. Also, the volcanic production of pyrophosphate has been detected recently and could have been a more abundant source under prebiotic conditions.[9]

An alternative possibility is that thioesters were the primeval source of energy for condensation reactions.[10] Thioesters are obligatory intermediates in a large number of ATP-supported condensation reactions in which one of the partners of the reaction is an acid. In such reactions, the step powered by ATP hydrolysis is the condensation of the acid with a thiol, usually pantetheine phosphate or coenzyme A, to form the corresponding thioester. The acid group is then transferred from the thioester to its acceptor. We saw in chapter 2 how such group-transfer reactions are involved in the formation of esters and some peptides. Many other important biological constituents are made by way of thioesters, including a large variety of fatty substances, cholesterol, several vitamins, parts of chlorophyll, and numerous metabolic intermediates.

What makes thioesters attractive is that they are energetically equivalent to ATP. The thioester bond is a high-energy bond. Thus, thioesters can support the assembly of ATP or be assembled at the expense of ATP hydrolysis equally well. The American chemist Arthur Weber, formerly of the Salk Institute in San Diego, now at the NASA Ames Research Center in Moffett Field, California, who has pioneered the study of prebiotic sulfur compounds, has shown that thioesters can, under very simple conditions, support the formation of inorganic pyrophosphate from inorganic phosphate by a mechanism similar to that of the present-day thioester-linked process whereby ADP and P_i are joined into ATP.[11]

We thus have the choice between two possibilities consistent with the congruence rule. Pyrophosphate was provided by the prebiotic setting and served as a condensing agent for the assembly of thioesters. Or, conversely, thioesters came first and supported the assembly of pyrophosphate. Or, of course, the two could have appeared independently and interacted later. Before we come to any conclusion, we must look first at the mechanisms whereby ATP is continually regenerated from its hydrolysis products in extant living organisms.

HOW THE WHEELS ARE KEPT TURNING

In living cells, ATP turns over very rapidly. It is continually consumed—that is, split hydrolytically—by the performance of chemical work (and many other kinds of work, as we shall see later), and it is regenerated equally fast from its hydrolysis products. Where does the energy required for this regeneration process come from? The answer to that vital question brings us back to the case of the missing hydrogen. The energy for the regeneration of ATP comes from electron flow.

We saw in the beginning of this chapter how electrons can be transferred from a reduced donor occupying a higher energy level to an oxidized acceptor occupying a lower energy level. A transfer of this sort releases energy in an amount proportional (per electron transferred) to the difference in the two energy levels. As a simple image, think of a waterfall. The energy released by the fall of a given amount of water is proportional to the height of the fall.

In all living cells, certain key "electron falls" are coupled to the assembly of ATP from ADP and P_i, the way certain waterfalls are harnessed to the running of a mill or to the generation of electricity. This universal mechanism is called oxidative phosphorylation—oxidative because the electron donor in the coupled reaction is oxidized; phosphorylation because ADP is phosphorylated, that is, fitted with an additional phosphate group in the process. It requires three conditions: (1) an appropriate source of electrons; (2) an outlet for the electrons situated at a sufficiently lower energy level for the amount of energy released by the electron transfer to cover the needs of ATP assembly (as a rule, one ATP molecule is assembled for each pair of electrons transferred); and (3) a coupling system—the equivalent of the waterwheel or turbine in the waterfall analogy—linking ATP assembly to electron flow.

Many different electron donors and acceptors are used in such reactions in nature. In organisms like ourselves, for example, the foodstuffs provide the electrons, and oxygen is the final acceptor. This is what really happens when we "burn" our food. Thanks to this mechanism, the energy released by such combustions is only partly given off as heat. Much of it is retrieved in the form of reassembled ATP. In illuminated green plants, the electrons are delivered at a high energy level by excited chlorophyll molecules and recovered at the lower level by the same molecules. In between, the electrons fall through coupled phosphorylation systems, as they do in our tissues. Electron donors and acceptors vary, but the energy-retrieval mechanisms are universal.

The most important such mechanisms are situated in membranes. We shall meet them later when we look at the first cells. We have agreed not to take them into consideration at this early stage in the origin of life. Some coupled phosphorylations do not depend on membranes and take place in the cell sap, the soluble part of living cells. Known technically as substrate-level phosphorylations, these mechanisms could qualify as prebiotic (with pyrophosphate being assembled instead of ATP). Interestingly, they involve thioesters as key intermediates. The reaction directly coupled to the energy-releasing electron transfer is the formation of a thioester, which then supports ATP assembly in the manner already seen.

Thioesters thus occupy a unique position in metabolism: They *bridge the two main forms of biological energy*—one linked to electron transfer and one linked to group transfer. In addition, we saw in the preceding chapter that thioesters could have played a key role in the generation of the first catalysts of emerging life. These facts, taken together with the probable abundance of acids and thiols on the prebiotic Earth, build a strong case in favor of thioesters being the primary energizers of the early biogenic process, perhaps preceding inorganic pyrophosphate. But there is an if—and oh what a big if, to quote Darwin![12] The assembly of the primeval thioesters would itself have required energy. So, we are back to square one.

There are several possible solutions to the problem of thioester assembly under prebiotic conditions. According to thermodynamic data, thioesters could form spontaneously from free acids and thiols in a watery medium, if this medium were very hot and acidic. Even so, however, the yields would still be slight. Nevertheless, the possibility is worth entertaining. Boiling acid may not be our idea of a cozy little niche. Neither is it a particularly favorable medium for a number of fragile biological molecules. Yet it is the choice habitat of certain bacteria known as thermoacidophilic, of particularly ancient origin.[13] A number of authors believe that life started in a hot environment. They have paid particular attention in this connection to the continuing recycling of water through deep-sea hydrothermal vents. One could imagine a scenario in which thioesters arising in the hot, acidic, sulfurous depths would be continually delivered to the prebiotic soup, where milder conditions prevailed and the energy-rich thioesters could carry out their function.

This is not the only possibility. Weber has described other plausible mechanisms for the formation of thioesters.[14] There is also the possibility, as yet unexplored, that thioesters may have assembled in the atmosphere from volatile thiols and acids. Finally, and perhaps most simply, there is the recourse to coupled electron transfer as a source of energy, as occurs in metabolism today.

What is known of this reaction indicates that it could have taken place under prebiotic conditions. The necessary ingredients were probably present, and primitive iron-sulfur complexes, ancestral to iron-sulfur proteins, could have catalyzed such a reaction. In some ancient bacteria, a process of this sort is indeed catalyzed by an iron-sulfur protein.[15]

As to the electron acceptor required by the thioester-generating process, an attractive possibility is ferric iron, the product of the reaction whereby ferrous iron serves to generate hydrogen from protons with the help of UV light. By serving as electron acceptor, ferric iron would return to the ferrous state, thus completing a cycle in which electrons are released from ferrous iron with the help of UV light and return to ferrous iron by way of a complex pathway completed by the sealing of thioester bonds. The net result would be the use of UV-light energy to close thioester bonds, which, in turn, could serve to meet all the energy needs of emerging life. Such a cycle would be entirely analogous to the water–oxygen cycle that supports much of the present-day biosphere—plants release oxygen from water with the help of visible light, and animals and other aerobic organisms use the oxygen as final electron acceptor and reconvert it into water—with the crucial difference that the water–oxygen cycle requires complex structures, whereas this iron cycle does not. The collaboration of iron and sulfur in such a cycle could have been a primeval manifestation of the contemporary fecund alliance between these two biologically important elements.[16]

THE THIOESTER WORLD

We have not uncovered the hidden trail of protometabolism, but we have found some telltale traces. These traces have been described in detail in this and the preceding chapter. A brief summary of what they have revealed may be useful. The main message comes out loud and clear: sulfur.

This element is, quantitatively, a minor component of living matter, but qualitatively a very important one. Two of the twenty amino acids that form proteins, cysteine and methionine, contain sulfur. So do several coenzymes. Sulfur is often found at the heart of the catalytic centers of enzymes. There is sulfur also in a number of structural macromolecules, for example, some of the main components of cartilage. Many of the most ancient bacteria live by metabolizing some sulfur compound. The prebiotic world was steeped in sulfur. It all amounts to a very strong case.

In present-day organisms, sulfur is derived mostly from the fully oxidized sulfate ion (SO_4^{2-}), which exists unchanged in a number of components, mostly structural, in which its role is mainly to provide negative charges to the molecules. Many of the most important biological functions of sulfur, however, require sulfate to be reduced to hydrogen sulfide (H_2S) and incorporated into organic molecules, mostly thiols and their derivatives. Hydrogen sulfide is also the form of sulfur that dominated in the prebiotic world. The traces we have detected unmistakably point to thiols.

In the prebiotic soup, thiols most likely were present together with a variety of amino acids and other organic acids, which are the main substances produced in Miller-type simulation experiments and found in meteorites. Thiols and acids readily join into thioesters, provided some means exist for the removal of the water molecule that must be extracted for a thioester bond to form. Several possible mechanisms whereby this could have happened exist. My main hypothesis is that, somewhere in the prebiotic world, conditions prevailed under which thioesters formed spontaneously. Granted this unproved but not implausible assumption, the way opens to a metabolism-like protometabolism supported by thioesters.

Thioesters provided protometabolism with two essential ingredients: catalysis and energy. The catalysts were peptides and peptide-like substances that prefigured present-day enzymes and guided the first building blocks of life into directions not too different from present-day metabolism. The energy was in a form that fitted within such pathways and could have served to usher in inorganic phosphate and the all-important pyrophosphate bond.

The high-grade electrons required for the first biosynthetic reductions could have been provided by ferrous iron with the help of UV light or by hydrogen sulfide with the help of ferrous iron. The first mechanism would have given rise to ferric ions, which could have served as the first electron acceptor in energy-yielding electron-transfer reactions coupled to the synthesis of thioesters and, in due course, of inorganic pyrophosphate. Together, the two processes would have closed an iron cycle whereby UV-light energy supported the assembly of thioesters and, by way of the splitting of the thioesters, the whole of protometabolism. In addition, iron associated with sulfide could have constituted the first electron-transfer catalyst.

This "thioester world," or, better said, "thioester-iron world," represents my hypothetical reconstruction, based on the few traces it has left, of the hidden trail that led from the first products of prebiotic chemistry to the RNA world, and that continued to support the RNA world during all the time it took emerging life to evolve from the RNA world to the RNA-protein world. This view of the trail is purely conjectural. Quite possibly, future findings will point to different pathways not thought of today. For my part, I would find it very suprising if these early pathways did not reveal glimpses of present-day metabolism.

Chapter 4

The Advent of RNA

EVEN IF WE accept the premises of a thioester world, we are no nearer to identifying the chemical pathways that led from the first building blocks of life to RNA. A possible experimental approach to the problem does suggest itself: Reproduce in the laboratory the primeval mixture of multimers and look for key catalytic activities in the mixture. According to my model, this is the heart of the problem. Protometabolism must have been channeled by the early catalysts the way metabolism is by enzymes today.

Several techniques now exist for making random peptides. One could even go back to the old Wieland procedure, which has the merit of actually using thioesters. Such experiments might not provide the selective conditions that led to the particular subset of multimers assumed by the model, but they would be a step in the right direction. The congruence rule, on the other hand, would help in choosing the kind of catalytic activities for which to search. I am, unfortunately, too far along my own trail to start such an approach. But other laboratories are becoming interested in it.

Meanwhile, we are left to conjecture, using present-day metabolism as a guide. A possible clue is provided by ATP.

THE ATP CONNECTION

ATP plays a key role in energy metabolism. It is also one of the four precursor molecules used in the synthesis of RNA. Here lies the connection. RNA molecules are constructed from nucleotides, which are combinations of phosphate, ribose, and one of four bases: adenine, guanine, cytosine, and uracil. AMP, the parent molecule of ATP, is one such nucleotide. The similarly constructed GMP, CMP, and UMP are the other three. Just as AMP can be phosphorylated to ADP and ATP, the other nucleotides can likewise be converted to GDP and GTP, CDP and CTP, and UDP and UTP, respectively. The pyrophosphate bonds in GTP, CTP, and UTP have the same properties as the pyrophosphate bonds in ATP; their splitting can similarly support energy-requiring processes and does so in certain cases. Even when this happens, the central role remains the prerogative of ATP. Whenever another energy donor is involved, it is regenerated at the expense of ATP splitting.

Why was ATP singled out in this manner? A possible answer to this question is that adenine happened to be there before the other bases. It is certainly the easiest one to make abiotically. Oró's celebrated synthesis of adenine from ammonium cyanide was mentioned in chapter 2.[1] Even though the relevance of this finding has not been established, it suggests that adenine may belong to the group of easily formed molecules that made up the early building blocks of life. In support of this hypothesis, traces of adenine have been detected on meteorites.[2]

The origin of the 5-carbon sugar ribose is obscure. Sugars arise readily from formaldehyde in alkaline solution, but as a complex mixture of different molecules. In metabolism, ribose is formed from the 6-carbon sugar glucose by a devious pathway that involves phosphate-linked intermediates. Work by the Swiss chemist Albert Eschenmoser, from the Federal Institute of Technology in Zurich, has shown that phosphate groups may, even in the absence of a catalyst, exert highly selective influences on the reactivity of sugar molecules.[3] Phosphate groups are also involved in the biological mechanism whereby ribose joins with bases, and they provide the phosphate component of nucleotides. Perhaps the early appearance of inorganic pyrophosphate—from natural sources or by way of the thioesters of my model—helped steer protometabolism in the direction of AMP formation. At present, one can only surmise.

Once AMP comes on the scene, an interesting possibility is offered by the thioester-world model. It is known that the thioester bond and the pyrophosphate bond are energetically equivalent; the hydrolysis of either one can support the water-releasing assembly of the other. In one such reaction, ATP hydrolysis to AMP and inorganic pyrophosphate (PP_i) serves in the condensation of acids with coenzyme A or pantetheine phosphate (two major thiol cofactors in metabolism). This is how the thioesters involved in the biosynthesis of esters and many other important biological compounds are assembled. The reaction is freely reversible. Now, imagine yourself in a thioester world containing thioesters and PP_i, at the time when the first molecules of AMP made their appearance. Through a reversal of the reaction just mentioned, AMP could have joined with PP_i, with the support of the energy supplied by thioester hydrolysis.[4] Had you been there, you would have witnessed one of the great events in the chemical origin of life: the birth of ATP, the universal purveyor of biological energy, destined eventually to replace in all its major functions the pyrophosphate from which it arose.

An interesting possibility is that ATP, in turn, served to usher in RNA: in other words, *information may have entered by way of energy.* What may have happened is the condensation of two ATP molecules to ATP-AMP, with the release of inorganic pyrophosphate. In this reaction, the second ATP donates an AMP molecule to the first ATP, and the bond between ATP and AMP is sealed at the expense of the bond that originally existed between AMP and pyrophosphate in the AMP-donating ATP molecule. Have another ATP donate AMP to ATP-AMP, and you get ATP-AMP-AMP. Repeat the reaction any number of times, and you get a chain of ATP-AMP-AMP-AMP--- of any length, called poly-A. This is not science fiction. Such a reaction actually occurs in many living cells, where it adds poly-A tails up to about 250 nucleotides long to many RNA molecules. Unlike the synthesis of true RNA, the formation of poly-A takes place without the supply of information; it consists merely of "dumb," repetitive assembly.

In the context of protometabolism, poly-A would have been no more than a storage form of AMP molecules, which would automatically have regulated the relative amounts of ATP and PP_i present. Imagine a situation where ATP was abundant and PP_i scarce. Because of the laws of chemical equilibrium, ATP would be driven to form poly-A, and PP_i would rise. In the opposite situation, excess PP_i would drive the formation of ATP from poly-A. The process may have been "dumb," as was the whole of protometabolism before RNA appeared. But the outcome was far from useless. It preserved AMP and adjusted the availability of the two forms of high-energy pyrophosphate bonds—PP_i and ATP—to their relative rates of consumption.

It is not known how the other three bases found in RNA appeared. Possible schemes for their abiotic formation have been suggested.[5] An origin from adenine is not inconceivable. Once they were present, their incorporation into nucleotides could have been brought about by AMP. Reactions are known by which one base replaces another in nucleotides. Thus, guanine could replace adenine in AMP, to form GMP; cytosine could similarly lead to CMP, uracil to UMP. In turn, these nucleotides could acquire phosphate groups from ATP, as they do in present-day metabolism, and give rise to GTP, CTP, and UTP. Finally, the same "dumb" reaction that caused poly-A to be formed from ATP could have incorporated GMP, CMP, and UMP into similar associations, which would have been the first RNA molecules, though still devoid of information, a mere jumble of letters. This scenario is conjectural but not unlikely. It seems reasonable to suppose that the chemistry came first, without information, and that information came later.

In the present-day world, RNA is assembled as described from ATP, GTP, CTP, and UTP, which donate AMP, GMP, CMP, and UMP units to a growing chain initiated by ATP or GTP. What the biological process has in addition, and the prebiotic process may have lacked, is a mechanism for the selection of whichever of the four available nucleotides is to be added at each step. The birth of this mechanism, which relies on simple molecular interactions, represents a true watershed in the development of life on Earth. It signals the transition from the age of chemistry to the age of information.

Before we ourselves effect this transition, we must take a look at an important group of catalytic components known as coenzymes, many of which may go back to the age of chemistry. Their existence raises interesting possibilities with respect to the functioning of the RNA world and the origin of RNA.

COENZYMES: CHILDREN OF THE RNA WORLD?

In metabolism, enzymes are often assisted by special molecules called coenzymes. These serve most frequently as intermediaries, or carriers, in transfer reactions. Imagine a process in which entity X is transferred from the X donor X-Y to the acceptor Z, to give X-Z and Y. In many cases, a carrier K mediates the transfer so that X is first transferred from X-Y to K, with the formation of X-K, which then donates X to Z, making X-Z and leaving K ready for another round. There are several advantages to this peculiarity. A major advantage is centralization and simplification of the transfers of the same entity. Suppose, for example, that X is to be exchanged between ten different donors and ten different acceptors. The number of individual reactions needed to allow all possible exchanges is one hundred. With K as common intermediary, only twenty reactions are required. It is like having a central currency, as against barter.

Two kinds of entities are exchanged in transfer reactions: electrons or chemical groups. We have seen the importance of electron transfers in biosynthetic reductions and in energy retrieval. As to group transfers, they represent the main mechanism of biosynthetic assembly reactions. Several examples were mentioned in chapters 2 and 3. RNA synthesis is another. In RNA elongation, the addition of, say, AMP is a transfer of the AMP group from ATP to the growing chain. Virtually every biological condensation proceeds by group transfer. Attesting to the importance of transfer reactions is the fact that more than 90 percent of the reactions that occur in living organisms are either electron transfers or group transfers. Most of these reactions take place with the help of a coenzyme carrier. There are thus two main categories of coenzymes: electron carriers and group carriers.

Some coenzymes may go back to early prebiotic days. We have already encountered iron-sulfur complexes as putative primitive electron carriers. Some thiols, such as coenzyme M or pantetheine phosphate, may be among the earliest group carriers. Many other coenzymes could be children of the RNA world. It is striking that all four nucleotides provide important group carriers participating in the synthesis of certain carbohydrate (derived from sugars) and lipid (fats) components. In addition, AMP is part of several other coenzymes, including coenzyme A, where it is linked with pantetheine phosphate, and several central electron carriers.

In a number of coenzymes, the active part is represented by a flat, ring-shaped, nitrogenous molecule chemically related to the bases found in the major nucleotides. Many of these special molecules are vitamins, that is, essential chemicals that the human organism is incapable of making and that must be supplied with food. Interestingly, some of these substances are engaged in combinations entirely similar to nucleotides. Such is the case of nicotinamide, an important vitamin known as vitamin PP (pellagra preventiva), whose deficiency causes pellagra, a severe nutritional disease that was once prevalent—and still is in remote areas—in many parts of Latin America. In living organisms, nicotinamide is linked with ribose and phosphate in a typical nucleotide, nicotinamide mononucleotide, or NMN. In association with AMP, NMN forms the two major electron carriers, NAD (nicotinamide adenine dinucleotide) and NADP (nicotinamide adenine dinucleotide phosphate).

Another vitamin, riboflavin, or vitamin B_2, is similarly engaged in a nucleotide-like combination (ribose is replaced by a related substance), flavin mononucleotide (FMN), which, combined with AMP, gives flavin adenine dinucleotide (FAD). Both FMN and FAD are also important electron carriers.

The fact that so many coenzymes are nucleotides is often quoted in support of the RNA-world model. These molecules are seen as going back to an RNA world entirely operated by ribozymes intimately connected with nucleotide coenzymes. In the words of the American biochemist Harold White III, the nucleotide coenzymes could be "fossils of an earlier metabolic state."[6]

The existence of molecules such as NMN and FMN raises the possibility that the first RNA molecules contained more than four kinds of nucleotides. The conditions that led to the abiotic synthesis of purines (adenine and guanine) and pyrimidines (cytosine and uracil) may well have spawned the formation of a whole array of related nitrogenous bases. A number of these could have formed nucleotides, which, in turn, could have become incorporated into RNA-like combinations. The four RNA components would then have been selected later by virtue of their unique ability to support information transfer. This intriguing possibility will be considered in the following chapter.

THE CHEMICAL FOUNDATION OF LIFE

The most important notion encountered in our attempt to reconstruct the early pathway from abiotic chemistry to the RNA world is that of congruence between protometabolism and metabolism. This notion runs contrary to the ideas generally accepted in the field. Here is how two pioneers in origin-of-life research, Miller and Orgel, summed up their views on the topic in 1973. Referring to the possibility that "metabolic pathways parallel the corresponding prebiotic syntheses that occurred on the primitive earth," they wrote: "It is not difficult to show that this hypothesis cannot be correct in the majority of cases. Perhaps the strongest evidence comes from a direct comparison of known contemporary biosynthetic pathways with reasonable prebiotic pathways—in general they do not correspond at all."[7]

What is a "reasonable" prebiotic pathway? No doubt, the very early reactions, such as we know have occurred on comets and meteorites and suspect occurred on Earth, are products of some basic reactions of organic chemistry, of the kind Miller calls "robust,"[8] that require only simple conditions to take place. A number of amino acids and other organic acids, perhaps adenine and some other nitrogenous bases, plausibly some sugars (although this is more problematical), could have formed in this way. These substances are what I call the building blocks of life.

However, the road from these simple molecules to the kind of chemical complexity required to generate and sustain an RNA world belongs to a different realm. There is no robust way of making ATP, let alone an RNA molecule. This, in itself, makes a strong argument in support of a catalyst-mediated protometabolism.

A further conclusion, which I believe to be compelling, is that protometabolism must have prefigured metabolism. I don't see how the RNA world could, by the progressive generation of protein enzymes, have given rise to a network of chemical reactions unrelated to those that produced and sustained it in the first place. Metabolism must have arisen congruently with protometabolism.

Thus, the basic chemistry that underlies life in all its forms was laid down right from the start, by a succession of steps ruled by the strictly deterministic factors that govern all chemical processes. This is true not only of the particular model I have proposed, but of any model. Whatever pathways emerging life followed on its way to the RNA world, the nature of the process as a multistep succession of chemical events precludes a significant participation of improbable accidents. The biogenic pathway involved reactions that were bound to take place under prevailing conditions.

Another conclusion is that the early phase of the biogenic process must have been fast, contrary to the commonly accepted notion that the emergence of life took a very long time.[9] With the kind of fragile chemicals involved in the construction of life, only a fast process could overcome the wear of spontaneous decay. In a primeval soup appropriately stocked with building blocks, sources of energy, and catalysts, RNA could be reached in a matter of years, if not less. A large number of such developments—many abortive for one reason or the other—could have been initiated in different parts of the world and at different times. Even the possibility of one day reproducing the process in the laboratory no longer belongs to the realm of science fiction.

PART II

THE AGE OF INFORMATION

Chapter 5

RNA Takes Over

LITTLE HAS BEEN SAID so far about information, the reason being that whatever led to the appearance of the first RNA molecules, it was not an anticipation of their informational role. Initially, RNA was a product of chemical determinism. Information arose as an emergent property. A few words concerning this property will be helpful before we proceed.

A GLIMPSE OF THINGS TO COME

All living organisms are constructed according to a blueprint that is transmitted from generation to generation. Plague bacilli beget plague bacilli, orchids beget orchids, mites beget mites, humans beget humans. For this reason, the blueprint is called genetic (the root *gen,* as in *genesis,* comes from the Greek verb meaning "to be born"). The genetic blueprint consists of units, or genes, which together make up the genome, or genotype, of the organism. Genes have two cardinal properties: (1) they can be *copied,* the condition of hereditary transmission; and (2) they can be *expressed* into properties of the organism, which, together, constitute its phenotype. The underlying phenomena are entirely chemical.

In all extant forms of life, the genetic blueprint is written into molecules of deoxyribonucleic acid (DNA), a material closely related to RNA and similarly constructed from four different kinds of nucleotides. The sequence of nucleotides determines the informational content of the molecules—just as the sequence of letters determines the informational content of words.

DNA is the stuff of our genes and, for this reason, deserves its preeminent position in the symbolism of life. Its function, however, is strictly limited to the storage of genetic information (and the replication of that information when a cell divides, so that each daughter cell has a copy). When it comes to expression of the information, DNA is invariably transcribed first into RNA. Transcription is not very different from replication, as RNA is chemically similar to DNA and is constructed, but for a single, minor difference, with the same four bases.

RNA is a much more versatile molecule than DNA. It can exhibit catalytic activities, as in ribozymes, and thereby expresses the information received from the transcribed DNA in the performance of chemical reactions. Among these reactions are changes affecting RNA molecules and—within the framework of a complex cell structure, called a ribosome, made of a number of RNA and protein molecules—the assembly of amino acids into proteins, a process of overwhelming importance in all living beings.

Such functional RNAs express only a small number of DNA genes. Most genes code for proteins, which, through their structural, regulatory, and especially catalytic functions (enzymes), are the main agents of phenotypic expression. Cells, by and large, and the organisms they serve to build, are the expression of their proteins. The sequences of amino acids in proteins are specified by the sequences of nucleotides in the DNA genes, though not directly but by way of RNA transcripts, which are the information "messengers" in this process. Because proteins are constructed with an "alphabet" of twenty amino acids, as opposed to the four-nucleotide alphabet of RNA (or DNA), the transfer of information from RNA to protein is called translation. The set of equivalences that govern translation forms the genetic code.

These relationships are summarized by the sequence:

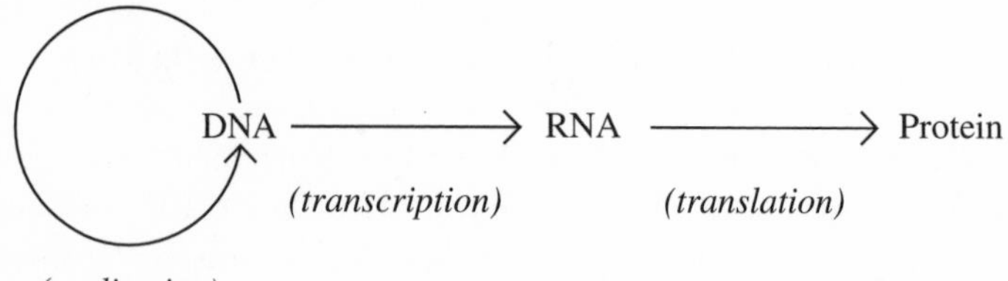

in which the arrows represent information transfers. Crick's Central Dogma[1] amounts to the amply verified statement that the last arrow (translation) is strictly unidirectional—reverse translation does not occur. This fact, together with the important catalytic role of certain RNA molecules in protein synthesis, explains why most investigators believe that RNA came before protein. But why RNA first, not DNA first?

The reason for this is that DNA is theoretically expendable, whereas RNA is not. All that is needed is for RNA to be replicatable:

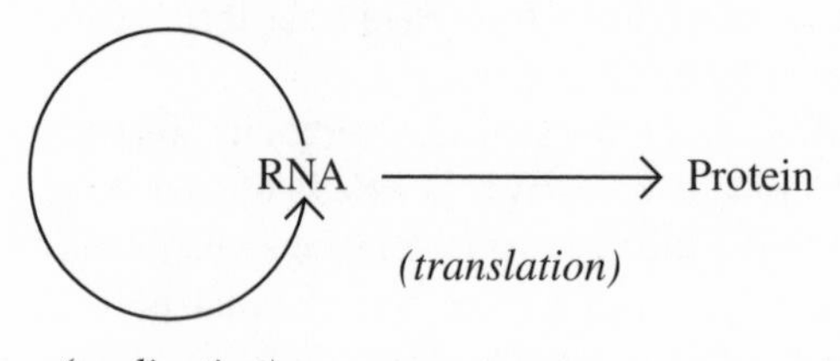

Such a process does not occur in normal cells, but it takes place in cells infected by certain viruses—the agent of polio, for example—that have an RNA genome. In these viruses, RNA is the replicatable storage form of genetic information; DNA is not involved. Among the proteins coded for by the viral RNA is RNA replicase, the enzyme catalyzing RNA replication.

Thus briefed, we may now attempt to reconstruct the historical events that led to the emergence of replication, protein synthesis, and translation, remembering that these developments came about through strictly chemical processes. Information did not beckon. It slipped in. As soon as it had done so, it became the central driving force of life, by making possible Darwinian selection.

THE MAGIC CIPHER

The first RNA molecules were probably random associations of nucleotides, which may, quite possibly, have contained other bases besides adenine, guanine, cytosine, and uracil. These bases are customarily represented by their initials. (In biochemical abbreviations, such as ATP or UMP, A, G, C, and U stand for the combinations of the bases with ribose; but in the informational—and informal—shorthand of molecular biology, these chemical nuances are dispensed with.)

What could have singled out the AGCU alphabet for information transfer is the existence of chemical complementarity relationships between A and U, on one hand, and G and C, on the other. These relationships, with T (thymine) replacing U in DNA, now dominate all forms of information transfer among nucleic acids, as well as the three-dimensional structures of these molecules.

In the late 1940s, the Austrian-born American biochemist Erwin Chargaff of Columbia University in New York, spurred by the rising interest in DNA, analyzed DNA samples of widely different origins for their content in their four constituent bases. He made the surprising observation that, within the limits of experimental error, the adenine content was always equal to that of thymine, and the guanine content to that of cytosine.[2]

It was left for Watson and Crick to discover the fundamental significance of these relationships. In the double helix, the two strands are linked by bonds between A and T, on one hand, and G and C, on the other. The two strands are complementary because one always has an A where the other has a T, a T where the other has an A, a G where the other has a C, and a C where the other has a G. If one knows the sequence of one strand, one can write the sequence of the other. Watson and Crick also noted that such base pairing could underlie replication.

This brilliant intuition has been fully corroborated by later experiments, and extended to RNA, where uracil replaces thymine as the base complementary to adenine. Thus, in their universal form, the "Chargaff equations" are now written:

A === T (or **U**) and **G === C**

These relationships express the ability of the bases concerned to join by their free edges—they are flat molecules—like two pieces of a puzzle, to form flat bimolecular associations stabilized by a special kind of bond known as the hydrogen bond. Hydrogen bonds govern many important interactions between and within biomolecules. The most fundamental of these interactions are those that cause base pairing in nucleic acids.

Two hydrogen bonds exist between A and U (or T). There are three between G and C, which thereby make up the stronger of the two pairs. Even so, hydrogen bonds are relatively weak with respect to the forces that tend to destabilize base pairs, namely thermal agitation and electrostatic repulsions between the negatively charged phosphate groups of the two chains. However, the links become cooperatively stronger as the number of adjacent bases engaged in pairing increases. Thus, any two nucleotide chains that contain complementary stretches of, say, three or more nucleotides can be cemented into relatively stable associations by base pairing.

These associations require the two chains to face each other by their bases, which implies that the chains are aligned in antiparallel orientation, with the phosphate end of one adjoining the ribose end of the other, and vice versa (see the structure on p. 23). Owing to the particular molecular anatomy of polynucleotide chains, the connected stretches are twisted into a helical coil, somewhat in the shape of a spiral staircase. The base pairs form the steps of the staircase, their planes perpendicular to the main axis and rotated horizontally with respect to each other. The phosphate-ribose backbones of the chains form the banisters of the staircase.

When two polynucleotide chains are complementary over their entire lengths, they join into a long, regular, helical, double-stranded thread stabilized by base pairing. This conformation was first discovered in DNA, which exists in nature mostly in double-stranded form. This is the famous double helix. Unlike Watson and Crick, nature discovered the double helix first in RNA. In contrast to DNA, most RNA molecules are single-stranded, except in certain viruses. However, single-stranded RNAs include many short nucleotide sequences that are complementary in antiparallel orientation. These stretches can join by base pairing, causing the slender RNA chains to form loops of varying complexity, generating structures that have been poetically described as clover leaves, flowers, and the like, but probably appear in reality more like hideously twisted knots. Whatever their aesthetic value, these shapes play important roles in determining the functional properties of the molecules.

Most likely, the first RNA molecules occasionally included such complementary stretches, which caused the chains to form loops. This may be how replication was initiated.

THE START OF REPLICATION

Consider a random RNA chain ending, for example, with the sequence GACU. Occasionally, an A unit in the body of this chain—one out of sixty-four, on average—will be followed by GUC. This AGUC sequence is complementary to the terminal sequence written in antiparallel fashion, and will cause the chain to double up as follows:

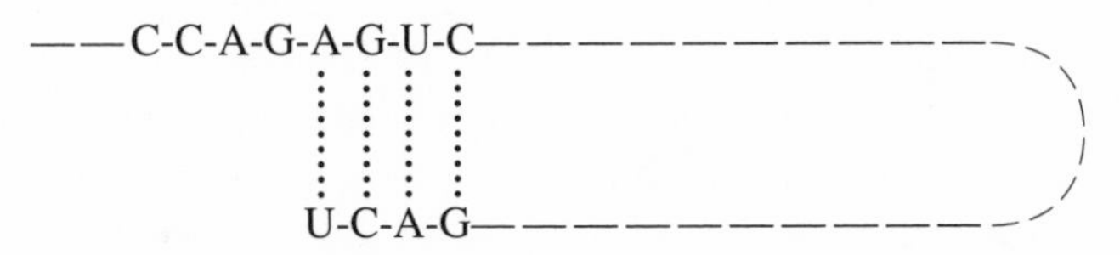

Assume now that this folded chain is subject to elongation, by the addition of new nucleotides, from right to left, to the U end. The presence of G next to the A paired with the terminal U is likely to favor the addition of a complementary C over that of the other three possible nucleotides. Repeat the process and you get U added opposite A, G opposite C, G again opposite the next C, and so on. What you get is the formation of a stretch complementary over all its length to the other end of the molecule:

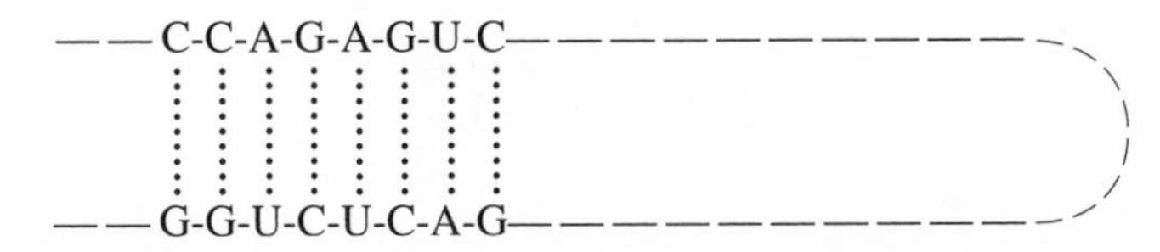

Cut off the loop at the C and G on the right and you get two entirely complementary chains. The same would happen if a short GACU stretch served as "primer" for elongation with a "template" molecule ending with AGUC (without a connecting loop). This, essentially, is what happens in biological RNA (and DNA) replication. In this process, a polynucleotide chain is assembled along an antiparallel, complete chain that serves as template. At each step, the nucleotide capable of base-pairing with the one facing it on the template is selected from the four kinds of available nucleotides. The mechanism is extremely simple. It comes down to choosing, from a total number of four, the piece of a puzzle that fits a model. A child of three could do it. Blind molecular agitation until a chance fit locks the correct molecule into place achieves the same result in nature without the child's observation power. The final outcome is a chain fully complementary with the template.

You might say that this is not replication. However, just repeat the process with the newly formed chain as template and you will obtain a copy of the first template. In other words, RNA (and DNA) replication takes place in the same way as photographic replication: A positive is made from a negative, and a negative from a positive. Double-stranded nucleic acids contain duplex copies of the same molecular (sequence) information, one in positive, the other in negative form. Their replication takes place in a reciprocal manner, with the positive serving as template for the assembly of a new negative, and vice versa, to give two identical duplexes as the final result.

This is the true significance of the Watson-Crick discovery. It is contained in a single sentence at the end of their 1953 note in *Nature:* "It has not escaped our notice that the specific pairing we have postulated immediately suggests a possible copying mechanism for the genetic material."[3] The American science writer Horace Judson has called this sentence "one of the most coy statements in the literature of science."[4] Purportedly written by Crick, the English member of the team, the sentence could also be described as one of the most famous British understatements. Just imagine the inner excitement of the two young men who were no doubt aware that they had just unlocked one of nature's most jealously guarded secrets.

The emergence of RNA was a truly revolutionary innovation in the development of life. By opening the way to molecular copying, it made possible for the first time a mechanism of evolutionary self-improvement by variation, competition, and selection. Henceforth, the historian trying to reconstruct the emergence of life is entitled to look for new explanations besides chemical determinism. A novel phenomenon had arisen to guide events: Darwinian selection.

DARWIN PLAYS WITH MOLECULES

Imagine the setting. The prebiotic soup has "cooked"—via protometabolism—to the point that ATP, GTP, CTP, and UTP (and perhaps other similar molecules) are all being made and combined into a variety of polynucleotide stretches. Among these stretches, some happen to possess appropriate complementary sequences allowing them to double up or join in a manner that permits their replicative elongation. Such favorable sequences will be replicated and will become more abundant than the others. For the first time, a mechanism of *selection by amplification* has arisen, thanks to replication. But this is not all.

RNA replication must have been a fumbling affair at first. Inevitably, many mistakes were made, resulting in the formation of many inexact complementary copies of the template sequences. Since these variants could themselves serve as templates, a further dispersion of the sequence information ensued. Not each variant, however, gave the same number of copies. Because of a particular feature of their sequence, some were replicated faster than others and were advantaged. Stability under the prevailing conditions was another favorable factor. Molecules combining these two assets—replicatability and stability—in optimal fashion produced the most progeny, which, being endowed with the same advantages, progressively crowded out less advantaged sequences. At the end of this process, a single sequence must have come to dominate the whole mixture, however complex the starting situation.

This scenario is not just a theoretical construction. It has been re-enacted many times in the laboratory, first in 1967, by the late American biochemist Sol Spiegelman, from Columbia University, a pioneer in this field.[5] What Spiegelman did was to mix in a test tube RNA extracted from a small virus called Qβ, the RNA-replicating enzyme (replicase) from the same virus, and the four substrates (ATP, GTP, CTP, and UTP) of RNA replication. After a brief incubation, during which RNA replication took place, he used the RNA formed to seed another similar mixture, repeating the procedure a number of times. In the end, the resulting RNA turned out to be very different from the viral RNA added in the beginning. It had been streamlined to retain only those parts of the molecule that were needed to ensure effective interaction with the replicating enzyme and to preserve stability. A different final product was obtained when the experiment was performed in the presence of an inhibitor that changed the conditions required for optimal interaction with the enzyme. Since these historic experiments were performed, they have been variously repeated many times, especially by Orgel[6] and by the German Nobelist Manfred Eigen and his coworkers at the Max Planck Institute for Physical Chemistry in Göttingen, who have also made a detailed theoretical study of the system.[7]

What is observed is no less than authentic Darwinian evolution at the molecular level. There is a gene, an RNA molecule of given sequence, able to undergo replication. This gene is subject to mutations. The mutants compete with each other for available resources—the limited quantity of nucleotides usable for replication. The winners are those that multiply fastest. The important point is that this result is achieved without any design or foresight. The mutations are caused by replication accidents, fortuitous events that bear no relationship to the production of better replicators. This is the essence of Darwin's theory. Natural selection operates blindly on material offered to it by chance.

At the end of the optimization process just described, the system settles into what is known as a steady state, a state of apparent stability in which replication and breakdown compensate each other and in which the optimized sequence maintains its supremacy thanks to continuing selection. Even in the optimized steady state, the RNA molecules are not all identical. Because replication errors continue to occur, the outcome is a continually shifting population of molecules, which Eigen has called a quasispecies. This population consists of perfect copies of the optimal sequence ("master sequence"), accompanied by a covey of variants generated by replication errors. It is believed that, in the origin of life, the master sequence of the first RNA quasispecies characterized the first gene. This was an utterly "selfish" gene—to use an expression coined by the British biologist Richard Dawkins[8]—geared only to its own replication.

Eigen has tried to draw the portrait of this primordial gene on the basis of all the information available from experiment and theory, somewhat in the way police artists draw the pictures of wanted criminals on the basis of testimonies.[9] His "identikit" picture of the first gene shows a surprising similarity to the image that can be deduced from comparative sequencing data for the common ancestor of all extant members of a special class of small RNA molecules, known as transfer RNAs, that play a key role in protein synthesis. Because of the many uncertainties affecting both kinds of reconstruction, no judge would yet pronounce a conviction on the strength of this identification. But it is highly suggestive. The possibility that the primordial gene might have been the ancestor of transfer RNAs carries implications of utmost importance and interest with respect to the manner in which RNA molecules first became involved in peptide assembly.

The first outcome of the kind of molecular selection just described may have been the *selection of the four bases that make up RNA today.* Molecules made exclusively of A, G, U, and C had the advantage that they could be replicated thanks to base pairing. Molecules containing other bases unable to engage in base pairing were weeded out.

THE BIRTH OF PROTEINS

With the appearance of the primordial gene, nascent life, having exhausted the possibilities of molecular improvement by Darwinian selection, must have worked itself into what looked like an evolutionary dead end, the only means of escape being a change of external conditions. Even then, freedom would not have lasted long. The selective pressure would have changed, but soon again the system would have run into another optimization blind alley, adapted to the new conditions. This would indeed have been so but for the occurrence of a new kind of reaction that initiated further progression.

According to the most likely scenario, it all began with the emergence of one or more RNA variants that *interacted with amino acids* in such a way that the amino acid became linked to the ribose end of the RNA molecule. This is how the primordial gene started the long evolutionary journey that eventually gave rise to transfer RNAs, its closest extant descendants according to Eigen's identification. For this is exactly what transfer RNAs do today. They join with amino acids in what happens to be the first step of protein synthesis.

Emerging life did not "know" that this interaction was opening the way to one of its most momentous developments: RNA-dependent protein synthesis. There must have been an immediate advantage whereby RNAs capable of interacting with amino acids were selected for preferential replication. The explanation could be simple. RNAs with amino acids hooked to their tails could have folded into a more compact conformation that protected them against breakdown. Or they could have served more efficiently as templates for replication—not an implausible possibility since the amino acids were attached to the end of the RNA molecules where reading of the template starts. The presence of the amino acid could have ensured that reading started at the proper place, or it could have facilitated the interaction of the template with the catalytic system in some other way. Thus, simple Darwinian selection at the molecular level could have provided the driving force, not only behind RNA evolution, but also behind the involvement of RNA in protein synthesis, one of the most crucial events in the history of life.

Attachment of amino acids to transfer RNAs requires energy, which, in nature, is supplied by ATP. In my model, the energy could have come from thioester bonds if the amino acids had reacted as thioesters. There are other possibilities, including a role for ATP, already present in the system at the time, and even a direct interaction beween RNAs and free amino acids.[10]

An interesting question is whether the interactions were specific. Did a particular kind of RNA join specifically with a particular kind of amino acid? Or did the same RNA react with several different amino acids, or the same amino acid with several different RNAs? In protein synthesis today, the interaction between transfer RNAs is highly specific. However, this specificity is ensured essentially by the enzymes that catalyze the association. These enzymes possess binding sites that specifically recognize a given amino acid and a corresponding transfer RNA, positioning the two molecules in such a way that they become linked together with the help of ATP. There is little evidence that transfer RNAs and amino acids recognize each other directly without the help of the joining enzymes, although this possibility cannot be excluded: Certain direct RNA–amino acid interactions have been observed.[11]

It is tempting to assume that the primitive process exhibited some specificity, either by direct interactions or through the mediation of some catalytic surface. This would explain the puzzling selectivity of protein synthesis, which uses only twenty distinct amino acids and leaves out many others that are available, including some that probably were abundantly present in the primeval soup. Also, there is the intriguing fact that nineteen out of the twenty proteinogenic amino acids—the twentieth, glycine, does not exist in two forms—are "left-handed." Molecular handedness—the technical term is "chirality," from the Greek *cheir,* hand—is a property of pairs of molecules that, like our two hands, are constructed identically except that one appears in space as the mirror image of the other. The two forms are designated D and L, the initials of the Latin words for right (*dexter*) and left (*leavus*). Proteins contain only L-amino acids. This strange preference of nature for left-handed amino acids is viewed by many scientists as one of the most intriguing mysteries surrounding the origin of life. It is conceivable that primitive transfer RNAs were instrumental, by their specificities, in selecting certain amino acids for protein synthesis.[12] Furthermore, it is difficult to explain the origin of translation and of the genetic code without assuming that the primordial couplings between amino acids and RNA molecules enjoyed a certain degree of specificity.

Once amino acid–carrying RNA molecules were roving around in increasing abundance, it is to be expected that they started to interact with each other. This is what happens today to amino acid–carrying transfer RNAs. In a first step, two such molecules confront each other in such a way that the amino acid of one joins with the amino acid of the other to form a dipeptide. Then, through a similar type of interaction, the dipeptide gains another amino acid provided by a transfer RNA and becomes a tripeptide. This act is repeated many times, until a particular polypeptide chain is completed. Proteins are assembled by this mechanism in the whole living world. It seems very likely that assembly of peptides was inaugurated by the primordial amino acid–bearing RNAs and that RNA-dependent protein synthesis was born in this way.

In nature, peptide assembly takes place on ribosomes, which are highly complex, compact particles made of several kinds of RNA molecules (ribosomal RNAs) and more than fifty different proteins. The protein-synthesizing machinery is completed by a thread of messenger RNA, which dictates which of the twenty amino acids is to be inserted at each step. However, this last mechanism need not yet concern us, as the code according to which it operated was not yet developed. We are dealing with an uninstructed form of peptide synthesis.

Even if we leave out the information aspect, the prominent role of RNAs in present-day protein synthesis remains striking. Impressed with this fact, Crick suggested in 1968 that the first protein-assembly machinery might have consisted exclusively of RNA molecules, without proteins.[13] This was not an unreasonable proposal since proteins could hardly have been available initially for making the machinery that was going to make them. The later discovery of catalytic RNAs gave a great boost to Crick's suggestion, which has become one of the main props of the RNA-world model. Even though the thioester-world model allows for the intervention of catalytic multimers in primordial peptide assembly, we cannot ignore the eloquent message from nature. It seems very likely that RNAs ancestral to ribosomal RNAs and, perhaps, to messenger RNAs were involved as structural and catalytic components of the primitive peptide-assembly machinery. The finding by the American investigator Harry Noller, of the University of California at Santa Cruz, that the ribosomal catalyst responsible for sealing peptide bonds may itself be of RNA nature, provides strong additional support to this hypothesis.[14]

What is not clear, however, is how RNAs catalytically active in peptide assembly came to be selected. Perhaps involvement in this process somehow favored the replicatability or stability of the molecules. But this explanation is not very convincing. In any case, there was soon to be a need for a new selection mechanism based on the usefulness of the assembled peptides. This will be the subject of the next chapter.

Chapter 6

The Code

WE HAVE REACHED a point in our hypothetical reconstruction of the age of information where the first peptides began to be assembled by an RNA machinery. We know what came next: translation and the genetic code. Two questions challenge the historian. First, by what succession of steps did translation and the genetic code arise? Second, what was the driving force that propelled such an extraordinary development? The two questions are intimately related, since no pathway can be considered that does not entail an explanation of its spontaneous emergence. Before we try to answer these questions, a new element needs to be introduced, namely, the concept of a primeval cell.

DARWIN NEEDS CELLS

The cell is the unit of life and figures at some stage in all attempted reconstructions of the origin of life. Some scenarios bring in cells early or, even, right at the start. Others begin with an unstructured soup and introduce cellularization later, sometimes postponing it to the last moment before it became indispensable for further progress. For reasons that will be explained in chapter 9, I have adopted the latter course. But a limit has been reached.

With the initiation of RNA-dependent peptide synthesis, if not before, emerging life had virtually exhausted the potential of molecular evolution. For further evolution to take place, less selfish criteria for selection—or, better said, less crudely selfish criteria—had to come into play. RNA molecules no longer had to be assessed solely on the strength of their intrinsic ability to survive and be replicated, but on the basis of their ability to *do something* that favored their survival and replication indirectly. But for this kind of selection to operate, the biogenic system needed to be parceled out into a number of discrete, semiautonomous, self-reproducing units—let us call them protocells—each containing its individual genome. Then, any useful mutation would benefit only the protocell in which it occurred, causing this protocell to reproduce faster than the others, together with its improved genome, and to progressively squeeze out the others by the ever-augmenting throngs of its similarly advantaged progeny.

In order not to break the thread of my narrative, I shall examine the mechanisms that led to the appearance of the first protocells in a subsequent chapter (chapter 9). I shall assume, for the time being, that cellularization has occurred and that the events we are about to consider took place in a population of protocells capable of individual growth and of reproducing by division.

Once protocells existed, selection could proceed on a wider basis and favor any replicatable RNA that, one way or the other, enhanced the capacity of its protocellular proprietor to grow and produce progeny. It is at this stage, and not earlier, that catalytic RNAs could have been retained and improved by selection, to the extent that their activities happened to be useful to the protocells concerned. In particular, there would have been a strong selective pressure in favor of more efficient parts of the peptide-synthesizing machinery if, as appears likely, the ability to make peptides was an asset in itself.

A considerable additional advantage would have been provided by anything that favored the making of useful peptides, as against useless or harmful ones. But for this to happen a *feedback loop* was needed whereby the useful peptides selectively promoted their own replication.[1] Direct copying of the useful peptides could conceivably have done the job, but there are good reasons for rejecting this possibility. Perhaps the best reason for doing so, unless evidence to the contrary becomes available, is that protein copying does not solve anything. We are still left with the problem of explaining translation.

ANATOMY OF TRANSLATION

Our clue, once again, comes straight out of present-day life. The protein-synthesizing machinery consists of several parts. First, there is the ribosome, which is the catalytic assembly bench. It is a small, dense particle, one-millionth of an inch in size, constructed of a large and a small subunit, each made of RNA and protein components in roughly equal proportion by weight. The ribosome automatically joins an amino acid to a growing peptide chain (a single amino acid to start with) when appropriately presented with these two molecules. In terms of information, the ribosome is illiterate. It acts blindly and links any two partners that are in the right chemical conformation and are properly aligned in regard to its catalytic center.

A second part of the machinery is messenger RNA. It runs like a tape between the two ribosome subunits, which it helps to keep together, and it provides the information that specifies the order in which amino acids are to be assembled in the course of peptide synthesis. In this translation from nucleic-acid language to protein language, the nucleotide sequence of the messenger RNA stipulates the amino-acid sequence of the corresponding peptide. This takes place in simple colinear fashion: Each consecutive triplet of bases (codons) in the messenger RNA specifies in the same order an amino acid of the polypeptide chain. Of the sixty-four different codons that can be constructed from the four different bases, sixty-one specify one of the twenty amino acids out of which proteins are constructed, and three are stop codons signifying the end of assembly. A special amino-acid codon doubles as the starting codon. These one-to-one relationships between a triplet of bases and an amino acid make up the genetic code. With minor exceptions, the same code is obeyed throughout the living world. It is a universal dictionary.

With its associated messenger RNA, the ribosome turns into an assembly bench that makes a *single type of polypeptide.* As a rule, ten or more ribosomes, busily reading the message and making the corresponding polypeptide chain, follow each other on the same messenger-RNA strand. Such strings are called polysomes. Tens of thousands of polysomes, assembling thousands of different proteins, are present at any given time in any given cell.

Amino acids are conveyed to these machineries by transfer RNAs. Properly aligned on the ribosome surface, two amino acid–carrying transfer RNAs interact so that one donates its amino acid to the amino acid attached to the other transfer RNA, which now bears a dipeptide. Subsequent lengthening of the chain takes place by what the late German-born American biochemist Fritz Lipmann has called "head growth."[2] At each step, the entire growing chain is transferred to the next amino acid brought in by its transfer-RNA carrier (see figure 6.1). Imagine a train being assembled, not by hooking on one car after another to the tail of the train, but by moving the whole lengthening train each time to attach it to the next head car, ending with the locomotive. In this way, the lengthening peptide chain continually remains attached to a transfer RNA by means of its most recently acquired amino acid until synthesis is completed, at which time a stop codon signals the detachment of the finished polypeptide chain from its final carrier.

The sites on the ribosome by which amino acid–carrying transfer RNAs are bound recognize features shared by all transfer RNAs, whereas the catalytic center joins chemical groups common to all amino acids. Discrimination is carried out entirely by the messenger RNA. In performing this function, the messenger RNA does not "see" the amino acids. It sees only the transfer RNAs. More precisely, it sees only a small part of that RNA, a triplet of bases, or anticodon, complementary to the codon. The location of the anticodon in the transfer-RNA molecule is such that it is correctly aligned in antiparallel fashion along the codon exposed by the messenger RNA when the transfer-RNA molecule occupies a binding site on a ribosome. The association between codons and anticodons depends on base pairing, with some leeway, or "wobble," at the third base of the codon, which allows a given anticodon sometimes to interact with more than one codon. (There are about forty transfer RNAs for the sixty-one amino-acid codons.) At each step of the assembly process, the ribosome and messenger RNA together shape a newly

FIGURE 6.1

The Main Steps in Protein Synthesis

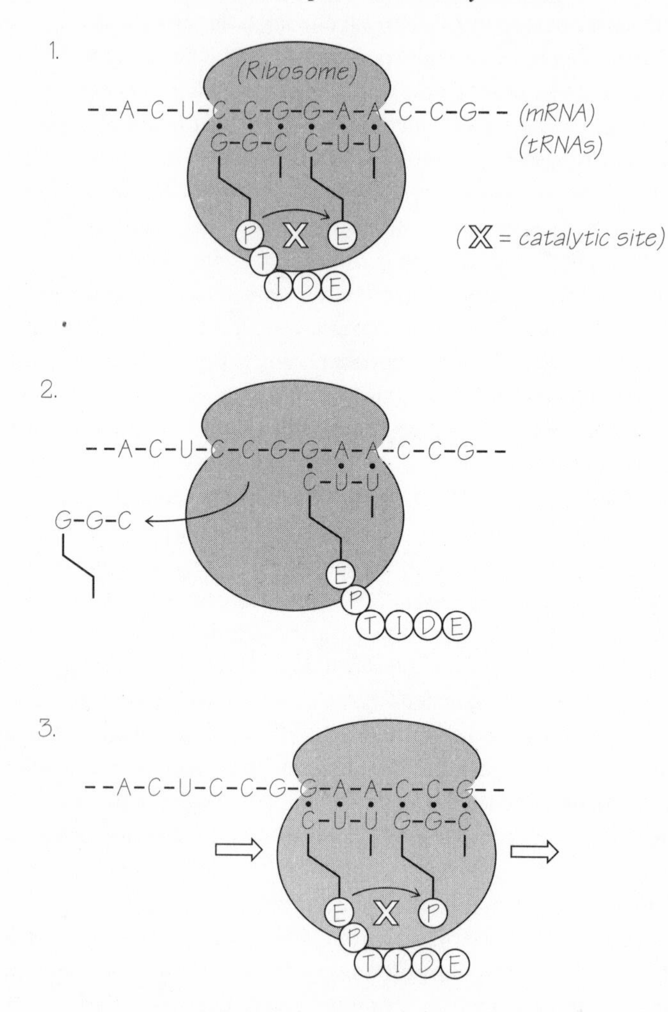

1. A transfer RNA (tRNA) bearing amino acid **E** is aligned on the ribosome next to a transfer RNA bearing the growing peptide.

2. The growing peptide is transferred from its transfer RNA (which detaches from the ribosome) onto amino acid **E** presented by the neighbor transfer RNA.

3. The ribosome has shifted along the messenger RNA (mRNA), and the next amino acid (**P**) is presented by a transfer RNA with a complementary anticodon.

Note that the peptide elongates by head growth.

opened site, within which only one among the forty-odd transfer RNAs involved in protein synthesis can fit correctly with its attached amino acid. This is how the message is read. It is the puzzle game all over again, but with forty pieces instead of the four used in RNA replication. It may take a child of five to do it successfully.

It is a characteristic of this process that the reading is done *entirely in RNA language* by means of base pairing between codons and anticodons. The translation step proper is carried out before assembly, by the enzymes that *attach amino acids to transfer RNAs.* These enzymes recognize both an amino acid and a corresponding transfer RNA. They are the only parts of the translation machinery that understand both "proteinese" and "RNAese," albeit only a single word of each language per enzyme. If a mistake is made by one of these enzymes and it attaches the wrong amino acid to a given transfer RNA, the assembly machinery has no way of detecting the mistake. It slavishly obeys the instruction provided by the anticodon of that wrongly loaded transfer RNA and adds the wrong amino acid to the growing chain.

Astonishingly, only about half of the "bilingual" enzymes that attach the appropriate amino acids to the appropriate transfer RNAs recognize the anticodons on these transfer RNAs.[3] The other half of the set of enzymes involved in translation recognize particular transfer-RNA features other than the anticodons and are unaffected by changes in the anticodon or even by its complete removal. They are thus bilingual by proxy, so to speak—their RNA specificity does not come straight out of the genetic dictionary. It relates instead to structural elements on the transfer RNAs distinct from the anticodon and sometimes separated from it by a considerable distance. This is a very puzzling fact, as it adds another link in the information chain and therefore increases the chance of error. It is difficult to understand why evolution would select such an unnecessary weak link, which seems likely to be a vestige of an earlier relationship that evolution failed to erase. I shall come back to the possible nature of this relationship. But first let us look at the development of translation itself.

THE ORIGIN OF TRANSLATION

To understand how a machinery as intricate as that involved in protein synthesis could ever have come into being, let us imagine a situation where no code existed and peptides were assembled in random fashion. You might expect that messenger RNAs, being meaningless, would not be needed in such a situation. This would not be so. Even in the present-day machinery, messenger RNAs, or their equivalent, would still be required because they play a *conformational* role in addition to their informational role. They help lock the two loaded transfer RNAs together with the two ribosomal subunits in the conformation required to transfer the growing peptide chain to the next amino acid.

This, I submit, explains the entry of messenger RNA on the peptide-synthesizing scene. Its precursor was part of the original catalytic RNA scaffolding on which the first peptides were assembled. In this scaffolding, the primitive RNA destined to become ribosomal RNAs provided the catalytic part, and the primitive RNA that evolved into messenger RNAs ensured that the amino acid–carrying and peptide-carrying RNAs were properly positioned, accomplishing this by the same kind of triplet interaction that now exists between codons and anticodons. For simplicity's sake, I shall use the terms "codon" and "anticodon" to designate these triplets, and I shall refer to the three RNA species involved by the names of their present-day descendants: ribosomal, messenger, and transfer RNAs.

Together, these three types of RNA made the first peptides, assembling amino acids in an order that, if not entirely random, was far from being as strictly imposed as it is with today's set of RNAs. Some of the peptides made in this way were useful. Therefore, *the mere ability to make peptides was an asset* to the protocells concerned, and any mutation of the RNAs that enhanced this ability conferred a selective advantage on the protocells. Thus, the various RNAs were jointly subject to Darwinian evolution and selection, with the efficiency of peptide synthesis as the screening criterion. The machinery might improve in this way, though not yet its products. There was no way for a good product to feed back positively on its production.

The seeds of such a feedback did, however, exist in the fact that codon–anticodon interactions were involved in the mechanism whereby messenger RNAs helped immobilize loaded transfer RNAs in the appropriate orientation for peptide synthesis. Messenger RNAs thus, from the start, *made a selection among the transfer RNAs at each step,* the degree of specificity of this selection depending on the number of different transfer-RNA molecules that carried the same anticodon. Such being the case, we may expect natural selection to weed out ambiguities and lead to a situation where each kind of amino acid becomes attached to a transfer RNA bearing a specific anticodon.

Imagine, for example, a case where two transfer RNAs carrying the amino acids glycine and alanine, respectively, have the same anticodon: GGC, that is, the sequence guanine-guanine-cytosine. Wherever the codon GCC (complementary to GGC in antiparallel orientation) appears in the messenger RNA, either glycine or alanine will be inserted in the forming peptide, with chance deciding between the two. Now let a chance mutation in the transfer RNA for alanine change the central G of the anticodon to C. Only glycine is now inserted in regard to codon GCC, whereas alanine is now called for by codon GGC. The system has gained in specificity. Let one or more of the new peptides turn out to be useful and the protocells that possess the mutated transfer RNA will enjoy a selective advantage and produce more progeny possessing these more specific transfer RNAs.

Should the same mutation affect the transfer RNA for glycine instead of the transfer RNA for alanine, a similar gain in specificity will be achieved, but the positions of the two amino acids in the peptides will be reversed. The protocells concerned may also be advantaged. The final outcome will depend on which of the two sets of peptides confers the greatest evolutionary advantage. According to the genetic code, the first set was better. GCC happens to be a codon for glycine, and GGC a codon for alanine.

This kind of scenario could be repeated with other amino acids and other transfer RNAs. Eventually, all twenty proteinogenic amino acids would progressively be pulled into the system. Evolutionary step by evolutionary step, *translation and the genetic code would emerge concomitantly as products of natural selection.* The proposed mechanism requires that each transfer RNA be specific for a particular amino acid. This is consistent with the view that transfer RNAs "fished" out the proteinogenic amino acids. Even if the original specificities had been relatively loose, natural selection would progressively make them stricter.

An interesting feature of the proposed model is that natural selection started screening peptides almost as soon as a primitive peptide-synthesizing machinery was set in place. The mutations that played an important role in the beginning were those that *affected the anticodons of transfer RNAs* and thereby changed the sequences of *all* the peptides containing the amino acid carried by the mutated transfer RNA. Whole sets of peptides with given positions occupied by either one or another amino acid were subjected to screening by natural selection. Later, as translation and an unambiguous code progressively emerged, mutations of the transfer RNAs became almost invariably lethal because the consequences were too widespread to be tolerable. Among the many peptides that were altered simultaneously, some were almost bound to become defective in the process. The motor of evolution shifted to *mutations in messenger RNAs.* These mutations led to the synthesis of only one kind of altered peptide, which could then be evaluated on the basis of its usefulness. Most often, the altered peptide will have been inferior to its nonmutated predecessor and the affected protocells eliminated in the competition. Occasionally, the mutation would bring improvement and give the affected protocells a selective advantage. This mechanism (with DNA eventually replacing RNA as the mutable storage form of the information) has become the central driving force of evolution.

There is a third possibility. The altered peptide was neither better nor worse than its predecessor. The mutation was neutral and carried along passively by what is called genetic drift. These are the mutations that allow us to reconstruct the tree of life by comparative sequencing.

STRUCTURE OF THE CODE

A central issue, still unresolved, is whether the structure of the genetic code is a product of chance or was imposed by deterministic factors. Put otherwise, if living organisms similar to those on the Earth existed elsewhere, would they use our code or another one?

The answer to this question would be straightforward if there happened to exist a direct structural correspondence between amino acids and their anticodons, that is, if the first transfer RNAs had actually fished out amino acids using their anticodons as hooks. The code would then be strictly deterministic. Many efforts have been made to uncover such relationships, almost invariably in vain. Although not entirely hopeless, the prospects of this line of research do not look encouraging.

The fact remains that primitive RNAs and amino acids must have "seen" something in each other. Why would they have come together otherwise? Furthermore, that something must have been different for each transfer RNA–amino acid combination, to account for the specificity without which the emergence of translation seems difficult to explain.[4] It is an intriguing possibility, so far unconfirmed, that the transfer-RNA features involved in these early recognitions may be related to the structural traits recognized today by those enzymes that ignore anticodons in the selection of the transfer RNAs to which they bind amino acids. This would explain the evolutionary retention of such features in some cases, and in other cases, their erasure and replacement by anticodons as the recognized features.

Be that as it may, the necessary recognition between primitive transfer RNAs and amino acids implies that the starting situation was far from random and could well be the same should the process be reproduced elsewhere. Further evolution through mutations affecting the anticodons would have to operate within those constraints. Perhaps amino acids would end up coded by the anticodons that specify them in present-day life, but this is far from certain.

Another important element is the historical factor. It is likely that the twenty proteinogenic amino acids were not all available initially. The code, therefore, must have started with a smaller number of amino acids—estimates vary between four and eight—and must have emerged progressively as more amino acids entered the circuit. Several different models have been proposed to account for this emergence.

All the models have in common that they impose limits on the kinds of anticodons likely to represent given amino acids. To take a simple example, the German chemist Manfred Eigen[5] has speculated, for reasons that we need not go into, that primitive transfer RNAs might be made of repetitive GXC triplets, with X being any one of the four available bases: G, C, A, or U. There is room for four anticodons in such a structure—GGC, GCC, GAC, and GUC—which, in antiparallel orientation, correspond to codons GCC, GGC, GUC, and GAC, respectively. In the present-day world, these codons specify the amino acids alanine, glycine, valine, and aspartic acid, respectively. These amino acids happen to be the most abundant among the proteinogenic amino acids formed in Miller's simulation experiments or found in meteorites. It is difficult to see this as mere coincidence.

Whether Eigen's scenario or another is the correct one, the important point is that chance and selection had to work within a severely constrained historical context. Amino acids were encoded in the order in which they became available for peptide synthesis, whereas the codons themselves were assigned in an order that most likely was not random but imposed by molecular exigencies of the RNAs involved. In other words, codons were distributed among amino acids—or amino acids among codons—on a mutual, "first come, first served" basis. It is impossible to estimate the stringency of these constraints, but their existence makes it very probable that the structure of the code is not purely accidental, as is sometimes claimed.

There is another aspect to the code that suggests a nonrandom origin. Its structure is remarkably regular. Codons coding for the same amino acid or for amino acids with similar properties are grouped together in such a way that the harmful consequences of mutations (due to chance replacement of one base by another in the triplet) are minimized. In many instances, the altered codon calls for the same amino acid or for an amino acid sufficiently similar to the original one for the properties of the altered peptide not to be significantly modified. This regularity suggests that the code was molded by natural selection during the long period when protocells were experimenting with different codon assignments and vying with each other for leadership in the evolutionary race.

In conclusion, it is not certain that aliens would understand our genetic language, but the odds that they would do so are far from negligible. To be true, evolution has played a few tricks with the genetic code since the code was first established, for example, in mitochondria, a characteristic part of eukaryotic cells. This, however, was a very late event that happened when less than a dozen genes were left in the mitochondria to be affected by the change. It tells nothing of the historical constraints that shaped the code in the course of early evolution.

METABOLISM REPLACES PROTOMETABOLISM

The development of translation and the genetic code only opened the way out of the RNA world. There followed a long period during which protocells progressively acquired new peptides. Let us imagine how this happened. Invariably, the first initiating step was the occurrence of a chance mutation in some RNA molecule. Remember, in the RNA world, RNA molecules served both as replicatable genes and as translatable messenger RNAs. Therefore, the mutation was hereditary and expressed as a new peptide. If this peptide gave the protocell in which the mutation occurred an advantage over the other protocells in the Darwinian "struggle for life," the protocell and its progeny would multiply faster than the others and progressively take over. The same series of events must have taken place hundreds of times in succession before a protocell emerged that was fully able to survive and multiply with the help of its newly acquired peptide armamentarium. Only then could this protocell population dispense entirely with whatever supported its ancestors in the RNA world.

What properties of the new peptides could have been useful to the protocells so as to induce their selection? In the majority of cases, catalytic properties must have been the main assets that singled out peptides for selection. These assets had to be appraised within the framework of existing protometabolism. A catalytic peptide, even of exceptional activity and specificity, that did not find a substrate to act on or that gave rise to an unusable product would have been no good to the protocell in which it arose and would not have been retained by selection. In contrast, a catalyst that fitted within the scheme of things would have been good material for selection, especially if it did a better job than the existing catalyst or if it extended the network of metabolic pathways into new directions that led to selective advantage.

This brings us to the important point emphasized in part I, namely, the need for congruence between early protometabolism and present-day metabolism. The network of chemical intermediates participating in protometabolism served as a powerful screen for the selection of the appropriate enzymes among the offered peptides. Protometabolism could gradually evolve into metabolism, and multimers give place to enzymes, without transgressing the sacrosanct Central Dogma. There was no need for peptide replication or for reverse translation of primitive peptides into the corresponding RNAs. All the required information was present in the metabolic network. Metabolic superhighways were not constructed independently of existing country roads, but came about by widening and resurfacing those country roads slowly over time.

Concurrently with the stepwise development of translation and the genetic code, enzymes manufactured by the RNA machinery, or rather their peptide precursors, progressively took over the jobs previously carried out by primeval catalysts. The transition was gradual, as it took a long time for translation to reach a stage where peptides were made reproducibly from well-defined, replicatable RNA blueprints. It also took a long time for protocells to acquire hundreds of enzymes, one by one, by mutation-selection. Protometabolism progressively gave way to metabolism during that time, but it could give over completely only after the last essential enzyme had been put into place.

During this transition, my hypothetical multimers, if they ever existed, became increasingly dispensable. However, the capacity to make multimers from thioesters may not necessarily have disappeared. That capacity could have been retained and perfected by mutation-selection if some multimers happened to carry out useful functions that none of the new peptides could accomplish. The synthesis of gramicidin S and other odd peptides from thioesters by certain bacteria could be a heritage from the ancient ancestral mechanism. It could, however, also be a recent evolutionary reinvention. Thioesters play so many important roles in all living organisms that their use for peptide synthesis could easily have emerged more than once.

Chapter 7

Genes in the Making

BETWEEN THE APPEARANCE of the first peptide, haphazardly assembled by interacting RNA molecules, and the inauguration of a fully integrated translation apparatus, complete with an unambiguous genetic code and a reliable set of functional RNAs and enzymes for enforcing the code, emerging life went through a long succession of tiny evolutionary jumps separated by more or less extended periods of random groping. An image that comes to mind is that of a surface of water slowly spreading over an irregular terrain. Fingerlets extend here and there, as local attractive forces battle with surface tension, until a minor breakthrough suddenly occurs in a given direction and all the pressure momentarily concentrates on a single rivulet. After that, groping soon resumes, sending out feelers until the next breakthrough.

Evolution did its groping by means of chance mutations resulting in the synthesis of altered peptide molecules; it achieved its directional jumps through the occasional emergence of an altered peptide product that conferred a selective advantage on the protocell concerned. As with spreading water, the outcome of such a process depends on the structure of the terrain. Without better knowledge of the prebiotic terrain, we cannot reconstruct this phase of evolution in any detail, but we can guess its eventual outcome with some measure of confidence. By the time this phase came to an end, most, if not all, of the twenty proteinogenic amino acids had been recruited for peptide synthesis, the genetic code had reached its present structure, except for possible minor adjustments, and translation of RNA messages into peptides was essentially unequivocal and reliable. What, then, were the next steps?

THE MODULAR GAME

Most likely, at this stage, genes still consisted of RNA. Those early RNA genes were short, no more than seventy to one hundred nucleotides long (the length of present-day transfer RNAs). This estimate follows from the rule, established by Eigen,[1] that the number of unit building blocks in a replicatable macromolecule cannot exceed the inverse of the error rate of the replication process. Otherwise, the information content of the molecule becomes irretrievably lost upon repeated replication. It is estimated that crude RNA replication had an error rate of one nucleotide wrongly inserted for every seventy to one hundred nucleotides added, depending on the base composition of the RNA molecule. Hence the estimate of seventy to one hundred nucleotides for the maximum length of the first genes.

It follows that the peptide products of the first genes cannot have been more than about twenty to thirty amino acids long—one amino acid for every nucleotide triplet—allowing for some noncoding parts in the genes. These peptides were retained by natural selection. Therefore, they had some useful function, most often a catalytic one. This tells us two things. First, peptides that short *can* display enzyme-like catalytic activities—a point of importance with respect to the multimers of my model. Second, enzymes *did* start as relatively short peptides.

This fact counters the argument, often proffered by creationists, allegedly proving that life cannot have originated by a natural process. Consider, it is said, a protein, such as cytochrome *c,* made of one hundred amino acids. Imagine that at each step in the synthesis of this protein the amino acid to be added is decided by the throw of a twenty-faceted die (one facet for each of the twenty proteinogenic amino acids). The chance that the right amino acid will be selected is one in twenty at each step. For the whole sequence of one hundred amino acids, the probability that the correct sequence will turn up is one in 20^{100}, or one in 10^{130}—for all practical purposes, zero. And cytochrome *c* is one of the shortest of several thousand proteins present in any cell. Hence the conclusion that life cannot have arisen by a natural process.

Let us, however, repeat the calculation for a peptide of twenty amino acids and let us assume further that only eight different amino-acid species are available for its formation, as might have been the case at an early stage of evolution. Each distinct possible sequence then has a probability of one in 8^{20}, or one in 10^{18}. It would take only one billion billion protocells, which could comfortably fit within a small pond if they were the size of bacteria, to try out all possible sequences. Even if all twenty proteinogenic amino acids were used, the number of protocells needed for a complete survey, which is on the order of 10^{26}, could still be accommodated in a small lake. In other words, if proteins started as short peptides, emerging life could have explored the totality of what is called the sequence space, leaving nothing to chance.

The reader may have detected a flaw in this reasoning. Granted that the number of possibilities stays within manageable limits for the first twenty amino acids, another eighty amino acids remain to be added in order to arrive at cytochrome *c.* Thus, the original objection still appears to be valid. Cytochrome *c* cannot have arisen by chance.

This would be true if the next eighty amino acids were added one by one. But they were not. The next phase in the evolution of proteins very likely took place by a combinatorial game using existing peptides (by way of their genes) as *modular construction blocks.* This fact completely changes the probabilistic outlook. Assume, for instance, that a set of one thousand peptides of twenty amino acids was selected during the first phase of protein evolution. With these peptides as building blocks, one can construct $1{,}000^2$, or one million, different peptides of forty amino acids. All possible combinations are readily tried and submitted to natural selection. Let something like one thousand such peptides eventually emerge and it will again be possible for all combinations of sixty and eighty amino acids to be screened. It thus appears that the whole gamut of present-day proteins could have been created through an exhaustive exploration of the sequence space, provided the expansion of this space by the lengthening of sequences was appropriately pruned by natural selection.

The importance of the historical factor in such a process must be underscored. At each stage, *evolution can work only with the materials that have survived from earlier trials.* Even if a previously rejected combination should turn out to be highly desirable at a later stage, there would be no way of retrieving it, except by chance mutations of existing combinations. The historical dimension of the evolutionary process is of very general significance. We shall encounter it many times in subsequent chapters. As evolution proceeds in a given direction, the range of available choices narrows, and its commitment becomes increasingly focused and irreversible.

Considerable evidence of modular construction exists in present-day proteins. The number of modular units out of which all existing proteins have been constructed has even been estimated. The indications, although still subject to considerable uncertainty, are that this number could be of the order of only a few thousand. If confirmed, such a figure would be extraordinarily suggestive: the whole variety of life created from permutations of a few thousand building blocks! I shall have more to say on this topic in chapter 24.

RNA SPLICING

The modular game was not played by peptides but by the genes that coded for them, most likely RNA genes. A new catalytic armamentarium was needed for this. The most important catalyst required was one that would join, or splice, two separate RNA chains into a single one. In modern jargon, this kind of activity is called *trans* splicing (*trans* means "on the other side of" in Latin). By itself, this activity often failed to create coherent messages because the coding parts of the spliced RNAs were not joined in phase—that is, without codon disruption—to allow continuous reading, or were separated by a noncoding stretch. In order to correct such defects, a different kind of activity was needed that would cut out a piece between two messages and splice them together again, in phase. This kind of splicing within a single molecule is called *cis* splicing (from the Latin word for "on the same side of"). Finally, proper insertion of the messenger-RNA molecule into the translation machinery may have required some trimming at the end of the molecule (the end of the tape must be cut off for proper fit).

All three of these processes occur in many extant organisms, though no longer in the course of RNA gene assembly—RNA genes were phased out long before the appearance of the common ancestor of all present life on Earth—but rather as a means of unscrambling at the RNA level a mysterious phenomenon of gene fragmentation at the DNA level. I shall discuss this phenomenon further in chapter 24. Let it simply be stated that many genes, especially in higher eukaryotes, are split into segments that are expressed, and therefore called exons, and into intervening segments, or introns, that are not. These split DNA sequences are transcribed in their entirety, and the resulting RNA molecules are subsequently processed in such a way that the introns are removed and the exons spliced together. Sometimes completed by end trimming, this processing gives rise to mature RNA molecules. These then either become part of some machinery, such as the protein-synthesizing system, or, more frequently, serve as messenger RNAs and are translated into proteins.

Split genes are virtually absent in bacteria; they are scarce in lower eukaryotes and more abundant in higher eukaryotes, where their frequency tends to increase with evolutionary advancement. Post-transcriptional RNA processing thus has the appearance of a late evolutionary acquisition. Whether this is so or not—we shall see later that this question is disputed—the processing enzymes could be very ancient heirlooms that go back to the RNA world. It is remarkable that all three of the activities involved—*trans* splicing, *cis* splicing, and end trimming—can be catalyzed by special RNA molecules without the assistance of any protein. Catalytic RNAs, or ribozymes, were discovered by the study of these processes. Proteins are invariably involved as well, but the fact that they are dispensable is viewed as highly significant. It suggests that the relevant activities were originally carried out by ribozymes alone. Thus, next to the translation machinery, a second major catalytic system arose from interactions among RNA molecules. This fact has supplied another powerful argument in support of the RNA-world model.

The interplay among RNA molecules that led to the emergence of RNA splicing presumably went through the usual combination of random mutation and then screening by natural selection. It is likely that protocells equipped with splicing ability gained a selective advantage from some of the longer peptides that were formed through the translation of spliced RNA genes. Replication of the spliced genes must have run into a problem, however, because their length exceeded the limit of seventy to one hundred nucleotides imposed by the error rate of replication. The solution to this problem was the development of more accurate replicating enzymes, which were a prize catch for selection to net. Until this happened, replication had to go on using the shorter genes as templates, which implies that protocells continued to rely on splicing to retain the longer, useful peptide products. Thus, any improvement in the specificity, accuracy, and reproducibility of RNA splicing was an advantage. This splicing process still plays a major role today, especially in higher eukaryotes, even though, with the advent of DNA, it is no longer the main mechanism generating variations for the evolutionary combinatorial game.

THE ADVENT OF DNA

As their genetic diversity and sophistication increased, protocells must have faced growing logistic problems. Picture the two complementary forms of hundreds of RNA "minigenes" and their splicing products vying for base pairing, replication, splicing, and translation, and you readily visualize the inextricable tangle in which protocells became increasingly snarled the further they progressed. There was only one way out: division of labor. Replication had to be separated from translation. DNA had to emerge. Nobody knows when this crucial development took place, but it seems likely that it happened at a time when the formation of larger RNA genes was already well advanced.

Chemically, DNA is a chainlike macromolecule very similar to RNA. It is likewise made of a large number of nucleotides chosen from four distinct species. There are two differences. The sugar ribose is replaced by deoxyribose, which is ribose from which an oxygen atom has been removed—hence the prefix "deoxy" and the name "deoxyribonucleic acid," DNA for short. The second difference concerns one of the four bases, uracil, which is replaced in DNA by thymine, which is uracil to which a methyl group (CH_3) has been added. This modification does not affect base pairing, so that the pair AT in DNA is equivalent to the pair AU in RNA. The pair GC is the same in both types of molecules.

Only minor metabolic innovations were required for the building blocks of DNA assembly, dATP, dGTP, dCTP, and dTTP—d stands for "deoxy"—to become available. When these molecules appeared, three key reactions became possible, all ruled by base pairing, as in RNA replication.

First came reverse transcription, the assembly of DNA on an RNA template. It is called "reverse" because it was discovered after transcription, the assembly of RNA on a DNA template, which is the main reaction linking these two information-carrying molecules in the contemporary world. Historically, however, reverse transcription most likely emerged first. It played a crucial role by allowing information stored in RNA molecules to be transferred to DNA molecules.

Storage without the possibility of retrieval would have been useless. Hence the need for transcription. The stored information could thereby be recovered in a form suitable for translation. This back-and-forth movement of information between a form (DNA) that is unavailable to the translation machinery and one (RNA) that is available to the machinery provided a valuable way of regulating the expression of genetic information.

Finally, DNA replication, or the assembly of new DNA on an existing DNA template, completed the installation of the new genetic machinery, by entirely dissociating the replication of information from its expression.

It is likely that these three functions were all performed at first by the same enzyme, which was also the catalyst responsible for RNA replication. The substrates and the templates involved in the four types of reactions were sufficiently similar for a crude catalyst not to discriminate efficiently among them. However, as soon as the use of DNA as a storage form of genetic information provided some selective advantage, evolution took over in its usual way, putting mutations to a test and letting natural selection retain whatever happened to be useful. Increasingly, specific enzymes thus arose from mutations of the original gene coding for the multifunctional catalyst of nucleic-acid assembly. Eventually, four distinct enzymes, each specific for a single type of reaction, emerged. They are, in current terminology: RNA replicase, DNA replicase (more commonly called DNA polymerase), transcriptase, and reverse transcriptase. (The suffix "ase" denotes an enzyme.)

Once the DNA system was firmly established, the two enzymes using RNA templates became useless, even harmful, as they could only confuse matters. It was more advantageous for the protocells to have an unambiguous chain of command, from DNA to RNA to protein, and to restrict replication to DNA. There was thus a considerable evolutionary pressure in favor of eradication of the RNA replicase and reverse transcriptase genes. These have, indeed, largely disappeared from the living world, except in certain viruses.

Viruses are infectious agents that can be reproduced only with the help of the chemical machinery of a living cell. Polio, rabies, smallpox, measles are caused by viruses that reproduce in animal or human cells. Viruses that infect plant cells, protists, or bacteria also exist. All viruses have in common a genome carrying their blueprint and the means to introduce this genome into a cell in a manner that allows reproduction of the virus by the cell. Some viruses have DNA genomes like the rest of the living world, but others have RNA genomes. RNA viruses come in two types.

In one class (for example, polio), the viral RNA is reproduced by direct replication, with the help of an RNA replicase encoded by a viral (RNA) gene. The viral RNA (or its complementary replica) also acts as messenger RNA in the expression of the viral genome.

When RNA viruses of the second type infect a cell, the RNA is first subjected to reverse transcription to DNA, with the help of a reverse transcriptase encoded by a viral gene. Transcription of the DNA then serves in expression of the viral genome and in its replication. Such viruses are called retroviruses. They include a number of cancer-causing viruses, as well as the dreaded human immunodeficiency virus (HIV), the causal agent of AIDS, or acquired immunodeficiency syndrome, the plague of the modern world.

It has been suggested that viruses are descendants of early forms of life that preceded cells. This cannot be so, however, since viruses cannot multiply without cells. They are viewed today as information-carrying remnants or fragments of cells, reduced to the bare minimum required for perpetuation with the help of other cells. Viruses are gypsy genes let loose from their original residences, equipped for wandering from cell to cell, and capable of refreshing and replenishing their stock at each passage. It is possible that some viruses started their wandering at a very early stage in the development of life. The RNA viruses, in particular, could go back to the days when protocells were getting rid of RNA replicase and reverse transcriptase. The viruses could thereby have saved these enzymes from total eradication. An alternative possibility is that these enzymes were "reinvented" at a later stage, for example, by mutation of some DNA replicase or transcriptase gene. Comparative molecular etymology may someday give us the answer to this intriguing question.

GENETIC ORGANIZATION

With DNA in charge, many important improvements to genetic organization became possible. First, genes could be stored in single copies or in the minimum number of copies needed to satisfy the requirements of growth. The need for multiple copies existed particularly for genes coding for structural or functional RNA molecules, such as ribosomal or transfer RNAs. In contrast, messenger RNAs could be generated from single copies of DNA, since translation provided an adequate means of further amplification.

As a second advantage, all the genes could now be kept as stable, double-stranded threads, from which one or the other strand, occasionally both, was selected for transcription through the mediation of special, strategically situated nucleotide sequences, called promoters, that control the interaction between the genes and the transcribing systems. As evolution proceeded, these sequences became the target of many regulatory interventions serving to turn transcription of given genes on or off. Transcriptional control of gene expression, which was destined to become an immensely important mechanism in adaptation and development, was thereby initiated. We shall encounter this mechanism on several occasions in subsequent chapters.

Because genes no longer had to serve as messengers, they could be joined together in strings of increasing length, which, in turn, offered the possibility of a synchronous and appropriately timed replication of all the genes of the string. This development was conditioned by improvements in the accuracy of replication. It is remarkable that much higher fidelity eventually came to be achieved for DNA than for RNA replication. Whereas the lowest error rate for RNA replication is on the order of one in a few tens of thousands—consistent with a maximum length of 20,000 to 30,000 nucleotides for viral RNAs—the error rate can be as low as one in one billion for DNA replication. Due to the existence of elaborate "proofreading" mechanisms, whereby wrongly added nucleotides are removed before they are sealed in by the next nucleotide in the growing chain, this remarkable accuracy has allowed all the genes of a bacterial cell, covering millions of nucleotides, to be strung in a single, circular chromosome, which is turned on for replication from a single commanding site, called the origin of replication.

It took evolution a great many steps to move from the tangled jumble of small RNA genes to the majestic orderliness of the bacterial chromosome. However, from the moment the first stretch of DNA was assembled, each step provided an incremental selective advantage. The whole succession followed the characteristic alternation of mutation and selection that is the *modus operandi* of evolution.

Chapter 8

Freedoms and Constraints

WITH THE APPEARANCE of the first RNA molecules, incipient life entered the era of molecularly encoded information and progressively built the DNA-RNA-protein triad that now rules the entire biosphere. Three key concepts were introduced in the wake of informational molecules: complementarity, contingency, and modular assembly.

COMPLEMENTARITY

Biological information transfer is based on chemical complementarity, the relationship that exists between two molecular structures that fit one another closely. Images such as lock and key, mold and statue, are often used to illustrate such a relationship. In the chemical realm, complementarity is a more dynamic phenomenon than these images suggest. The two partners are not rigid. When they embrace, they mold themselves to each other to some extent. Furthermore, the embrace leads to binding. Its degree of intimacy is such that electrostatic interactions and other short-range physical forces act strongly enough to prevent the association from being disrupted by thermal jostling.

Base pairing, the support of the genetic language, is the most spectacular manifestation of chemical complementarity in biology. But it is only one of many. Every facet of life depends on molecules that "recognize" each other. Self-assembly, the phenomenon whereby complex structures are formed from a number of parts, rests on complementarity relationships between the parts, as did the assembly of furniture in the old days, except that chemical parts even provide their own glue.

Take the immune system and its astonishing versatility and specificity. What makes us resistant against polio or diphtheria—as a result of a previous attack or vaccination—is the presence in our blood of special protein molecules, antibodies, that specifically bind to some component, termed antigen, of the polio virus or of the diphtheria bacillus. The cells that recognize a grafted heart or kidney as foreign, and reject it, do so through the mediation of surface molecules that join with some surface component peculiar to the intruder. The white blood cells that stalk invading microbes and gobble them up recognize their prey by a similar mechanism.

Hormones, drugs, poisons, and every other chemical that exerts a biological effect owe this property to their ability to interact with a receptor molecule on their target. This kind of relationship is now exploited on a vast scale in research. Endorphins, which are natural inducers of pleasurable sensations, were discovered through the morphine receptor.

Enzymes offer another fundamentally important example of complementarity. Most enzyme-catalyzed reactions take place in three interconnected steps. First, the molecule or molecules on which the enzyme is to act—its substrate or substrates—become physically bound to special binding sites on the enzyme surface. This binding is such that the molecules are offered in appropriate spatial orientation to the catalytic site of the enzyme. Catalysis is the second step, followed, in a third step, by detachment of the products, so that the cycle can start again. As a simple analogy, imagine a welder immobilizing two pieces of metal in a vise, then proceeding with the welding, and finally removing the welded product to start a new operation. Alternatively, a single piece of metal could be similarly immobilized prior to sawing or filing.

In enzymatic reactions, there is no workman to select materials. The process is self-powered and depends on molecular affinities between binding sites and substrates. Thanks to these affinities, enzymes can "fish out" their substrates from highly complex mixtures. In any living cell, hundreds, if not thousands, of different substances coexist, all at very low concentration, as they might in a prebiotic mixture. Enzyme specificities, as defined by the affinities of the substrate-binding sites of the enzymes, determine the chemical pathways the molecules follow.

This relationship can work both ways. Just as receptors may fish out hormones, substrates can select their enzymes, either directly by protective binding—many enzymes are more resistant to degradation when linked to their substrates—or indirectly through the enzymes' activities. This is what I believe happened when the RNA machinery started delivering peptides. The catalytic peptides that fitted within the protometabolic network were retained. In this sense, protometabolism already contained information. It provided the blueprint for metabolism through the mechanism of enzyme selection.

CONTINGENCY

Contingency entered the history of life on Earth with the onset of replication and the inevitable mutations that perturb this process. Thus was set in motion the process of Darwinian evolution that has governed the history of life on Earth. Genetic information is accidentally altered. The modified message is replicated and expressed. The ability of the modified phenotype to perpetuate the modified genotype by means of offspring is evaluated by natural selection, which weeds out harmful mutations with poor reproductive success, favors useful mutations that enhance survival and reproduction, and lets neutral mutations simply drift along. As soon as replication started, this process was initiated, first at the molecular level, then at the level of the protocell.

Because mutations are accidental, no two RNA worlds, even in one billion or more, can have exactly the same microscopic history. But what about their macroscopic history? No two streams follow exactly the same course down a mountain, but all may end up in the same valley.

We have no way of answering this question with certainty, but the likelihood is that, in a large number of cases, incipient RNA worlds will issue into an RNA-protein world similar to ours. My main reason for stating this lies in the stringency of the selection factors that came into play at each stage. Behind these factors lurks a great deal of chemical determinism, often based on complementarity.

If my proposed reconstructions are correct, the four RNA bases were selected among a number of related products on the strength of their pairing ability, which allowed amplification of the corresponding RNAs by replication. Molecular selection, based on optimal replicatability-stability, next led to a reproducible RNA master sequence, as in the Spiegelman-Eigen type of experiment. Chemical interactions between RNA and amino-acid molecules, once again determined by chemical complementarity, then selected the proteinogenic amino acids and the corresponding transfer RNAs. All very reproducible and leaving little to chance.

Also important was the historic factor, which severely channeled the emergence of translation and of the genetic code, itself shaped further by the condition of lowest harm to the organism by mutations. Finally, the existing protometabolic network acted as a unifying screen for the first enzymes produced by the machinery. The order in which the enzymes appeared might have varied with the vagaries of mutations, but the end result would, in each case, be a metabolism largely copied from protometabolism.

Later evolution may also have been subject to more deterministic factors than is often surmised, in spite of the increasing role of contingency. Quite possibly, when DNA originated from RNA, driven by the advantages of a separate storage form of genetic information, there were not too many different possible modifications of the RNA molecule that could have ensured specificity, while at the same time saving information transfer between the two molecules.

MODULAR ASSEMBLY

A third lesson we learn from our reconstructions is the importance of modular assembly. This is a recurring theme in the history of life. Evolution works with pre-existing modules—RNA minigenes at the stage we have considered—which are modified and combined in different ways into larger assemblages that are then screened by natural selection. Implicit in this mechanism is the possibility of extensively exploring the available sequence space at each step, thus further reducing the role of chance.

In conclusion, a number of RNA worlds might abort—and perhaps did on our planet—because chance did not provide a necessary mutation. But those that mature would probably lead to a form of life supported by the same basic metabolic processes and ruled by the same DNA-RNA-protein triad and, perhaps, the same genetic code that characterize our own form of life. Furthermore, because of the limited size of the sequence space available to incipient life, the success score is likely to be high.

PART III

THE AGE OF THE PROTOCELL

Chapter 9

Encapsulating Life

FOR A FULLY operational genetic system to develop, emerging life had to become partitioned into a population of protocells capable of multiplying by division, so that protocells, not simply molecules, henceforth were subjected to natural selection. So far, we have been content to assume that this partitioning took place. Let us now turn back in time and look into the mechanisms of cellularization and into the new properties that the confinement of life within boundaries both allowed and required.

THE TIMING OF CELLULARIZATION

There are two conflicting views of the time at which the first cellular structures appeared. A number of scientists, impressed with the fact that microscopic aggregates or vesicles of various kinds, crudely reminiscent of living cells, can be observed to form under relatively simple conditions, believe that the formation of primitive cells was the seeding event in the origin of life. Extensive laboratory investigations—by Alexander Oparin in Soviet Russia,[1] Alphonse Herrera in Mexico,[2] and Sidney Fox in the United States,[3] to mention only the most prominent—have been devoted to such artificial "cells," though without disclosing any plausible pathway for the progressive "vitalization" of the structures. Other scientists have defended early cellularization on the grounds that a membranous structure was required for the initial trapping of sunlight energy.[4] Yet others find unacceptable for theoretical reasons the possibility that life could have originated in an unstructured "soup."[5]

The opposite view is also defended by many. It has been pointed out that the "primeval soup" need not have filled the whole of oceans. Coastal areas, lagoons, ponds, even puddles, could have provided appropriate sites for the soup to thicken and evolve chemically. The hindrances an enveloping structure might have posed to the free circulation of biogenic substances are also mentioned in objections to cell-first theories. Eigen, for example, believes for this reason that "organization into cells was surely postponed as long as possible."[6]

The thioester-based metabolic model I have proposed is not readily compatible with early cellularization and fits better with the concept of an initially unstructured soup. It is suggestive in this respect that some metabolic systems generally considered most ancient, for example, the system involved in the fermentation of sugar to alcohol, which uses a thioester-linked energy-retrieval mechanism, are situated in the cytosol, or cell sap, the unstructured part of the cell. Thus, the primeval soup, energized by thioesters, may be seen as progressively developing into a sort of extended protocytosol.

In the early stages, the need for free exchanges would have given the unpartitioned protocytosol a clear advantage over separate entities subject to the constraints of a peripheral boundary. For encapsulation to take over, the advantages of confinement must have outweighed the drawbacks. This implies that the isolated systems both enjoyed enough autonomy to survive on relatively simple exchanges with their environment and derived a clear benefit from being enveloped. This situation was reached, at the latest, when the RNA machinery for peptide synthesis began to come together, since further evolution of this machinery made the existence of a large number of competing protocells mandatory. How did the first cell boundaries form, and from what materials? In order to answer these questions, let us again look for clues in present-day organisms.

CELL BOUNDARIES

All living cells are surrounded by an unbroken, filmy envelope, called the plasma membrane. Many cells are also partitioned by internal membranes. The universal fabric of biological membranes is the lipid bilayer, a tenuous, double molecular leaflet, about one five-millionth of an inch thick, usually made largely of phospholipids. Qualified as amphiphilic, or amphipathic ("with two loves"), these molecules are characteristically composed of two parts with opposite affinities: a hydrophilic (water-loving) head and a hydrophobic (water-hating) tail, also called lipophilic (fat-loving).

Hydrophilicity depends on the attractions that exist between electric charges of opposite sign. The water molecule has no net electric charge, but it is electrically polarized. It has a negative pole situated on the side of the oxygen atom, which tends to appropriate more than its share of electrons, and a dual positive pole made up by the two hydrogen atoms, which protrude asymmetrically as partly naked protons on the same side of the molecule. As a consequence of this structure, water molecules bind by either their positive or negative pole to any oppositely charged or polarized molecules or chemical groups. Water molecules also bind to each other for this reason. Were this not so, water would be liquid only at very low temperatures; the Earth would be dry, lifeless, and forever barren.

Hydrocarbons, the main constituents of petroleum, and all other substances made entirely or mostly of carbon and hydrogen, being uncharged and nonpolar, are hydrophobic. Many such substances exist in the living world. They are grouped under the term of lipids, which comes from the Greek word for fat. Hydrophobic molecules do not really hate or repel water; they are excluded by it owing to the strong tendency of water molecules to join by electrostatic attractions. In the presence of water, hydrophobic molecules are thus driven together by crowding water molecules. The formation of such allocations is itself facilitated by hydrophobic–hydrophobic interactions mediated by short-range forces, weaker than electrostatic forces and known as van der Waals forces, from the name of the Dutch chemist who discovered them.[7] Each thus keeps to itself. Oil and water don't mix.

The heads of phospholipid molecules owe their hydrophilic character to a negatively charged phosphate group, often associated with other charged or polar groups. Two long hydrocarbon chains make up the hydrophobic tails of the molecules. In the presence of water, phospholipids satisfy their two contradictory loves by forming bilayers. In these structures, each of the two layers consists of closely packed molecules, lined up perpendicularly with respect to the plane of the layer (like the bristles of a brush) and oriented in such a way that the hydrophilic heads all face one side and the hydrophobic tails the other. Each layer is thus a very thin sheet one molecule thick, with a hydrophilic face and a hydrophobic face. In bilayers, the two sheets are sandwiched by their hydrophobic faces held together by van der Waals forces, whereas the hydrophilic faces are directed outward in contact with water. Such bilayers thus interpose an oily film between two watery milieus.

Phospholipid bilayers are very fluid and flexible. They form a sort of two-dimensional liquid, within which the constituent molecules easily slide along each other within the plane of the bilayer. Because of this property, bilayers can mold themselves around any kind of surface and readily adapt to changes in the conformation of the surface, as often happens with cells. Phospholipid bilayers are always continuous and self-sealing, and therefore always form closed sacs. In this respect, they resemble soap bubbles, with which they share a number of physical properties. In particular, they can join (fusion) or be split (fission) without loss of continuity. Two phospholipid vesicles may fuse into a single one, like two soap bubbles that bump into each other. Conversely, a single vesicle may divide into two, as sometimes happens to a soap bubble caught in an air drag.

A last important property of phospholipid bilayers is their ease of formation. No more than vigorous mechanical agitation, by means of ultrasound, for example, is needed to turn a mixture of phospholipids and water into a suspension of small vesicular bilayers. A whole industry has been built around this phenomenon. Artificial phospholipid vesicles, called liposomes, have found many applications as carriers for cosmetics, drugs, vaccines, genes, and other agents.

Phospholipid bilayers are impermeable to most water-soluble (hydrophilic) molecules. This property makes bilayers excellent boundaries that allow cells to maintain an internal composition different from that of the surrounding medium. But cells cannot survive sealed off from the outside. They must be able to take up nutrients, get rid of waste products, and respond to environmental signals. These functions are carried out by proteins inserted into the bilayers.

The sequences of membrane proteins are characterized by one or more transmembrane stretches of about twenty to thirty largely hydrophobic amino acids, typically coiled into a helical rod called an α-helix. These rods pass through the bilayer, in close contact with the hydrophobic parts, with which they establish links stabilized by van der Waals forces, and serve to position the proteins within the membrane. The other parts of the protein molecules protrude on the outer and inner faces of the membrane.

Most cells in both the prokaryotic and eukaryotic worlds are surrounded by peripheral structures external to the plasma membrane, ranging from a fluffy down to massive, rigid walls. These structures serve to support and defend the cell. They act as molecular filters and may fence off an intermediary space, called the periplasmic space, between the cells proper and their environment. A variety of substances, including proteins, lipids, complex carbohydrates, and special constituents of unique chemical composition, participate in the building of these outer structures.

MECHANISMS OF CELLULARIZATION

Phospholipids are complex molecules that could hardly have been available in the primeval soup. But they could have arisen through the development of protometabolism and been present in the soup at the time encapsulation became advantageous. It would then have needed no more than some violent storm for vesicular bilayers to form spontaneously in such a soup, the way artificial liposomes arise today in phospholipid–water mixtures exposed to ultrasonic vibrations. Primitive cells could have been born in this way, but only to die almost immediately of starvation because their phospholipid envelopes would not have let through even the simplest of nutrients.

It is conceivable, however, that the empty ghosts of stillborn cells provided anchoring points for some metabolic systems and offered a harbor for hydrophobic peptides. Progressive curving of this structure could give rise to a double-membranous cup, which could further close into a double-membranous pouch once the structure had acquired the necessary systems of transmembrane communication. According to this model, which has been proposed by the German-born American cell biologist Günter Blobel, of the Rockefeller University in New York, the first cells would have been bounded by a double membrane.[8] This happens to be a characteristic feature of gram-negative bacteria (so called because they react negatively to a test devised by a Danish bacteriologist named Gram). It has, indeed, been suggested that gram-negative bacteria may have preceded gram-positive organisms, which have a single membrane. The British biologist Thomas Cavalier-Smith, who champions this idea, has adopted Blobel's model for this reason.[9] However, the outer membrane of gram-negative bacteria is very different in structure from the inner membrane, which represents the true cell boundary, or plasma membrane.

A possible alternative is that the first boundary was not made of phospholipids but of peptides and other multimers of largely hydrophobic character, which could have formed a looser and more permeable network than lipid bilayers. This is a plausible possibility, as hydrophobic multimers must have been abundant right from the start, considering the nature of many of the available building blocks. Phospholipids could have come later to plug the holes in the boundary and expand it into a more flexible and versatile membrane, as needed communications became established.

Whatever their nature, the mechanisms that led to the encapsulation of the first protocells must have been intimately associated with the creation of appropriate passageways allowing the necessary molecular traffic between the protocells and their environment to take place. There are unfortunately no clues to the long succession of molecular events that determined this progressive tightening of barriers around increasingly sophisticated means of crossing them. We can only look at the finished product and speculate about its origin. Let us first consider construction.

THE ASSEMBLY OF MEMBRANES

Membranes grow by accretion, that is, by the addition of components to a pre-existing membrane.[10] Thus, *de novo* synthesis of a membrane needed to occur only once in the history of life, and all subsequent membranes could have arisen from this ancestral membrane by expansion followed by fission. We don't know whether things happened this way, but it is an intriguing possibility. At least, membranes develop in this manner in the living world today.

Once the first membranes arose, any innovation that facilitated the insertion of new components into them was advantageous. For lipids, the simplest and most effective innovation was to have them synthesized right in the membranes, which provided an excellent milieu for housing the hydrophobic building blocks used. Thus, a number of enzyme systems involved in the synthesis and assembly of lipids, especially phospholipids, became associated with membranes. Today, CMP, the cytosine-containing constituent of RNA molecules, is heavily implicated in these processes as a carrier of several key building blocks. If historically significant, this fact suggests that phospholipid membranes came with or after the RNA world, in agreement with the hypothesis of late cellularization.

In the case of proteins, adaptations were of a more subtle kind, as the ribosomes on which protein assembly took place were situated in the soluble compartment of the protocells. Homing of proteins to membranes was achieved by means of certain amino-acid sequences, called signal or targeting sequences, typically present in membrane proteins. These sequences were specifically recognized (bound) by certain membrane components that served as docking areas for the proteins carrying the right address tag. Consequent to this binding—another typical example of complementarity—the proteins carrying the tag became inserted into the fabric of the membranes. Two main variations on this theme developed. In one, the targeting sequence occupies the initiating end of the nascent polypeptide chain and joins with the membrane as soon as this end emerges from the ribosome. Called cotranslational because it takes place while translation is still going on, this transfer is revealed by the observation of ribosomes closely apposed to the inner face of bacterial cell membranes. The second, posttranslational, kind of protein transfer occurs after completion of the polypeptide chain and depends on targeting sequences that may be situated anywhere in the chain.

THE CONSTRUCTION OF OUTER DEFENSES

The construction machineries considered so far played an important role in the functional enrichment of the first membranes but contributed little to their structural strength. Phospholipid bilayers, even reinforced by proteins, are flimsy fabrics. They are easily torn or damaged by physical or chemical agents and offer virtually no resistance to osmotic swelling, a phenomenon induced by the inflow of water that occurs when cells are exposed to a medium in which dissolved substances are less concentrated than they are inside the cell. This fragility of their surface boundary severely curtailed the ability of the protocells to withstand outside aggressions and to adapt to different environments. Then, an event happened that exerted an enormous influence on the prospects of life on Earth. Protocells "learned" to build rigid extracellular structures from carbohydrate building blocks.

This historical event probably started with the appearance of mechanisms for joining sugar molecules together into chains, or saccharides, of various lengths that served mainly as reserve substances. What we call sugar in everyday language is actually a disaccharide made of two elementary sugar molecules, glucose and fructose. Starch is a polysaccharide made entirely of glucose. One readily sees how the ability to store energy-rich foodstuffs as large molecules that could not escape through the surrounding boundary provided the protocells with enough advantages to favor selection. It is interesting, and possibly suggestive, that the main carriers involved in saccharide synthesis today are derivatives of UMP, or occasionally of AMP or GMP, that is, typical RNA constituents. Thus, together with phospholipids, polysaccharides could also be products of the RNA world or of the post-RNA world.

The next decisive step was initiated by the formation of a new kind of sugar carrier, derived from a substance called dolichol, anchored in the membrane by a long hydrophobic tail. Sugars or saccharide chains were transferred from their nucleotide carriers to the membrane-bound carrier and thereby made to stick closely to the inner face of the membrane. By an intriguing flipping phenomenon, these bulky, highly hydrophilic masses came to be translocated across the hydrophobic barrier of the phospholipid bilayer and to pop up on the outer face of the membrane. There they could be handed over to protein molecules or to other acceptors. In this way, the surface of the protocells became progressively bolstered and defended by carbohydrate parts, which greatly augmented the survival potential of the protocells concerned.

The building of outer defenses involved a remarkable molecule that is still present in a large part of the bacterial world today and has all the hallmarks of a surviving fossil. Called murein, this molecule consists of sugar molecules and of short heterogeneous peptides that could, according to their structure and content in both D- and L-amino acids, have come straight out of the primeval multimer mixture. These parts are interlocked into a single, huge, meshlike molecule that entirely envelops the cell within a sort of organic coat of mail. Called the cell wall, this structure is remarkably strong and resilient while being sufficiently porous not to impede molecular passage.

Murein is broken down by lysozyme, an enzyme that plays an important role in the defense of organisms against invading bacteria. The naked cells, or protoplasts, that are stripped of their wall by lysozyme usually burst osmotically unless the medium composition is such as to prevent the influx of water. On the other hand, the miracle drug penicillin owes its unique therapeutic virtues to its ability to block the building of murein and thereby prevent the growth and multiplication of sensitive bacteria. As it happens, lysozyme and penicillin were both discovered by the same scientist, the Scottish microbiologist Alexander Fleming, at a time when nothing was known about the chemistry and synthesis of the bacterial cell wall.[11]

The wall was further strengthened by the thickening of the murein layer or by the coating of this layer with a membranous skin constructed from special lipopolysaccharide molecules and rendered permeable to small molecules, but not to proteins, by inserted, tunnel-shaped protein molecules called porins. As mentioned, there is a possibility that this second membrane, which characterizes gram-negative bacteria, may be a legacy of an early encapsulation stage in which protocells were enveloped by a double membrane.

THE NECESSARY INLETS AND OUTLETS

The first, inescapable condition of survival in confinement was the possibility for the protocells to take in food from the outside and get rid of waste material. The simplest way in which fully enveloped protocells could fulfill this condition was by means of pores, mere holes kept open in lipid bilayers by some kind of inserted protein framework. The porins, just mentioned, are an example of such proteins.

Next came transport facilitators, which are transmembrane proteins that act as molecular turnstiles for certain specific substances. Like simple turnstiles, facilitators are passive systems. They open in either direction and give in to the side from which the pressure is greatest. That is, they let substances flow in the direction leading from a higher to a lower concentration. But they do this with a certain degree of chemical discrimination. Many cells, for example, contain a transport facilitator that provides specific passage for glucose molecules.

A more sophisticated kind of molecular turnstile is the gated channel, analogous to some of our controlled admittance devices. Gated channels, like facilitators, merely let certain substances of given chemical specificity move through passively, but they are unidirectional and regulated by a gate that needs to be unlocked by some chemical or electrical signal.

The next improvement in the building of molecular transport systems was active transport, hooked to a source of energy, usually the splitting of ATP, so that the spontaneous direction of flow could be reversed and substances could be forced uphill, from a lower to a higher concentration. For the protocells involved, such acquisitions meant that they could now fish out rare but essential substances from their surroundings or, alternatively, rid themselves of toxic refuse even in a highly polluted environment. Although there was an energy bill to pay, the gain in survival potential was high enough to tilt the direction of natural selection.

Among the substances that could be actively transported into or out of protocells, a number were ions, that is, electrically charged entities. In many cases, the displacement of ions in one direction is linked to an equivalent displacement of ions of the opposite charge in the same direction, or of ions of the same charge in the opposite direction. The membrane boundary remains electrically neutral. Sometimes there is no such compensation and the forced transport of ions creates an imbalance of electric charges, or membrane potential, between the two sides separated by the membrane. Such pumps—a name often used for ion-transporting systems—are termed electrogenic.

A particularly important electrogenic pump uses the energy supplied by the splitting of ATP to drive sodium ions (positively charged) out of cells and replace them partly (two against three) by potassium ions (also positively charged), with the consequent building of a membrane potential positive to the outside. In eukaryotic cells, this potential has come to play a role of exceptional importance, as the basis of all bioelectric phenomena, including the functioning of the nervous system in animals.

The origin of the sodium-potassium pump is obscure. It is of possible significance that the principal positively charged ion of sea water—and also of animal blood—is sodium. It is conceivable that fenced-off life had to defend itself very early against excessive sodium. Interestingly, some of the most ancient bacteria, known as halophiles, are particularly effective in coping with external sodium. They thereby manage to survive—in fact, to thrive—in concentrated brine.

Another kind of electrogenic pump of central importance forces protons across membranes. Proton pumps powered by the hydrolysis of ATP serve in a number of instances to raise the proton concentration, that is, the acidity, of certain intracellular or extracellular regions (think of the acid produced in the human stomach). By far the most important function of proton pumps is in energy transfer, which I shall examine in the next chapter.

CELL DIVISION

Whatever the mechanism of encapsulation, it had to include the possibility of turning growth into division. Without such a link, advantageous mutations could not have turned into selective assets; they would even have been self-defeating. This point is easy to understand. Consider a spherical cell. As it grows, its mass and, therefore, its maintenance and repair needs increase as a function of the third power of its radius. On the other hand, the surface area it has available for the import of nutrients increases only as a function of the second power of the radius. Growth of such a cell must necessarily reach a point where import just suffices for maintenance and repair. Further growth is impossible, unless the cell becomes asymmetrical, forms a bud, for example. Burgeoning of the bud would eventually lead to its falling off as a free entity, especially if a self-sealing membrane surrounded the whole system. Thus, any surface property that favored asymmetric growth and budding of the first protocells would, if hereditarily transmissible, automatically have been selected.

With the development of outer structures, division became a more complex process, dependent on a progressive constriction of the wall into a deepening circular furrow. Little is known of the mechanisms controlling this phenomenon, but a link exists between membrane and wall. Bacteria unable to build a wall as a result of exposure to penicillin and protected against rupture grow bigger but do not divide.

For division to be of any use, each daughter protocell had to include all that was needed for autonomous survival and proliferation, in particular a full set of genes. At first, this condition was probably satisfied on a statistical basis by the random mixing of the protocell components within their membranous envelope. When the genetic material became centralized into a single, circular chromosome, a relationship was established between DNA replication and division, such that each daughter inherited a chromosome. This partition was facilitated by anchoring of the chromosome to the plasma membrane. Upon initiation of DNA replication, the complex of enzymes and ancillary factors needed for this process was assembled around the anchoring point, and the chromosome was gradually reeled in through this complex, exiting in duplicated form. After the two resulting chromosomes became disentangled, each ended up anchored to a different site of the membrane. The furrow initiating division then formed between the two sites, thus ensuring that each daughter cell inherited one of the duplicated chromosomes.

Chapter 10

Turning Membranes into Machines

ENCAPSULATION was a slow, progressive process, punctuated by many evolutionary acquisitions. By necessity, the earliest of these acquisitions concerned mostly means of ensuring vital exchanges with the environment. Soon, however, the scope widened. Once phospholipid bilayers were formed, this new fabric turned out to be much more than a convenient boundary. It presented burgeoning life with numerous opportunities for useful innovation. A whole new class of proteins emerged, fitted with one or more hydrophobic sequences that allowed insertion within membranes. Thus immobilized, the proteins could participate in a variety of novel functions that were sufficiently advantageous to favor the evolutionary selection of the mutant protocells that made the proteins. By far the most important development of this sort was the putting together of a machinery coupling downhill electron transfer reversibly to proton extrusion. Emergence of this machinery was a truly revolutionary advance in the ability of life to derive energy from environmental sources.

PROTONMOTIVE ELECTRON TRANSFER

Imagine the following scenario. It may not have happened as depicted, but the scenario is plausible and tells in a simple fashion how emerging life may have hit upon the invention that completely transformed its future—made this future possible, in fact.

Owing to some mutational event, a protocell acquires an electron-carrying molecule constructed so as to fit within the fabric of the protocell's membrane. What makes this carrier useful, and favors its selection, is that it can serve as a bridge for electrons across the membrane, between an internal donor and an outside acceptor to which the membrane is impermeable. Access to this acceptor is, thus, the protocell's gain from the mutation.

The lunch is not free, however. There is a price to pay: The carrier transports electrons in the form of hydrogen atoms. This means that if the transaction between internal donor and external acceptor involves naked electrons, protons are necessarily translocated together with the electrons. The carrier must pick up protons from inside the protocell—one proton for each electron—to make the electrons transportable as hydrogen atoms, and it must discharge the same number of protons outside when it delivers electrons to the external acceptor. Thus, *electron transfer is obligatorily coupled to proton translocation,* and vice versa. One cannot take place without the other. What this amounts to is a reversible, electron-driven proton pump.

For such a pump to be of any use, the membrane needs to be impermeable to protons. Otherwise, the translocated protons immediately diffuse back inside. If proton leaks are plugged, the coupling between electron transfer and proton translocation becomes an energy link. As electrons are transferred across the membrane, the accompanying protons create a rising imbalance, or proton potential, which, depending on circumstances, is manifested in the form of an excess of external over internal proton concentration; of a membrane potential, positive outside; or of a combination of both. Whatever its physical form, the proton potential tends to oppose the further translocation of protons. The higher the potential, the stronger the opposing force. The coupled process grinds to a halt when the amount of work required to push more protons against the existing proton potential becomes equal to the amount of energy released by the electron transfer. This amount of energy is itself a function of the difference between the energy levels at which the donor gives out the transferred electrons and the acceptor takes them up.

Could such a liability be turned into an advantage? Yes, in several ways. Survival in an acidic medium is an attractive possibility. By definition, acids are hydrogen-containing substances that, when dissolved in water, tend to release free protons (the rest of the molecule being left as a negatively charged ion). The higher the proton concentration created in this way, the stronger the acidity (and the lower the pH, an expression swimming-pool owners will understand). From the tangy sourness of lemon juice (citric acid) or vinegar (acetic acid) to the metal-biting causticity of nitric acid, it is all a matter of proton concentration.

As discussed earlier, there are reasons for suspecting that life started in or near an acidic environment. Some of the most ancient microbial species belong to the group of thermoacidophiles, which inhabit a very hot and acidic milieu. We saw in chapter 3 that such a milieu could have been conducive to the formation of the first thioesters. Also, it would have released inorganic phosphate (or pyrophosphate) from its insoluble combinations and allowed this essential ingredient of many biomolecules to enter primitive metabolism. There is a difficulty, however, with early life actually developing in such an environment because a number of metabolic intermediates, including several critical phosphate compounds, are extremely sensitive to hot acid. The existence of volcanic springs and the recent discovery of deep-sea hydrothermal vents suggest a possible way out of this quandary. Thioesters could have arisen, and phosphate dissolved, in hot, acidic subterranean waters and come to the surface with pressurized jets that transferred them to milder conditions. Protocells developing at the edge of such a source could have invaded increasingly acidic waters by acquiring the means to drive out protons and so to keep an appropriately mild internal milieu against the pressure of a strong external proton potential that would take advantage of any weak spot in the membrane to push protons in.

The coupling between downhill electron transfer and proton extrusion could be put to use in an alternative manner if the outside proton potential were strong enough to reverse the flow of electrons across the membrane, that is, to force the electrons to move in the uphill direction, from the reduced form of the outside acceptor, turned into donor, to the oxidized form of the inside donor, turned into acceptor. For this to happen, the protocells would need a proton "sink," that is, a metabolic system capable of consuming the protons that enter through the pump turning in reverse.

The same evolutionary advantages would be associated with the acquisition of an ATP-driven proton pump (see the preceding chapter). Whichever pump came first, it would have conferred a substantial benefit to protocells occupying an acidic medium. It is tempting, though admittedly speculative, to take the emergence of proton pumps as another clue pointing to an acidic cradle for life.

A dramatic change, no longer linked to outside acidity, occurred when the two kinds of proton pumps—one electron-driven and the other ATP-driven—turned up together in the same protocell membrane. Imagine the scene. The two pumps start by acting in concert, joining efforts to build a rapidly rising proton potential. Because the two pumps are not likely to be of exactly equal strength, a stage will be reached where the weaker one stops, while the stronger one goes on driving out protons, raising the proton potential above the weaker pump's limits. When this happens, the weaker pump starts running in reverse. The proton potential built with one source of energy mediates the replenishment of the other source of energy. If the electron-driven proton pump is the stronger of the two, downhill electron transfer supports the assembly of ATP from ADP and P_i. If the ATP-driven proton pump is the stronger, ATP hydrolysis supports uphill electron transfer, from a lower to a higher energy level. A new form of reversible coupling between electron transfer and ATP assembly, based on protonmotive force, is born.

The importance of this event can hardly be overestimated. Before it happened, the reassembly of ATP (or pyrophosphate) with the help of electron-transfer energy took place entirely by the thioester-dependent mechanism of substrate-level phosphorylation (see chapter 3). Today, probably less than one molecule of ATP in one million is reassembled by this mechanism (which nevertheless remains universal and vitally important). The membrane-bound mechanism of carrier-level phosphorylation now dominates biological energy retrieval. Without it, we could not cover our energy needs by the combustion of foodstuffs. Nor would plants be able to harness the energy of the sun.

The evolutionary advantage of the new energy-retrieval mechanism was immediate. As soon as its most primitive seed was planted, every improvement in the efficiency and versatility of protonmotive coupling was strongly favored by natural selection. The crowning achievement of this long evolutionary development is the electron-transfer chain—also called the respiratory chain, because, in all aerobic organisms, the electrons are collected at the end of the line by molecular oxygen, which is itself made available by respiration. Such a chain consists of a number of electron carriers, arranged within the fabric of a membrane in a manner that has been compared to an electron bucket brigade or to an electron cascade. The bucket-brigade image underscores the participation of carriers in the flow of electrons along the chain. The image of a cascade makes clear that the pathway followed by the electrons is downhill and includes steps where the electrons fall down a substantial difference in energy level. One or more of these steps—three in all most advanced systems—are obligatorily coupled with the extrusion of protons and can serve to power the assembly of ATP by way of protonmotive force. Thus, when electrons fall down the cascade, ATP molecules are assembled. Conversely, electrons can be forced up the cascade with the expenditure of ATP or with the help of protonmotive force generated by electrons flowing down the lower part of the chain.

A number of important membrane-embedded molecules participated in the construction of electron-transfer chains. We have already encountered in chapter 3 the group of iron-sulfur proteins, built around iron-sulfur clusters, which operate by way of ferrous/ferric oscillations. Also dependent on the same oscillations are a number of membrane-bound, red-colored substances called cytochromes, which are members of the larger group of hemoproteins, of which the prototype is the red blood pigment, hemoglobin (from the Greek *haima,* blood). The active part of hemoproteins is a complex, flat, dish-shaped organic molecule, made of carbon, nitrogen, and hydrogen atoms, belonging to the porphyrin group. In the center of the dish is a hole occupied by an iron atom. In cytochromes, this iron atom alternates between the ferrous and the ferric form, and thus accounts for the electron-carrier function of the molecule. Cytochromes are found in membranes as members of electron-transfer chains. The hemoproteins also include a number of soluble substances, in which the iron remains permanently in the ferrous or ferric form. The ferrous hemoproteins mostly act as oxygen carriers, like blood hemoglobin. The ferric ones exert some enzymatic activity involving hydrogen peroxide.

In addition to members of these two groups of iron proteins, electron-transfer chains also include cuproproteins, with copper as the electron carrier; flavoproteins, with FMN or FAD (see chapter 4) as the electron carrier; and electron-carrying quinones, highly hydrophobic organic molecules composed of carbon, hydrogen, and oxygen atoms. Altogether, as many as fifteen distinct carriers may be associated in a given chain, physically organized in decreasing order of energy level, so that each carrier is strategically positioned with respect to the carriers with which it transacts direct electron transfers.

THE ATTAINMENT OF AUTONOMY

It is commonly believed that early life drew its building blocks from preformed organic products of abiotic syntheses. As to its energy needs, they could also have been covered by preformed energy-rich molecules, such as inorganic pyrophosphate or polyphosphates, or, as in my model, thioesters. Alternatively, the breakdown of preformed organic molecules could have supplied the necessary energy through some coupled process, such as the thioester-generating electron-transfer process envisaged in my model.

If such was the case, life started in heterotrophic form. The term "heterotroph" (Greek *heteros,* other; *trophê,* food) designates organisms that, like ourselves, feed on products made by other organisms, by contrast with autotrophs (Greek *autos,* self)—plants, for example—which manufacture their constituents from mineral building blocks. Early heterotrophy did not rely on autotrophs, of course, but on the celestial manna of abiotic chemistry.

By the time the manna became exhausted, some form of autotrophy had to be developed. We don't know when this happened, but it is not likely to have been before protocells appeared, unless some unknown kind of mechanism was involved. All known autotrophs depend on membrane-embedded electron-transfer chains. It is likely, therefore, that such chains started by supporting heterotrophic processes and became converted to autotrophy later. How did this conversion take place?

To answer this question, we must look at the anatomy of the biological electron cascade. It consists of four distinct "chutes," separated by five energy levels—say, in the order of decreasing altitude: A, B, C, D, and E. Each chute is high enough to support the assembly of ATP from ADP and P_i, at the rate of one molecule of ATP per pair of electrons falling down. The A-B chute is the oldest; it involves water-soluble components and depends on thioester-linked, substrate-level phosphorylation. The B-C, C-D, and D-E chutes involve membrane-embedded components and depend on protonmotive, carrier-level phosphorylation. For the cascade to operate, electrons must be fed into it and collected from it. Feeding occurs optimally at level A and collection at level E, but electron inlets and outlets also exist at intermediate levels.

This cascade is not a pre-existing feature of nature. It is the product of evolution, which, guided by natural selection, succeeded in putting the A-E span to optimal use, within the limits imposed by the energy requirement of ATP assembly. In brief, when two electrons fall from level A to level E, enough energy is released to power the assembly of four ATP molecules altogether. The cascade, with its four chutes, each harnessed to a separate machinery, exploits this possibility to the full. Our problem is to explain how this masterpiece of natural selection came into being and how it paved the way to autotrophy.

There is a simple answer to this question. It may not be the correct one, but it will do for our purpose. In terms of the image of an electron cascade, it can be summarized as follows. Life started at the top of the cascade and harnessed first the fall of electrons from level A to level B or below. This harnessing took place by way of thioesters and evolved into the mechanism of substrate-level phosphorylation. Abiotic syntheses, perhaps helped by a source of high-grade electrons (see chapter 3, "The Case of the Missing Hydrogen"), supplied the appropriate electron donors, some of which still fulfill this function 3.8 billion years later. For example, pyruvic acid, which readily arises from lactic acid or alanine—two characteristic abiotic products—is one of the major electron donors in substrate-level phosphorylations today. As for the collection of electrons, it seems safe to assume that the prebiotic world offered a choice of mineral electron acceptors operating at level B or below. Even organic molecules, produced by either abiotic syntheses or protometabolism, could have done the job, as we know from present-day metabolism.[1]

According to my scenario, emerging life was supported in this way up to the stage of genetically independent protocells capable of making proteins and competing with each other on the strength of their protein innovations. A turning point was reached with the appearance of the first membrane-embedded, two-pump machinery capable of using the energy released by falling electrons for ATP assembly by way of protonmotive force.

Most likely, this machinery received electrons at level B, which is the level at which the majority of metabolic electron donors feed electrons into electron-transfer chains today, mostly by way of NAD (see chapter 4). Entry at level C is not excluded, since a few metabolic intermediates—succinic acid, for example, a typical product of abiotic syntheses—deliver electrons at this level. Oxygen, the universal final electron acceptor of present-day life, was not available in the prebiotic world to collect the electrons at the bottom of the chute, but other acceptors could have been present, for example, ferric iron (see chapter 3), which, like oxygen, accepts electrons at level E.

As a means of satisfying energy requirements, the new machinery was far superior to the pre-existing thioester-dependent machinery, since level-B electron donors are much more numerous than level-A donors. But this is only a trivial advantage compared to another, truly life-saving consequence of the new development: The thioester-dependent machinery could now be reversed with the help of ATP provided by the protonmotive machinery. Electrons could be lifted from level B to level A—which is the key level for biosynthetic reductions. The assimilation of carbon dioxide, for example, requires electrons delivered at level A. The term "reverse electron transfer" designates the energy-dependent lifting of electrons from a lower to a higher energy level.

Once initiated, this kind of "bootstrapping" of electrons could continue. A second protonmotive chute, fed at level C and unloading electrons at level D or below, could have appeared, thereby opening the possibility of lifting electrons from level C to level A—first from level C to level B by reversal of the first protonmotive machinery, and then from level B to level A by reversal of the thioester-linked machinery. In all such cases, a lower part of the cascade provides power for lifting electrons to the upper part of the cascade.

This, I submit, is how autotrophy arose and emerging life was freed of its dependence on abiotic chemistry. All that was still needed for authentic autotrophy was for the environment to provide a suitable mineral electron donor delivering electrons somewhere between levels B and C. Then, in the presence of an appropriate acceptor at level D or below (oxygen today, perhaps ferric iron in prebiotic times), downhill electron transfer in the lower part of the cascade could cover all the ATP needs, whereas uphill (reverse) electron transfer in the upper part of the cascade, powered by ATP assembled in the lower part, could provide the high-grade electrons required for the reduction of mineral building blocks, such as carbon dioxide, nitrogen or nitrate, sulfate, and so on. This mode of life characterizes a number of autotrophic bacteria, called chemoautotrophs, which use mostly sulfur derivatives such as hydrogen sulfide or elementary sulfur—likely components of the prebiotic environment—as electron donors.

Once the basic skeleton of autotrophy was in place, it is easy to see how further improvements could have arisen, up to the formation of a complete, three-chute, protonmotive, B-C-D-E electron cascade, such as is found today in all the most advanced organisms, be they autotrophic or heterotrophic.

The ultimate improvement in electron-transfer chains occurred when a porphyrin variant arose in which a magnesium atom came to replace the iron atom in the central hole of the dish. Thus, presumably, was born chlorophyll (from the Greek *chlôros,* green, and *phyllon,* leaf), the green pigment of photoautotrophic, or more simply, phototrophic, organisms (Greek *phôs,* light).

The immediate benefit of chlorophyll sprang from its ability to garner energy from sunlight and to undergo what is known as an electron delocalization in the process. That is, one of the electrons in the chlorophyll molecule is displaced by the absorbed light energy from its resting level to a level of higher energy. The molecule is said to be excited by light. This phenomenon occurs with many colored substances (which are colored for the very reason that they absorb part of the light they receive), but is most often short-lived. The displaced electron promptly falls back to its resting level and the absorbed energy is given out as heat—not a useless commodity, as we know, but one that cannot be used effectively for the performance of biological work.

In the case of the excited chlorophyll, this dissipation of the light energy is avoided. Thanks to an intimate association of the molecule with a membrane-embedded respiratory chain—it was born from a member of such a chain, remember—the delocalized electron is diverted and led to fall productively through this chain. The light energy thus supports the generation of protonmotive force and, through this force, the assembly of ATP.

When they are used for ATP assembly, the electrons boosted by light energy eventually return to the chlorophyll molecule at their resting level, ready for another light-driven lift up. The electrons can thus cycle endlessly—up with light, down the cascade—and support ATP assembly in the process (cyclic photophosphorylation). In addition to this cyclic process, a noncyclic process also exists whereby the boosted electrons are used for the many biosynthetic reductions required for autotrophic life. In this event, chlorophyll recovers the diverted electrons from some outside source.

The evolution of phototrophy went through two main stages. First to appear was photosystem I, which picked up electrons from mineral donors, such as certain sulfur compounds, and boosted these electrons, with or without the additional help of thioester-dependent reverse electron transfer, to the energy level (A) required for biosynthetic reductions. Then came photosystem II, which has the ability to remove electrons from water molecules, with the release of molecular oxygen. Photosystem II lifts the electrons up to an intermediate level of energy from which they can be taken up further by photosystem I and pumped to the top. This development freed autotrophic life from its dependence on an external electron donor. Henceforth, life needed only light and water to turn air and a few water-borne minerals into a luxuriant green mantle, which could, in turn, support many heterotrophic forms of life. This momentous event did, however, also signal the appearance of molecular oxygen on our planet, with long-term effects of dramatic magnitude (see "The Great Oxygen Crisis," pp. 135–36).

Chapter 11

Adaptation to Life in Confinement

WITH CELLULARIZATION, for the first time life became a property of discrete, autonomous, individual units capable of diversification. Darwinian competition was the most immediate consequence of this development, as well as the main driving force of its further evolution. In addition, cellularization permitted a number of acquisitions that further enhanced the capacity of the protocells to survive and proliferate as individual units, and, therefore, offered likely catches for natural selection to net. In most cases, these adaptations were the result of modifications of the cell membrane, which, in addition to its functions as boundary, site of controlled passage, and protonmotive machinery, evolved into a sensitive interface capable of exchanging many signals with the environment and initiating appropriate responses. A brief review, not necessarily chronological, of these aids to protocellular life follows.

SENSING

Protocells probably first "learned" to explore their environment chemically, by "tasting" it, so to speak. Here is how it may have happened. It is known that membrane proteins often have one end sticking out on one side of the membrane, and the other end on the other side, with the hydrophobic transmembrane segment in between. Imagine now that the outer part of such a protein possesses a site complementary to a given substance and, therefore, has the ability to bind this substance. Imagine, further, that this binding induces a conformational change—a coiling, for example, or an uncoiling—of the transmembrane segment, such that the inner part of the protein, in turn, undergoes a change in shape. Imagine, finally, that this alteration of the inner part of the protein causes a specific effect, for example, the opening or closing of a channel, or the activation or inhibition of an enzyme. What you are witnessing, with the appearance of such a protein, is the creation of a link that allows an outside chemical to influence internal events without entering the protocell, clearly a valuable acquisition if the response turns out to be adaptive—which decision will be left for natural selection to make.

One can think of such a protein as a switch controlled by a chemical trigger. The triggering substance is called an agonist, or active agent. The outside part of the switch that binds the agonist is termed a receptor. The internal responding part is the effector. The switch is off (effector inactive) when the receptor is vacant, and on (effector activated) when the receptor is occupied by the agonist. There are usually many switches of any given kind on the surface of a single cell. This allows for a graded response between all switches off and all on. The extent of the response depends on how many of the total receptor sites are occupied by agonist molecules. This proportion depends on the abundance of agonist molecules in the environment and on the avidity with which receptor sites fish them out. With a properly tuned system, a fine adaptation of the response to the amount of agonist present may obtain.

As a simple example, imagine a protocell equipped with a channel allowing the entry of a substance that is useful or even essential in small amounts, but becomes harmful in excess. Such a protocell would lead a precarious existence, strictly dependent on finding the right concentration of substance in the environment. Let the protocell now acquire a receptor that binds the substance and is connected to an effector that closes the channel when the receptor is occupied. As the substance's concentration in the outside medium increases, the number of receptor sites occupied, and therefore that of channels closed, increases. Fewer channels are open, but each lets in more molecules of substance per unit of time, so that the total number of molecules of substance entering per unit of time remains approximately the same over a wide range of concentration. For the protocells concerned, acquisition of such a protein means the ability to adapt to considerable fluctuations of the substance in the environment, a powerful selective advantage.

The transmembrane receptor-effector combination has enjoyed an enormous evolutionary success, linking an immense variety of substances to a great diversity of responses, including the seeking and catching of food, the avoidance of noxious substances, the triggering of cell division, the stimulation of secretion, and many others. The outcome is especially intriguing when the agonist is produced by another cell. The receptor then makes it possible for the agonist-producing cell to influence events in the receptor-bearing cell. What takes place is *communication between cells by means of chemical signals,* a phenomenon that has played increasingly important roles in later evolution, especially in eukaryotes. As you read this sentence, billions of cells communicate with each other through chemical signals in your brain to make this array of printed symbols intelligible.

There is no way of knowing when protocells first acquired transmembrane proteins with useful receptor-effector combinations, or what these first transmembrane proteins were. It is likely, however, that such acquisitions were made almost as soon as the degree of structural and functional sophistication of the protocells allowed, since the resulting selective advantages would have been considerable.

MOTILITY

Transmembrane proteins can also serve to transmit chemical signals from the interior of a cell to its surface. One particularly intriguing such molecule has an energy-consuming inner part connected to a mobile outer part in such a way that the energy spent inside is converted into mechanical work outside, like an arm connected to an oar or an engine to a propeller. The source of energy could be the splitting of ATP or, alternatively, protonmotive force. In the bacterial world, the ultimate machine of this sort, named flagellum (plural, flagella; Latin for whip), is a rotating helical rod that emerges on the surface of the cell and is connected to an internal "turbine" driven by protonmotive force. The shaft of this rod traverses the cell membrane and the cell wall through special, tight-fitting "bearings." No doubt, this elaborate engine was preceded by simpler motors, possibly constructed with ATP-splitting proteins that bend when they bind ATP, and straighten when the bound ATP molecule is split. Eukaryotic motor systems, including our own muscles, are all built with proteins endowed with this ability.

Properly arranged on the cell surface, such proteins could cause the cell to move with respect to the surrounding water. This sort of motility started as a random walk. Cells moved for a while in a given direction and then "tumbled," resuming movement in another direction. There was little advantage to such a property, which was as likely to bring the cell to a less favorable as to a more favorable environment. Things changed when the motor systems became coupled with chemical receptors. The coupling was primitive, the variable affected being merely the frequency of tumbling. Receptors sensitive to useful substances came to inhibit tumbling, so that the cells continued to move toward the substance for a longer time. In contrast, receptors sensitive to harmful substances precipitated tumbling, thus shortening the time of progress of the cells in the wrong direction. This mechanism—known as positive and negative chemotaxis, that is, the ability to seek useful substances and to avoid harmful ones—has remained to the present day, at least in the bacterial world. Even though it acted simply by modulating random events and was only marginally effective for single cells, this mechanism was very powerful at the population level. Its acquisition entailed a considerable evolutionary gain.

PROTEIN EXPORT AND THE BIRTH OF DIGESTION

In their wanderings, protocells left traces of their passage in the form of discharged waste products and other chemicals. Such traces could conceivably alert protocells to stay away from or congregate around each other. A more specific mode of discharge arose through a modification of the mechanism whereby nascent proteins are targeted to membranes (see chapter 9). Through the combinatorial vagaries of gene assembly, targeting sequences came to be added also to proteins other than those carrying the appropriate hydrophobic sequences needed for nestling within lipid bilayers. A number of soluble proteins were fitted with the right tag and became embroiled cotranslationally or posttranslationally with membranes. It took only minor changes in the machineries involved for insertion to make way for complete translocation. The outcome was protein export, or secretion.

Among the proteins that came to be discharged out of protocells in this way, a particularly favored group consisted of enzymes that split with the help of water the chemical bonds whereby building blocks are linked in natural macromolecules. These hydrolytic enzymes, also known as hydrolases, fragment proteins into amino acids; nucleic acids into nucleotides; nucleotides into sugars, bases, and phosphate molecules; saccharides into individual sugars; phospholipids into their constituents; and so on. Hydrolases play havoc with vitally important substances, and their acquisition must have carried major risks. The protocells in which such enzymes emerged in fully active form were promptly eliminated until chance provided the harmful proteins with a tag that caused them to be exported as soon as they were made. Then the evolutionary disadvantage suddenly turned into an advantage. Enzymes released into the environment by living protocells could break down organic debris left around by dead protocells, and the resulting small molecules could serve to support the nutritional requirements of the secreting protocells. Digestion was born and, with it, the possibility for a living organism to exploit the synthetic activity of another; in other words, the possibility of heterotrophy at the expense of autotrophy. This event had far-reaching consequences. It freed the organisms concerned from the heavy burden of making their own building blocks and gave them greater scope for innovation. Notably, it has been a key step in the emergence of the whole animal world, including the human species. We live, directly or indirectly, on the products of plant photosynthesis, which we digest in our stomach and intestine with the help of extracellularly secreted enzymes.

Protocells surrounded by a simple murein wall permeable to protein molecules—such protocells would correspond to present-day gram-positive bacteria—had to reside within a stagnant and confined environment in order to benefit from the secretion of digestive enzymes. Otherwise, the enzymes would have been washed away before they could act. In contrast, protocells equipped with a second membrane—corresponding to gram-negative bacteria—kept the secreted enzymes within the periplasmic space intercalated between the two membranes. There, the enzymes were able to act on such molecules as were let through from the outside by the porins of the outer membrane.

Whether membrane-bounded or not, the space surrounding primitive heterotrophic cells may be viewed as the first digestive pocket in the history of life. The first stomach, so to speak, except that it was not situated inside the organism, but the other way round. We shall see later that internalization of this stomach may have played a decisive role in the conversion of an ancestral prokaryote into the first eukaryote (see chapter 16).

A TOUCH OF SEX

One last surface acquisition must be mentioned. It consists of long, slender filaments, up to several times the length of the cells themselves, that adorn the surface of many bacterial cells. Appropriately called pili (singular, pilus; Latin for hair), these structures serve mainly as anchors allowing the cells to attach to some support. They are neither motile nor connected to effectors. But they have some chemical specificity and therefore can act as receptors. For example, pili endowed with the right specificity could immobilize wandering cells in the neighborhood of a rich food supply or cause cells to congregate in colonies. It is easily seen how selection would favor the emergence of pili with useful specificities.

Certain pili specific for cellular surface components allow cells to touch each other in a special way. To call such contacts between bacteria fondling would no doubt be too anthropomorphic. The fact remains that the development of such pili did lead to sex and thereby unleashed one of the most, if not the most, powerful forces of diversification to drive evolution. Cells possessing sex pili—called male—use such filaments as some sort of molecular penis to copulate with female cells, that is, cells devoid of sex pili. In the course of such conjugation, as it is termed, male cells introduce into female cells a small, satellite circular piece of DNA, called a plasmid, on which, among others, the genes coding for the sex pilus proteins are situated. Not infrequently, a duplicate of a lesser or greater part of the male chromosome is injected into the female cell together with the plasmid. Subsequent recombination of the injected DNA with the recipient cell's DNA then brings about the formation of hybrid chromosomes made of genes contributed by the two parents.

Until this development occurred, the mutations that provided natural selection with materials from which to choose were mostly base replacements due to replication errors or to chemical injuries, as well as insertions, deletions, inversions, and duplications of certain DNA (or RNA) stretches. Thanks to conjugation and recombination, a whole new gamut of hereditary variations involving entire genes or clusters of genes was offered to selection. The evolutionary game became immensely richer and more innovative. It is amusing that, in this earliest form of sex, males did their job but females were the real beneficiaries of the evolutionary advances. If you are tempted to derive some sort of sexual moral from this fact, note that the first thing males do upon conjugation is to give the females genes (coding for pilus proteins) that turn them into males.

Chapter 12

The Ancestor of All Life

THE EVIDENCE that all known living organisms are descended from a single common ancestor is overwhelming. We cannot exclude the possibility that unknown or poorly known organisms of different origin exist in some remote environment that has remained isolated for a very long time. However, no discovery suggestive of a major break with "our" way of life has yet been made. Until proven otherwise, the hypothesis of single ancestry holds true.

In this chapter, I shall try to reconstruct the profile of our common ancestor and to retrace its emergence historically, paying special attention to the shape of its hidden roots. Did the universal ancestor arise as a single shoot along a highly deterministic pathway that left little to chance? Or was it one of many branches, a branch that simply happened to spread faster, smothered all others, and ended up filling the world with its progeny?

RECONSTRUCTING THE DISTANT PAST

The universal ancestor is defined as the organism that existed just before the tree of life divided into two separate branches that have spread out extensions to the present day. This definition distinguishes the universal ancestor from the more primitive ancestral forms that came before; it also leaves open the possibility of earlier branchings that have left no extant progeny. In principle, drawing a portrait of this ancestral organism is simple: just put together all the properties that are shared by all living organisms. In practice, three caveats complicate matters. First, we must subtract from our picture such shared properties as could have been acquired separately in individual branches after the first forking of the tree of life. Second, we must also subtract properties that appeared solely in one branch and were later acquired by the other branch or branches by some mechanism of gene transfer.

Finally, we must add to the picture the properties that are lacking in certain organisms, perhaps in all, because they were lost in the course of evolution.

The first caveat refers to what is known as evolutionary convergence. This is an important phenomenon in later evolution, illustrated, for example, by the development of flight independently in insects, pterosaurs, birds, and bats, but it is probably of minor relevance in the kind of molecular evolution we are mostly concerned with here. It is very unlikely, for instance, that a molecule such as cytochrome *c*, which shares more than fifty out of one hundred or so amino acids in all species investigated, could have arisen independently in two or more branches.

The second caveat is more serious. Horizontal gene transfer[1]—so named in opposition to "vertical" transfer from generation to generation—is believed to be a common phenomenon in the bacterial world. It seems likely that primitive organisms exchanged genes at least as readily as do present-day bacteria. However, to account for the occurrence of the same gene in *all* extant organisms, horizontal transfer of the gene must have occurred very early in evolution, when very few branches (most likely only two) existed in the same or closely connected niches. In addition, members of the transformed branch that did not acquire the gene must have been eradicated.

As to the third caveat, it is obviously relevant and must be applied with careful discrimination. Fortunately, enough universally shared features are left to make the main picture fairly clear. Uncertainties concern a few additional properties that are not found in all forms of life but could conceivably have been present in the common ancestor and been lost subsequently in some of its descendants.

Even with the above caveats, such a wealth of biochemical information has become available on all major forms of life that one would expect the reconstruction of the common ancestor to be fairly straightforward. This would be so but for what has come to be known as the rooting problem.

In the late 1970s, the American microbiologist Carl Woese,[2] from the University of Illinois, dropped what amounted to two simultaneous bombshells on the scientific world. First, he announced, on the basis of the comparative sequencing of RNA molecules found in the ribosomes of all living beings, that extant bacteria are not members of a single family, as was generally assumed, but fall within two groups that must have separated at the dawn of cellular life. He elevated these two groups to the rank of kingdom and named one archaebacteria, because of a number of characters he believed to be particularly archaic (from the Greek *arkhaios*, ancient), and the other, eubacteria (from the Greek *eu*, good). The two kingdoms are grouped together under the name prokaryotes (Greek *karyon*, kernel), which indicates that neither possesses a true nucleus, in contrast to the eukaryotes, which encompass all protists, plants, fungi, and animals. Woese[3] has more recently promoted the two kingdoms to an even higher rank, for which he has proposed the term "domain," and has renamed them archaea and bacteria to emphasize their differences. This proposal, however, is not yet generally accepted. I have not followed it in this book because the familiar term bacterium has become so much a part of everyday vocabulary that redefining it is likely to be confusing to most readers. On the other hand, I have adopted Woese's original classification, which met with some resistance at first but is now almost unanimously accepted.

Woese's second bombshell was even more stunning. Eukaryotes, which were commonly believed to have detached from the (single) prokaryotic trunk some time around one billion years ago or later, are almost three billion years older. They originated from a line that branched from the tree of life virtually at the same time that archaebacteria and eubacteria separated.

Thus, the common ancestor lies at the root of a trifurcation. However, bifurcations, not trifurcations, trace the development of an evolutionary tree. There are thus three possibilities: (1) The first bifurcation separated archaebacteria and eubacteria, and eukaryotes then branched off the archaebacterial line. (2) Prokaryotes separated first, as above, but eukaryotes branched off the eubacterial line. (3) The first fork separated eukaryotes from prokaryotes, which later subdivided into archaebacteria and eubacteria. Hence the rooting problem. In possibilities 1 and 2, the common ancestor must have been a prokaryote, ancestral to eukaryotes by way of archaebacteria in the first case and by way of eubacteria in the second. In the third possibility, the common ancestor could have been anything between a eukaryote and a prokaryote.

Unfortunately, available sequencing data not only have failed to provide an unambiguous answer to the rooting problem but have yielded conflicting answers. These questions are highly technical and concern the interpretation of the data at least as much as the data themselves. To sum up a complex situation, most authors favor a prokaryotic root. Woese's proposal of a common ancestor related to the most thermophilic (heat-loving) archaebacteria, which sequencing has identified as particularly ancient, is widely accepted. Even an ancestor more closely resembling eubacteria is believed to have been adapted to a high temperature, since the most thermophilic eubacteria are also the most ancient of the group.[4] The origin of the eukaryotic line remains uncertain. Eukaryotes share many properties with archaebacteria, but also a few with eubacteria. I shall examine in chapter 14 the various explanations that have been proffered to account for these discrepancies.

Not all researchers accept a thermophilic prokaryotic ancestor. The French investigator Patrick Forterre[5] has argued strongly that thermophily cannot go back to the common ancestor because a primitive system is unlikely to have withstood the harsh conditions imposed by a very high temperature. He believes instead that adaptation to a high temperature is a later development that was achieved through simplification. According to Forterre, the common ancestor was a primitive eukaryote, and prokaryotes were born by a streamlining process associated with the invasion of an increasingly hot environment. The prokaryotic way of life, instead of being primitive, as generally assumed, would thus have arisen as a secondary adaptation to heat. Once it had arisen, this type of organization would have proved enormously successful and invaded all niches presently occupied by bacteria.

An even stranger hypothesis has been advanced by Mitchell Sogin,[6] an expert in the field of comparative sequencing from the Marine Biological Laboratory at Woods Hole, Massachusetts. According to Sogin, the common ancestor was a primitive cell—a progenote, to use a term coined by Woese—straight out of the RNA world, and DNA appeared only after the first forking of the tree of life, in the line that was to lead to prokaryotes. The RNA line went on developing into a large cell resembling eukaryotes in several respects, but devoid of a nucleus and lacking the entire machinery involved in DNA synthesis, replication, and transcription. This cell then allegedly acquired a nucleus and its machinery by engulfing a prokaryote, probably of archaebacterial lineage.

I shall come back to these proposals later, when examining the early evolution of prokaryotes and eukaryotes. For the time being, I shall stick to the commonly accepted hypothesis.

PORTRAIT OF AN ANCESTOR

The common ancestor of all life was a single-celled organism of prokaryotic type, that is, resembling present-day bacteria in lacking a fenced-off nucleus and having only a rudimentary internal organization. This appears as the more probable possibility in the present state of our knowledge, although alternative descriptions have been proposed.

What did this organism look like? There is no clue to this question. The familiar picture of the blunt, rod-shaped *Escherichia coli,* the main resident of our gut and the most extensively studied specimen of the bacterial world, is a misleading stereotype. Bacteria come in all shapes—spherical, cylindrical, and filamentous. Some microbes recently isolated from deep-sea hydrothermal vents even look for all the world like miniature, flat, sharp-edged rectangular tiles. Some ancient microfossils are long, thin, threadlike structures. But this is no more than a hint. The choice of ancestral shape is entirely open.

Although rare wall-less bacteria exist today, it seems likely that the ancestral cell was surrounded by a solid wall. Wall-less forms are very fragile, and the ancestral cell probably could not have survived without some outer protection. Also, we know from microfossils that organisms encased by a wall existed as early as 3.5 billion years ago.

In all likelihood, the plasma membrane of the ancestral cell was built on the universal theme of the lipid bilayer with inserted transmembrane proteins. Granting the existence of a typical membrane, the question arises as to which of the many specialized systems mentioned in the preceding chapters were already incorporated within it and which came later.

A valuable clue is provided by the fact that the ancestral cell almost certainly used protonmotive force. This major mechanism of energy retrieval is too widespread to be a later product of evolution. So are several of the main components of membrane-bound electron-transfer chains, including iron-sulfur proteins, hemoproteins, flavoproteins, and perhaps others. The most elaborate respiratory chains were probably yet to come, but some of their main components were already in place. On the other hand, it is likely, although this point is debated, that the ancestral cell lacked the ability to use light energy for the generation of protonmotive force.

Be that as it may, the use of protonmotive force indicates that the ancestral cell membrane was impermeable to protons and other ions, and therefore to most of the molecules that must move in and out to satisfy the metabolic requirements of a cell. Therefore, the membrane must have possessed the minimum number of transport systems needed for metabolic exchanges with the environment, and these systems must have been sufficiently sophisticated to operate without letting protons through.

The ancestral membrane must also have possessed the various enzymes and insertion systems needed for its own construction. In addition, the membrane must have included the translocation systems needed for the extrusion and assembly of the surrounding wall constituents. Quite possibly, the ancestral cell was able to secrete proteins and carried out extracellular digestion. The widespread distribution of the mechanisms involved in these activities and their close molecular similarities throughout the living world strongly support these contentions.

We do not know to what extent the ancestral cell was equipped with surface receptors, or whether it possessed motile or sensing structures, including sex pili. These possibilities are by no means excluded. There can be little doubt that the ancestral plasma membrane and its associated elements already displayed many of the structural and functional attributes that characterize bacterial cell membranes today.

Metabolically, the ancestral cell carried out all the reactions needed for the construction and breakdown of its constituent molecules and for the support of its energy requirements. It did so by proven pathways now operative in a wide variety of prokaryotic and eukaryotic organisms. It had available for this purpose many of the coenzymes found in present-day cells, and used ATP as the main purveyor of energy. Some details of the ancestral metabolism must remain conjectural, as they depend on the kind of environment the cells occupied. We have seen that because the most ancient among known bacteria live in a hot environment, it is widely believed, though not unanimously, that the common ancestor occupied such an environment. Like present-day thermophilic organisms, it may have used some sulfur compounds as a final electron acceptor, or perhaps ferric iron, as I have suggested.

The question has often been raised whether the ancestral cell fed on pre-existing organic molecules, as heterotrophs do today, or shared with extant autotrophs the ability to build organic molecules from simple mineral precursors. The heterotroph theory was popular in the days when the ancestral cell was viewed as a resident of the primeval soup. Indeed, such must have been the case for the first protocell. The ancestral cell, however, is the product of a long evolutionary history, in the course of which complex electron-transfer chains were assembled, and autotrophy, therefore, could very well have developed. Also, the abiotic supply of organic molecules must have dwindled progressively and could well have died out long before the ancestral cell appeared. Finally, autotrophy is sufficiently widespread to allow the hypothesis that it was already an attribute of the ancestral cell. I consider it likely, therefore, that the ancestral cell was autotrophic, though not necessarily phototrophic. It more probably resembled present-day chemoautotrophs and, like these organisms, relied on mineral electron-transfer reactions to satisfy its requirements for energy and high-grade electrons, though not with oxygen as the final electron acceptor, but rather ferric iron or some other mineral substance.

Last but not least, it is highly likely that the ancestral cell had DNA genes, probably strung together in a single, circular chromosome; that it transcribed these genes into RNA; and that it translated most of these transcripts (except those that served catalytically or structurally) into proteins according to the universal genetic code. The ancestral genes and the corresponding proteins had reached the length and complexity that they have in all living organisms today. The proteins were assembled on typical ribosomes with the help of the complex machinery that, with minor variations, is universally operative.

In conclusion, we may take it that the ancestral cell was a fairly typical prokaryote, which we might well mistake for some contemporary bacterium if we should happen to meet it today. Exactly which kind of modern prokaryote the ancestral cell would have resembled most is, however, beyond our knowledge, because several blanks unfortunately remain in the portrait we are able to draw. Following are some of the major questions awaiting a clear answer: Was the ancestral cell wall made of murein or other constituents? Were the ancestral membrane lipids of the type chemists define as ether lipids, or were they of the ester type? Was the ancestral organism phototrophic or simply chemoautotrophic? Were the genes of the ancestral cell split by introns or were they continuous? These uncertainties exist because we are faced in each case with clear-cut differences, obviously of very ancient origin, among major groups of extant organisms. I shall address these questions in due time when examining the early evolutionary history of the ancestral cell's progeny.

Some readers might be dismayed by the many blanks in the portrait I have painted. I would rather have them marvel at the details that have been gathered—all in the lifetime of this writer—about a minuscule entity of enormous complexity that existed some 50 million human lifetimes ago.

Chapter 13

The Universality of Life

ALL EXTANT LIVING organisms are descendants of a single ancestral form of life. So much is clear. But why is it so? There are several possible answers to this question.

First, we could, with the proponents of an extraterrestrial origin of life, identify the ancestor with the immigrant germ that seeded the Earth four billion years ago.

A second explanation is that no other ancestral form was possible. It is single because it is unique.

A third possibility is that the ancestral form arose among a number of competing forms by a process of Darwinian selection.

Or, alternatively, there were several forms to start with, but all the other lines are extinct.

Finally, there is the possibility that a mere accident caused the ancestral form to emerge among several that were equally possible.

Having agreed to disregard the first possibility, we are left with the other four. The central issue here is the old dichotomy between chance and necessity. How much in the common ancestor was due to contingency, how much to determinism? We have no solid clues to answer this question, only surmises based on what we know of the nature of life, and suspect of its origin.

IS LIFE UNIQUE?

Given the physical-chemical conditions that prevailed on our planet 3.8 billion years ago, a protometabolism leading to RNA-like molecules was bound to arise along well-defined, reproducible chemical lines. Such is the unambiguous conclusion I have drawn from a consideration of the mechanisms involved. Because of the congruence rule, this conclusion extends to all features of today's metabolism that were prefigured in protometabolism, including such key elements as electron transfer, group transfer, thioester-dependent substrate-level phosphorylation, the central role of pyrophosphate bonds with, most likely, a privileged position for ATP, and perhaps the participation of several major coenzymes, such as pantetheine phosphate, coenzyme A, and NAD. Life is very much constrained by its early chemistry, which was itself ruled by deterministic factors.

What about other kinds of "life," based on a different kind of chemistry, born under different physical-chemical conditions, and adapted to a different environment? I cannot reject such possibilities outright, but I find it unprofitable to raise them on the pretext of leaving no stone unturned, as long as not the slightest clue to their reality, or even plausibility, is available. The properties we most intimately associate with the concept of life depend on versatile macromolecules that all chemists agree could not be built with other than carbon frameworks. Even silicon, carbon's closest relative, would not do. Water is uniquely suited as a medium for life. No other liquid is known that has a comparable combination of favorable physical properties. In addition, water provides two indispensable elements for the construction of carbon-containing molecules, hydrogen and oxygen. The irreplaceability for life of nitrogen, sulfur, phosphorus, and other biogenic elements has likewise been emphasized. Add to these considerations the predilection of interstellar chemistry for compounds of these elements and you end up with a strong case in favor of a life uniquely constructed according to the same kind of "organic" chemistry.

Whether this kind of chemistry could, under a different set of conditions, develop into life-generating worlds different from the RNA world is a question that must be left open. Much contemporary research in organic chemistry aims at "imitating" life with artificial molecules. Even if such efforts should one day be successful, the question of the artificial process ever occurring under natural conditions would still have to be answered. Until this happens, if it ever does, let us be content with the life we know. It is wondrous enough to make recourse to other hypothetical lives unnecessary.

Within the straightjacket of its chemistry, could emerging life have evolved a different genetic system? I addressed this question in part II and reached the conclusion that only secondary details—perhaps in the genetic code, although even that is far from certain—could have been different. Contingency no doubt played a role in shaping the exact evolutionary history of the RNA world, but the stringency of the selection factors ensured that the end result could hardly have been different, including the formation of protocells, a mandatory condition of further evolution at a certain stage.

There remains the long pathway from protocell to common ancestor. Chance played a role at every step of the pathway by providing an appropriate mutation, thus offering enormous scope for diversity. There was a major bottleneck, however: the need to develop autotrophy before the supply of abiotic products became exhausted. By that time, any heterotrophic line that might have existed was perforce extinct. The survivors either had available a direct supply of electrons at level

A from some mineral source—a very unlikely occurrence according to our knowledge of the mineral world—or had converted their thioester-dependent machinery to the reverse transfer of electrons from level B to level A, with the help of energy supplied by the hydrolysis of ATP. This implies that they had developed an alternative machinery for generating ATP and had available an appropriate mineral source of electrons with, if necessary, the means of bringing these electrons to level B.

These conditions may have left little leeway for contingency, especially if, as I suspect, circumstances—such as a high environmental acidity—put a high selective premium on the development of energy-driven proton extrusion. Under such circumstances, harnessing protonmotive force could well have been the only means of getting through the autotrophy bottleneck. If not, it was most likely the most efficient and, perhaps, the easiest to attain, considering that the required cofactors, such as the flavin derivatives FMN and FAD and, perhaps, even porphyrins may have been present as products of the carbon-nitrogen combinatorial chemistry that generated the RNA world. There may have been some competition at the entrance of the bottleneck, but little choice at the exit. If my reconstruction is correct, therefore, some cell resembling the common ancestor in its main characteristics may well have been the obligatory outcome of the biogenic process set off on the Earth some 3.8 billion years ago.

In the introduction to this book, I argued on theoretical grounds—remember the thirteen spades—that the emergence of life must have involved a very large number of steps, most of which had a high probability of occurring under the prevailing conditions. But I left open the possibility that there might be more than one pathway compatible with this exigency. My conclusion, after a consideration of the underlying chemistry, is that, given the opportunity, the development of life is very likely to take the course it actually took, at least in all essential aspects.

EXTRATERRESTRIAL LIFE

Is there life elsewhere in the universe?[1] The two Viking spacecraft (launched in 1976) carried equipment designed to answer this question by probing for traces of life on Mars. The results, unfortunately, were negative or at best "ambiguous." But Mars is only our nearest neighbor. What about other solar systems? There are about 100 billion stars in our galaxy alone, and there are billions of galaxies in the universe. How many of the trillions of existing stars have planets? How many of those planets would have a geological history comparable to that of planet Earth? On how many of those would the physical-chemical setting that fostered the birth of life on Earth be duplicated? And, finally, on how many of the planets exhibiting these conditions would life in fact emerge, and how closely would that life resemble life on Earth?

Nobody knows the answers to these questions, but they have been much in the limelight since a memorable day, November 1, 1961, when a number of scientists met at the National Radio Astronomy Observatory in Green Bank, West Virginia, to launch the search for extraterrestrial intelligence (SETI). On this occasion, they devised the so-called Green Bank equation, which includes as relevant parameters the number of stars capable of having planets in the universe and the probability of a planet's being capable of harboring life. Some of the greatest experts on cosmology have pondered these problems in the light of available evidence. Although quantitative estimates vary widely, the consensus is that the history of the Earth is probably not unique. The figure of about one million "habitable" planets per galaxy is considered not unreasonable. Even if this value were overestimated by several orders of magnitude, it would still add up to trillions of potential cradles for life. If my reading of the evidence is correct, this means that trillions of planets exist that have borne, bear, or will bear life. The universe is awash with life.

Unfortunately, interstellar distances are such that we may never get confirmation of this, unless extraterrestrial life somewhere reached a stage of development that rendered it capable of sending out messages that we can receive and decode. Hence the interest of the Green Bank participants in the search for extraterrestrial intelligence. In the meantime, we must rest content with the message we get from life down here, and that message is that there must be plenty of life out there.

ARTIFICIAL LIFE

This term has nothing to do with life in a test tube, although this may happen some day, but rather with life in a computer.[2] Ever since the famous Hungarian-American mathematician Johannes von Neumann devised the first "cellular automaton" in the late 1940s, theoreticians have evinced considerable interest in the mathematical modeling of such typical properties of living organisms as complexity, self-organization, development, reproduction, and evolution. They have paid special attention to the spontaneous emergence of these properties as a result of interactions among different variables intended to represent catalysts and their reagents, or genes and their products.

These models have highlighted conditions under which order can arise out of disorder through fluctuations that take place randomly until the system becomes caught in a network of interactions that drives it toward a dynamically organized configuration in which it settles. What is depicted is the stochastic exploration of a "space," eventually leading the system to fall into a "basin." Stuart Kauffman,[3] a pioneer in the field and a prominent member of the Santa Fe Institute, which has become the mecca of artificial life, prefers the opposite images of a "rugged fitness landscape" with "adaptive peaks" separated by "valleys." Somehow, I find the image of falling into a basin more representative of reality than that of climbing to the top of a peak.

Irrespective of the imagery, what the new computing methodologies have revealed is how a system of multiple interacting variables may become stranded in a basin (or on a peak), and how, depending on the structure of the landscape, it may escape and land in another basin (or on another peak), according to a saltatory, non-linear kind of process, similar in its mode of unfolding to what evolutionists call punctuated equilibrium. This process includes prolonged periods of pseudostability during which little changes connected to each other by occasional jumps induced by some chance event.

According to Kauffman, the condition for this process is a kind of restricted instability separating chaos from fixity. "Life," he concludes, borrowing a phrase from his colleagues Norman Packard and Christopher Langton, "adapts to the edge of chaos."[4] Darwinian evolution operates within the confines of a landscape and can be understood only in relation to the configuration of this landscape.

"Artificial life" studies fit within the current interest in dissipative structures, complexity, chaos, catastrophe, turbulence, and other phenomena that obey non-linear relationships such that very small changes may precipitate major events, the so-called butterfly effect: A butterfly fluttering in Rio unleashes a storm over Chicago. Many natural phenomena owe their relative unpredictability to this kind of intermingling of stochastic and deterministic factors. Viewed in a certain perspective, life would seem to be a particularly striking example of such an occurrence. The possibility that the spontaneous origin and development of life could be accounted for through the "freezing" of some fortuitous configuration of matter has obvious appeal, especially among those who reject deterministic explanations.

The modeling approach, which uses elegant mathematical procedures, has thrown much valuable light on the intrinsic properties of self-organization and self-regulation that characterize all living systems. It has also illuminated some important aspects of biological evolution. But the term "artificial life," applied by analogy with "artificial intelligence," could be misleading. Life is a chemical process. If it is ever to be created artificially, it will be by a chemist, not by a computer.

PART IV

THE AGE OF THE SINGLE CELL

Chapter 14

Bacteria Conquer the World

THE COMMON ANCESTOR of all living things most likely was a bacterium, or prokaryote. Were it not for one line—which entered the long, complex, and mysterious pathway that led to eukaryotes—all of its progeny today would consist exclusively of bacteria. Even though they are no longer alone, bacteria still make up the larger part of the living world. Their evolution from the common ancestor illustrates the astonishing durability and versatility of prokaryotic forms of life. These qualities have allowed them to adapt to all sorts of different environments and to establish themselves and flourish in almost every kind of habitat. The diversity of bacteria is staggering and still incompletely inventoried. The reason for this success is simple: Bacteria are built to *grow and multiply as fast as materially possible.* They epitomize life at its rawest, with no frills.

THE SECRET OF BACTERIAL SUCCESS

A bioengineer attempting to construct a cell designed to proliferate as fast as possible could not come up with anything better than a bacterial cell. The bacterial genome is "streamlined" for fast replication. Genes are not split by introns and are crammed in the chromosome with hardly any space left for "junk" DNA. The chromosome itself is loosely structured, offering little impediment to the replication process. Furthermore, bacteria hardly ever stop duplicating their DNA and they manage to transcribe their genes and build all the RNAs and proteins they need for growth while they go about this activity. Some even start a second round of duplication before the first one is finished. As soon as two copies of their genome are available, they divide. As a result, it takes the average bacterium no more than twenty to thirty minutes to go through a complete growth and division cycle, as opposed to some twenty hours for the average animal or plant cell.

The bioengineer responsible for this feat of design was natural selection. We are therefore led to ask what evolutionary advantage could have driven the process. It cannot have been mere rapidity in producing progeny. A single bacterial cell would, by unrestricted exponential growth, cover the whole surface of the Earth with offspring in less than two days. It would take a eukaryotic cell little more than two months to achieve the same result. Clearly, lack of available resources will soon curtail multiplication in either case. There seems little advantage in beating the generation clock.

No, the main advantage bacteria gain from their fast multiplication rate lies in the enormous number of mutants they can offer to natural selection. By the time two eukaryotic cells have arisen out of one, a bacterial cell can produce up to one trillion cells, among which several billion mutations due to replication errors alone would be distributed. (A bacterial genome contains about three million base pairs, and the minimum replication-error frequency is of the order of one wrongly inserted base in one billion.) Many of these mutations will be neutral, that is, they will have no effect on the proliferating ability of the cell. Many others will be deleterious and will be weeded out by natural selection because the affected cells cannot multiply, or multiply more slowly than the others. But the odds are that an occasional mutation will turn out to be advantageous, especially if the environment changes. That is why the fight against disease-causing microbes never ends. Whatever new antibiotics may be discovered, some resistant mutant is likely to arise and to proliferate preferentially in the presence of the drug. This no doubt happened to bacteria countless times during their long evolution. Whenever conditions changed, some mutant was there that could take advantage of the new conditions. This versatility has allowed bacteria to invade every possible ecological niche. Bacteria do indeed cover the whole surface of the Earth with multiple layers of thriving life. They are the great survivors.

The evolutionary strategy of bacteria illustrates a statistical form of adaptability that is foreign to our intuitive understanding of this term. We tend to think of adaptation in terms of individual responses—conscious and deliberate or unconscious and automatic—to changing circumstances. We react to cold by putting on more clothes. The pupils in our eyes react to strong light by contracting. If we were bacteria, the individual would not count. We would let most of us perish frozen or blind, and we would rely on the odd person who happened to be warmly covered or to have naturally narrow pupils to rapidly replenish our stock with similarly adapted individuals. Should the temperature go up again or the light diminish, the same strategy would quickly bring back a population of scantily clad or wide-eyed individuals. Human beings could not behave that way, even if unhampered by any sort of respect for the individual, because it would take them too much time to recover. Bacteria can behave that way because of their rapid proliferation rate. Note, however, that the bacterial strategy has also been followed by more complex organisms, but much more slowly and gradually, over eons of time; variation screened by selection is the mainspring of Darwinian evolution.

Bacteria do have some built-in mechanisms for individual adaptability. For example, certain bacteria, when transplanted into a medium containing galactose (milk sugar) as the sole source of carbon, respond by producing the enzymes they need to utilize this sugar. This is a famous case in the history of science because it suggested at first that galactose somehow "instructs" the cells to make the appropriate enzymes. The French investigators François Jacob and Jacques Monod gained a Nobel Prize for demonstrating that the bacteria have the ability to make the enzymes all the time but fail to make them in the absence of the sugar because the corresponding genes are blocked by a protein, called a repressor, that prevents their transcription into RNA. A derivative of the sugar unlocks the genes by binding to the repressor in such a way that the repressor can no longer exert its blocking action.[1] This mechanism is understandable in the context of the general streamlining evolutionary strategy of bacteria. Don't waste time and energy making something until it is needed.

The operation of the immune system in humans and higher animals offers another intriguing example of an apparently instructional mechanism. When the organism is exposed to a foreign macromolecule, known as an antigen, it responds by making the corresponding antibodies, which are proteins that specifically neutralize the antigen. If the antigen is borne by a virus, a microbe, or a foreign cell, from a transplant, for example, the response includes the development of killer cells specific for that particular antigen. Cancer patients may even mount an attack against their own tumor cells in this manner. The cells responsible for manufacturing antibodies and killing foreign cells are called lymphocytes. They are generated in the bone marrow and circulate in the blood. When, in the early 1950s, these mechanisms began to be elucidated, it seemed obvious that the antigens must be instructing the lymphocytes. How else could one explain the fact that virtually any antigen could elicit a specific response? Only the Australian immunologist Macfarlane Burnet thought otherwise. He proposed the view, which appeared fantastic at the time but turned out to be correct, that small numbers of lymphocytes capable of fighting almost every possible kind of antigen exist preformed in the organism and that when any such cells come into contact with "their" antigen, they proliferate to form a clone (a population of identical cells) of antibody-making or killer cells specifically shaped to recognize the foreign antigen.[2]

Known as clonal selection, this mechanism recalls the bacterial strategy of having mutants waiting in the wings, so to speak, for a wide variety of occasions. But it is of a much higher degree of sophistication. The lymphocyte "mutations" are not accidental but programmed. During their maturation, lymphocytes undergo complex genetic rearrangements that create the millions of different genes—perhaps as many as one billion—responsible for their diversity. In this mechanism, which recalls the primeval modular game whereby the first large genes were assembled, a piece of the gene—let us call this piece A—is taken at random from a set of different A pieces and joined with B, C, D, and E pieces similarly taken from corresponding sets. Thanks to this process and to some additional causes of diversification, each lymphocyte ends up with a different ABCDE combination, which is translated into an antigen-recognizing protein of different specificity.

Also much more elaborate than the passive proliferation of the better-adapted bacteria is the mechanism whereby lymphocytes are specifically triggered to multiply upon contact with the appropriate antigen. The whole problem of growth control—and of its derangement in cancer—is posed here. Once we have stopped growing, most of our cells no longer divide. Exceptions include the cells that must be renewed because they have a short life span, as do many blood cells, or because they are lost by sloughing off, as are the cells that line our skin and mucosae. Most of our cells do, however, retain the capacity to multiply when appropriately triggered, for example, in wound repair. Enormous advances have been made in recent years in our understanding of these mechanisms, which involve a number of receptors and cell type–specific "growth factors" that activate the receptors. Many cancerous transformations are due to alterations of one of the genes that code for these receptors and growth factors. Known as oncogenes (cancer genes) for this reason, these genes all belong to the normal growth-control machinery.

In the clonal selection of lymphocytes, exposure to a given antigen selectively triggers the division of the cells that recognize the antigen, thus building an army specifically directed against the enemy. There is a disadvantage to all this sophistication, namely, slow proliferation. It takes lymphocytes a couple of weeks to build a clone that bacteria would generate in a few hours. In addition, there is the danger of derailment of the growth-controlling machinery, leading to such deadly diseases as lymphoma and leukemia, which are "lymphocyte cancers."

Even though modern bioengineering may not have invented the streamlined, ultrafast multiplier, it is now making full use of it. In gene cloning, a piece of foreign DNA is introduced into a bacterial host by genetic-engineering techniques, and the engineered cell is then given the opportunity for unrestricted clonal growth. The next day, appreciable amounts of the inserted gene, which was duplicated with the bacterial DNA at each generation, are available for sequencing and other uses. If the gene has been inserted in such a way that it is actually expressed, its product can similarly be obtained in essentially limitless amounts. This is how human insulin is now being made industrially by bacteria. Even slow multipliers are now used as factories. Lymphocytes rendered able to multiply indefinitely by fusion with a cancer cell are cloned on a large scale for the production of specific antibodies, called monoclonal for this reason.

THE FIRST FORK

Some time around 3.8 to 3.6 billion years ago, some members of the primordial ancestral cell population became separated from the bulk of the group by some geological or climatic phenomenon that brought them into a different, less favorable environment. This would have been their undoing but for a rare mutant—remember the bacteria exposed to an antibiotic—that happened to be adapted to the new circumstances and proliferated. We know this because both the parent line and the offshoot have grown, evolved, and diversified into a wealth of different varieties that are now found, often side by side, in every part of the world. Yet unmistakable proof of their kinship within one or the other group—either archaebacteria or eubacteria—and of the deep, primeval rift between the two groups remains inscribed in the structures of some key constituents, in certain characteristic metabolic reactions, and, especially, in the sequences of nucleic acids and proteins. And so, amazingly, events that happened eons ago in microscopic entities that have left no decipherable trace in the fossil record have been uncovered and reconstituted in some detail thanks to the magic of modern molecular biology.

What was the environmental change that triggered the rift? We don't know but we can hazard a guess. If, as is likely, though not certain, the universal ancestor was a prokaryote of archaebacterial type adapted to a high temperature, a plausible hypothesis is that the dissident group found itself in a cooler environment where heat resistance turned into an impediment. There is some support for this possibility. Whereas archaebacteria exist that thrive at temperatures of up to 110°C (230°F)—with pressure sufficient to prevent the water from boiling—no thermophilic eubacteria are found at a temperature higher than 80°C (176°F), and, according to comparative sequencing, those that live at this temperature are among the most ancient.

If this hypothesis is correct, what aspect of heat adaptation would have become unfavorable to survival in a cooler environment and could have been offset by an appropriate mutation? Extremely thermophilic archaebacteria survive and multiply in their inhospitable habitat thanks to their possession of heat-resistant proteins and other constituents. Most proteins unfold and lose their specific conformation irreversibly when heated to temperatures on the order of 50° to 70°C (122° to 158°F). The coagulation of egg white is an example. The usual enzymes become inactive under such conditions. The proteins of thermophilic organisms are much more resistant to heat denaturation, a property that is now attracting great interest on the part of those who want to use enzymes as industrial catalysts.

Also characteristic of thermophilic archaebacteria,[3] of all archaebacteria, in fact, are special membrane lipids, known as ether lipids, that form particularly strong bilayers. (Ethers arise from the joining, with loss of water, of an alcohol molecule with another alcohol molecule.) In the most extreme thermophiles, the bilayer is further welded into a rigid structure because the hydrophobic ends of the lipids are joined chemically into single chains. In eubacterial membrane lipids, the rigid ether bond is replaced by the more flexible ester bond (between an alcohol molecule and an acid molecule), and the two layers of the bilayer can slide freely.

Consider now what could have happened to highly thermophilic bacteria suddenly exposed to a (relatively) cooler environment. The possession of heat-resistant proteins could hardly have been a drawback, especially one correctible by a single mutation, unless one specific protein became locked in an unfavorable configuration below a certain critical temperature. In contrast, the rigid ether lipids could have been a major handicap. In order to understand this, think of lard, butter, and salad oil. Each changes from solid to liquid at a given temperature. Salad oil is liquid at room temperature but congeals in the refrigerator. Butter is solid but melts in hot weather. Lard melts at an even higher temperature. All three natural fats consist of similar substances known as triglycerides. Secondary chemical differences among the triglycerides account for the differences in melting temperature.

The same is true for membrane lipids. Depending on their chemical composition, their melting temperatures may differ as much as do those of lard and salad oil. In particular, ether lipids generally melt at higher temperatures than do the corresponding ester lipids. Furthermore, there is a clear correlation between the melting temperature of a cell's membrane lipids and the temperature of the environment occupied by the cell. This is understandable. Membranes can accomplish their functions only if their lipid bilayers are kept fluid. On the other hand, excessive fluidity of the membranes may endanger cellular stability. So, each cell type has membrane lipids that are fluid at the normal surrounding temperature but congeal about 10° to 15°C (18° to 27°F) below this temperature.

With this information, it is easy to visualize what would have happened to the highly thermophilic archaebacterial ancestors transferred by some climatic or geological change from, say, 110°C to 80°C. Their ether membrane lipids would congeal, the cells would become sluggish, their exchanges with the environment would come to a halt, and the cells would literally freeze to death, albeit in water we might still consider scalding. Only a mutation leading to the formation of membrane lipids that remain fluid in the new environment could save the cells from this sorry fate. This is what I suggest may have happened, the mutation being one that replaced ether lipids by ester lipids in the mutant cell's membranes. The price the mutant cell and its progeny paid for their rescue was that they could no longer return to their boiling cradle. But the bounty was immensely greater. The whole world was theirs to invade. The first eubacteria were born.

This story of the genesis of eubacteria is hypothetical. The alternative possibility, that ester lipids were inherited from the common ancestor and ether lipids acquired by archaebacteria, cannot be excluded. My choice is based on the assumption that the ancestral cell occupied a hot environment and possessed the better-adapted lipids. This view is shared by many scientists, but not by all.

Some members of the eubacterial family have kept a predilection for a hot environment—though not as hot as that occupied by the most thermophilic archaebacteria—or have returned to such an environment at a later evolutionary stage. You can see some at work creating their favorite surroundings by their own metabolism in the steaming compost heap at the back of your garden. Most eubacteria, however, are adapted to milder temperatures. Some even thrive in the icy waters around the polar caps. In contrast to archaebacteria, which, with rare exceptions, have remained confined to their original hot niches and to a few specialized environments, eubacteria are found everywhere; they are by far the dominant prokaryotes. Among them are all the bacteria that cause diseases and the many additional harmless ones that we harbor in our gut and elsewhere on our body, but also a multitude of other invisible organisms that reveal their existence by fermentation, food spoilage, rotting of organic matter, and other natural manifestations. Fungi, which are eukaryotes, also play an important role in these processes.

OUTLANDISH COLONIES

While eubacteria were conquering the world, the archaebacteria may have long remained confined to the boiling waters of their birth, to which they were particularly well adapted. At this stage, these organisms presumably lost the ability to synthesize the characteristic cell-wall constituent murein, which is found exclusively in eubacteria. Murein, with its content of both D- and L-amino acids and its other structural irregularities, has the characteristics of a primeval substance. Therefore, the ability to make murein was more likely lost by archaebacteria than acquired by eubacteria. Most archaebacteria do have a cell wall, but made of protein and carbohydrate constituents different from murein.

Eventually, some archaebacteria ventured outside their original medium and succeeded in invading other habitats.[4] A particularly flourishing group developed—or quite possibly retained, as we may be dealing with a primitive metabolic attribute—the ability to use hydrogen for the anaerobic conversion of carbon dioxide to methane, and to support all their energy requirements with the help of this reaction. Methane is a highly flammable substance, the most volatile component of natural gas.

In line with their presumptive origin, the most ancient methanogens, as these organisms are called, are thermophilic. Later forms acquired the capacity to grow at lower temperatures while retaining ether lipids in their membranes. They are now established in almost every site where organic matter is decomposed anaerobically with the generation of hydrogen. They are found in the digestive tracts of animals, especially of cattle, which have become significant producers of atmospheric methane and, thereby, participants in the greenhouse effect (see chapter 30). Methanogens are also abundant in marine and freshwater sediments. From such muddy depths, they send up the bubbles that break the silence of swamps by their muffled plopping, and fuel the will-o'-the-wisps that flit on the surfaces of marshes at night.

Other archaebacteria have succeeded in colonizing waters of very high salinity, even the saturated brine of drying seas. They are the only living things that still inhabit the Dead Sea and the Great Salt Lake. Among these remarkable salt-loving organisms, or halophiles, is found the only known archaebacterial phototroph: *Halobacterium halobium.* Unlike all other phototrophs, this organism does not depend on chlorophyll to catch light. Instead, it relies on a purple substance known as bacteriorhodopsin. This substance is a membrane-bound protein linked with a carotenoid, a relative of vitamin A, which acts as the light-catching part of the complex.

Carotenoids are found everywhere in the living world, including the specialized membranes of phototrophic organisms. But bacteriorhodopsin offers the only known example of a substance of this family actually effecting the conversion of light into usable energy. This example is also unique in another respect. The absorbed light is used directly to generate protonmotive force, without the participation of electrons. Unlike chlorophyll, bacteriorhodopsin is not a light-powered electron pump; it is a light-powered proton pump.

It is interesting, and possibly revealing, that the closest chemical relative of bacteriorhodopsin is the light-sensitive purple pigment of animal eyes. This is the original rhodopsin, a name that combines the Greek roots for rose and vision. Because of this relationship with the eye, carotenoids are also known as retinoids. In vision, however, the excited rhodopsin does not fuel an energy-converting mechanism; it triggers a series of signals along nerves leading from eye to brain. It is, however, tempting to assume that this vital pigment of our eyes is a descendant of some remote, ancestral bacteriorhodopsin.

THE GREEN REVOLUTION

In chapter 10, I related how some red cytochrome turned into a green chlorophyll through the evolutionary appearance of a variant porphyrin molecule in which the central hole came to be occupied by a magnesium atom in lieu of an iron atom. This event most likely took place in the eubacterial line after this line separated from the archaebacterial line, as no organism endowed with chlorophyll is known among archaebacteria and only a small number of eubacterial species are phototrophic. It is for these reasons that I have assumed that the ancestral cell was not phototrophic. The alternative possibility, that the ability to make chlorophyll is an ancestral heirloom that was lost by all archaebacteria and by many eubacteria, seems much less likely. The appearance of chlorophyll had consequences of the utmost importance, first for the bacteria concerned and, eventually, for the whole living world and for the planet itself.

As mentioned earlier, photosystem I appeared first in the conquest of solar power by emerging life. When energized by light, photosystem I can draw electrons from mineral compounds, sometimes also from organic substances, but not from water. A number of phototrophic bacteria dependent on photosystem I are found in the world today.

The next major step in the green revolution was the development of the water-consuming, oxygen-producing photosystem II, presumably through evolutionary modifications of photosystem I. The two systems operate with related but different chlorophylls. In present-day phototrophs, photosystem II is always associated with photosystem I, which gives the electrons lifted from water by photosystem II the extra boost they need to participate in biosynthetic reductions.

In the bacterial world, the association of the two photosystems is found in a large and abundantly distributed class of bluish microorganisms, originally called blue-green algae because they tend to associate into multicellular chains resembling some primitive seaweeds (algae). However, authentic algae are eukaryotic organisms. To avoid confusion, the prokaryotic phototrophs possessing the two photosystems are now called cyanobacteria (from the Greek *kyanos,* blue).

When did these important events take place? According to many experts, at least 3.5 billion years ago, perhaps as early as 3.75 billion years ago. The most solid evidence comes from stromatolites,[5] those layered rocks that originate from superimposed bacterial colonies. In extant colonies of this type, the top layers are occupied by cyanobacteria, which serve as essential food providers for the deeper heterotrophic layers. The oldest known stromatolites date back 3.5 billion years. If the bacterial colonies from which these rocks arose were anything like their present-day counterparts, they were topped by cyanobacteria-like organisms, meaning that photosystem II, as well as photosystem I, is at least 3.5 billion years old.

This estimate is supported by microfossil traces of the same age. The internationally known microfossil expert William Schopf,[6] from the University of California at Los Angeles, has identified authentic traces of at least seven distinct cyanobacteria-like organisms in rocks situated in northwestern Australia and accurately dated between 3.46 and 3.47 billion years. Many of the traces appear like chains of up to several tens of walled cells almost indistinguishable from the morphology of some present-day cyanobacteria.

If Schopf's identification is correct—he is the first to admit the uncertainty of purely morphological criteria—the phototrophic production of oxygen started at least 3.5 billion years ago. Yet all the available evidence indicates that molecular oxygen did not start rising in the atmosphere until about 2.0 billion years ago and reached a stable level only around 1.5 billion years ago. A likely explanation of this discrepancy is that, for the first two billion years of oxygen-producing phototrophy, enough oxygen-avid minerals were present to trap the oxygen produced and that atmospheric oxygen started rising only after these oxygen "sinks" were saturated. A major such sink could have been ferrous iron, believed to be very abundant in the early oceans. The reaction of ferrous iron with oxygen could explain at least in part—there are other possibilities—the large-scale generation of the mixed ferrous/ferric oxide deposits (magnetite) that are the main constituents of the banded iron-formations mentioned in chapter 3. It is suggestive and possibly significant that the deposition of banded iron-formations dates back at least 3.75 billion years—the geological record does not go further. It continued uninterrupted until oxygen started appearing in the atmosphere, then declined progressively to come to a halt about 1.7 billion years ago.

Note that if banded iron-formations attest to the presence of oxygen-producing phototrophic organisms, the universal ancestor must have arisen some time before 3.75 billion years ago, since one must account for the separation of eubacteria from archaebacteria; for the appearance of chlorophyll in a given kind of eubacteria; and for the evolution of phototrophy to the point where oxygen was produced. Thus the time span left for the emergence of life, up to the common ancestor, would be limited to a maximum of some 200 million years, the interval between the time when the Earth first became livable, four billion years ago, and the minimum age of the common ancestor. Such a time span was once considered too short for such a complex event as the origin of life. As I have pointed out before, there is no valid reason for such a view. Two hundred million years is really an enormous duration, more than twenty times the time it took an ape to become human. During such a stretch, life as we know it could have arisen and disappeared many times.

The inauguration of phototrophy was a fateful event in the spread of life on Earth, as it enabled living organisms to plug directly into the huge energy reservoir of the sun for raising electrons to the high energy level required for the construction of biomolecules from mineral building blocks. Earlier forms of life may have done so via the UV-supported production of hydrogen at the expense of the ferrous-ferric conversion described in chapter 3. This latter mechanism had the advantage of simplicity—an enormous asset at the time life was taking its first faltering steps—but it was no match for the chlorophyll-dependent process once the membrane-embedded infrastructure needed for this process was in place.

Another advantage of phototrophy is that it freed autotrophic life from its dependence on the specific environmental provision of appropriate electron donors and acceptors. Especially after the acquisition of photosystem II, virtually the whole surface of the Earth could be invaded. Our planet became green and its stores of carbon, nitrogen, and other bioelements became increasingly tied up in this green mantle. This situation, in turn, gave a tremendous boost to the development of heterotrophic organisms capable of feeding on the biomolecules manufactured by others. Thus was inaugurated the great coalition that, starting with the first stromatolite colonies, has come to link us, all other animals, all fungi, and many bacteria, to green plants and phototrophic microbes, within a planetary superorganism, the biosphere, whose metabolism is manifested by the continual recycling of the major biogenic elements.

Perhaps the most important consequence of phototrophy in relation to evolution was the rise of atmospheric oxygen. The crucial time, in this connection, irrespective of the date of the first appearance of photosystem II, was the period between 2.0 and 1.5 billion years ago, which witnessed what was probably the greatest ecological disaster—and the most far-reaching adaptive reaction of living organisms—in the history of life.

THE GREAT OXYGEN CRISIS

Until the development of photosystem II, the world had remained virtually oxygen-free. We tend to think of oxygen as a vital element, which indeed it is for us and for all other aerobic organisms, that is, those "living in air." For the early forms of life, however, oxygen was a redoubtable poison, as it still is for those bacteria known as obligatory anaerobes, which can survive only in the absence of oxygen. The toxicity of oxygen is due to its ready conversion in the presence of living systems into highly reactive chemical species, with names such as free hydroxyl radicals, superoxide ions, and hydrogen peroxide, that can severely damage vital cell constituents, including DNA and lipid bilayers.

When oxygen made its appearance, life had no defense against these poisons, and a major holocaust threatened. Fortunately, the process was slow and there was plenty of time for the main strategies of evolutionary adaptation to come into play. Victims probably were legion, but a few survivors emerged to people the world with new forms of life, thus turning an impending catastrophe into a major source of innovation.

The first to adapt to oxygen were its producers. In itself, the key mutation that gave rise to photosystem II was lethal. It may have happened many times, until chance associated it with another genetic change that protected the cells against the toxic effects of the oxygen they generated. This change could have been the ability to make large amounts of substances, called antioxidants or free-radical scavengers, that can mop up the injurious reactive chemicals formed from oxygen. Among such antioxidants are ascorbic acid, or vitamin C, a number of thiols, and tocopherol, or vitamin E. Or the cells could have acquired some protective enzyme, such as superoxide dismutase, which inactivates the superoxide ion, or catalase, which destroys hydrogen peroxide. The need for such protective adaptations could explain the delay that may have separated the appearance of photosystem II from that of photosystem I.

When other forms of life became exposed to oxygen, their first reaction was retreat. However, retreat from oxygen also meant retreat from the majority of phototrophs, the principal food supply for heterotrophs. In addition, oxygen was all-pervasive. It invaded every crack in the soil and, being soluble in water, reached the depths of the oceans. Sheltered areas capable of supporting anaerobic life soon became scarce. There was thus considerable pressure on existing anaerobes to develop protective mechanisms similar to those that had allowed the original phototrophs to produce oxygen without suffering any harm. Through random mutations, many bacteria, both autotrophic and heterotrophic, became able to survive in the presence of oxygen.

Most did not stop at mere survival. Thanks to relatively simple mutational events, they acquired the ability to transfer to oxygen the electrons exiting from their electron-transfer chains, with the consequent formation of water. The function known today as respiration was thereby initiated. This was a major evolutionary development, which, by returning oxygen to water, inaugurated the planetary water/oxygen cycle. The ubiquitous oxygen replaced the special mineral electron acceptors to which the organisms were chained; electrons flowing through phosphorylating chains could fall right down to the lowest level of energy, thus maximizing the potential energy yield of the process. Spurred by these great benefits, many bacteria evolved to the point of becoming obligatorily dependent on their erstwhile foe. Some succeeded in exploiting this dependence to the full by acquiring respiratory chains that allowed energy retrievals close to the maximum possible yield.

Some archaebacteria also acquired the ability to detoxify and exploit oxygen when it started appearing in the atmosphere. The halophilic bacteria are aerobic. So are the thermoacidophilic bacteria, which are residents of what we would view today as the most inhospitable of all ecological niches: waters that are very hot, very acidic, and reeking with the stench of hydrogen sulfide. However, except for its oxygen content, this sort of medium may well have been the cradle of life and the favorite haunt of the universal ancestor.

In spite of all these adaptations, oxygen has retained remnants of its life-threatening properties. Our white blood cells kill microbes with flashes of toxic oxygen derivatives (free radicals). The same derivatives sometimes also form accidentally in our tissues, where they may participate in the aging process, cause genetic lesions, or initiate cancerous transformations. There is considerable interest in the administration of antioxidants as protection against such injuries.

The events I have recalled would have been difficult to reconstruct were it not for the rich world of bacteria, where forms representative of almost every intermediate stage in the development of phototrophy and aerobic life can still be found. Their ancientness, as revealed by comparative sequencing and other evidence, generally corresponds to their proposed position in the chain of events. But their story goes further. Some of the participants in the great oxygen saga played a fundamental role in the remarkable process whereby a prokaryotic precursor turned into the ancestor of all eukaryotes, including ourselves. Thus, our origin, and that of all the plants and animals that surround us, is intimately rooted in the major events of eubacterial evolution, as the next three chapters will show.

Chapter 15

The Making of a Eukaryote

SOME 3.5 BILLION YEARS AGO, while bacteria were beginning to cover the world with triumphantly successful colonies, an obscure offshoot started to evolve in a strange direction that would most likely have appeared to an extraterrestrial visitor as totally aberrant in the context of life on Earth as it was then, and leading nowhere. In reality, "nowhere" fanned out, more than two billion years later, into the immensely varied groups of protists, plants, fungi, and animals, including humans—virtually the whole visible part of the biosphere. The way to this extraordinary diversity of living forms passed through a new type of cell entirely different from any known bacterium, past or present. Called eukaryotic because it possesses an authentic nucleus, this type of cell has many other characteristics that distinguish it clearly from the bacteria, or prokaryotes.

THE PROKARYOTE-EUKARYOTE TRANSITION

The nature of the obscure originator of the eukaryotic line is uncertain. Most of the evidence points to a prokaryote that detached from the archaebacterial branch after the first forking of the tree of life. In seeming conflict with this possibility, eukaryotes possess a few traits that appear to be derived from an ancient eubacterium. These anomalies have been variously attributed to convergent evolution, to horizontal gene transfer, and even, as proposed by the German investigator Wolfram Zillig,[1] to a primeval fusion event between an archaebacterial and a eubacterial partner. The apparent mixed ancestry of eukaryotes has also been used in support of more radical proposals that picture the primitive eukaryotic ancestor as a cell anterior to prokaryotes and endowed with either DNA genes or even RNA genes.[2]

In this chapter, I shall assume that the eukaryotic branch issued from an ancestral prokaryote, presumably of archaebacterial nature with some admixture of eubacterial traits acquired one way or another. This hypothesis enjoys the greatest favor and agrees with most known facts. The problem facing us, therefore, is to retrace a pathway from the prokaryotic to the eukaryotic type of organization. This pathway must be mechanistically plausible, consistent with available evidence, and causally explainable by natural selection.

At first sight, the task appears daunting. Place an average bacterium and even the most primitive of eukaryotes side by side and you observe differences of such magnitude as to make the conversion of one into the other almost unimaginable. Fortunately, one valuable piece of information is at hand. It is now established with a high degree of certainty that certain parts of eukaryotic cells, including mitochondria, chloroplasts, and, perhaps, peroxisomes (see chapter 17)—all three of which are membrane-bounded granules about the size of bacteria—are, in fact, the descendants of bacteria that were taken up by some eukaryote ancestor about 1.5 billion years ago and adopted permanently by this cell as endosymbionts (from Greek roots that add up in reverse to "living together inside").[3] Thus, the history of the eukaryotic cell may be divided into two separate eras: the pre-endosymbiont era—between 3.5 and 1.5 billion years ago—and the post-endosymbiont era—from 1.5 billion years ago to the present.[4]

Reconstructing the second era is no major problem. Cells take up bacteria all the time—our white blood cells do nothing else when we suffer an infection—and we know in detail the mechanisms involved in this uptake. To be sure, the bacteria taken up are usually killed and broken down by their captors, or, alternatively, attack their captors and kill them. A number of cases are known, however, in which this conflict leads to a stand-off that is solved by peaceful cohabitation. Thus, we have plenty of information to draw on for our reconstruction.

The first era, which covers the transition from the ancestral prokaryote to a cell capable of capturing bacteria and adopting them as endosymbionts, is more mysterious. But we have a few clues. First, if bacterial uptake took place in those remote times the way it does today, we can identify from present knowledge a number of properties that must have been gained in the course of the transition to provide the cells with the ability to capture. Also, a set of valuable clues are provided by sequencing results and other biochemical data that allow us to trace the prokaryotic ancestry of a number of eukaryotic structures. Finally, and most informative, a few primitive unicellular eukaryotes are known that date back to the pre-endosymbiont era and that may give us some idea of what eukaryotic life was like at that time. Let us take a look at such a "living fossil."

GIARDIA, A LIVING FOSSIL

The most ancient eukaryotes known are the diplomonads, a group that includes *Giardia lamblia*, a parasitic microorganism responsible for a number of severe intestinal infections in humans and some animals.[5] According to sequencing results, *Giardia* is at the end of a line that may have branched from the main eukaryotic trunk more than two billion years ago, that is, before oxygen started appearing in the Earth's atmosphere. The organism thus had an extremely long time to evolve and may bear little resemblance to the remote forebear it shares with other eukaryotes. It no doubt has changed, but probably not beyond recognition. It has so many features found also in more recent eukaryotes that we may safely attribute most of these features—there is always the odd possibility of evolutionary convergence, the inevitable caveat—to a common pre-endosymbiont ancestor. Therefore, *Giardia* should provide us with useful information about this ancestor, leaving as gaps to be filled properties that were lost in evolution, for example, in the conversion of the organism from an autonomous to a parasitic way of life.

Giardia is a single-celled, pear-shaped organism, about one-thousandth of an inch in size, which makes its volume more than 10,000 times bigger than that of an average prokaryote. We are definitely dealing with a giant cell in comparison with bacteria. It is not encased within a solid wall, as are bacteria, but coated only by some fuzzy material without any rigidity. The cell nevertheless maintains its characteristic shape thanks to a number of inner props forming what is known as a cytoskeleton. One face bears a disk-shaped structure that serves as a sort of sucker enabling the cell to adhere to the intestinal wall, a feature no doubt acquired in the course of adaptation to its particular habitat.

Giardia is highly mobile, propelled by four pairs of long, undulating flagella.[6] These organelles are totally different from the motor appendages of bacteria bearing the same name; they are built of slender, hollow rods of protein nature, called microtubules, combined with many other proteins to form a long flexible shaft endowed with the ability to bend in a wavy movement, using energy provided by the splitting of ATP. Flagella share this basic structure with cilia, which are short, rapidly beating appendages usually present in large numbers on the surface of the cells they equip. The two kinds of motors never exist together on the same cell, as witnessed by the taxonomic distinction between flagellates and ciliates.

Flagella and cilia are widely distributed throughout the eukaryotic world. We each owe our existence to the proper functioning of a flagellum that propelled a sperm cell emitted by our father into an egg cell presented by our mother. It is quite impossible that a structure as complex as a eukaryotic flagellum could have arisen independently twice by convergent evolution. *Giardia* thus tells us that the distant eukaryotic ancestor had already acquired all the main proteins that take part in the construction of flagella.

It is particularly informative to watch *Giardia* feeding. It engulfs extracellular objects.[7] It does what the ancestral cell that first came to harbor endosymbionts is known to have done. And it does so by a mechanism virtually identical to that whereby our own white blood cells and innumerable other eukaryotic cells eat bacteria or other solid objects. Called phagocytosis (from the Greek roots for eating and cell), this process is so complex that we may assume, once again, that it existed already in the pre-endosymbiont ancestor. We don't have to search further to account for the manner in which endosymbionts entered their host cells. Just as we would have guessed solely from our knowledge of the present, these important guests were taken up by a typical phagocyte endowed with the basic attributes of similar extant cells. This is an invaluable piece of information. Lynn Margulis, a University of Massachusetts biologist known for advocating the endosymbiont theory at a time when there was little evidence to support it, has postulated aggressive invasion by a "fierce predator" to account for endosymbiont entry.[8] In my opinion, *Giardia* makes this hypothesis unlikely. Uptake by phagocytosis seems a more probable explanation, supported by all that we know.

Let us look at *Giardia*—or one of its free-living cousins—on the hunting trail. See it accidentally brushing against a bacterium or, perhaps, moving toward it in a seemingly purposeful manner, attracted by a chemical signal, as would our white blood cells in a similar situation. Whether chance or chemistry caused the encounter, the consequence is that the contacted bacterium remains stuck to the cell's surface, like a fly to flypaper. Triggered by this contact, the organism goes into action, slowly sucking in its hapless victim, which progressively vanishes from our sight. Soon, no trace of this dramatic gobbling remains on the cell's surface, which has regained its smooth, unruffled appearance.

Not all bacteria are caught in this way. This is because sticking requires the presence of receptors on the captor's surface that recognize certain components on the bacterial cell wall—another case of a complementary lock-and-key arrangement. If the lock or the key is missing, the bacterium simply bumps away after the collision and escapes. Some bacteria actually take advantage of such a defect to elude capture, a fact that has played a tremendous role in the history of science. It so happens that infectious pneumococci, the redoubtable agents of bacterial pneumonia, differ from their harmless relatives by lacking the gene coding for a cell-wall component that is specifically recognized by our white blood cells. They can't make the lock, and so the blood cells' key does not find a fit. In 1928, Fred Griffith, a medical officer in the British Ministry of Health, closely followed by the American Martin Dawson working at the Rockefeller Institute for Medical Research (now the Rockefeller University) in New York, found that the genetic ability to make the cell-wall molecule that renders microbes catchable and thereby harmless could be transferred from dead, noninfectious bacteria to live, pathogenic ones. Sixteen years later, three Rockefeller scientists, Oswald Avery, Colin MacLeod, and Maclyn McCarty, announced to an initially incredulous world that they had purified the "transforming factor" and identified it beyond reasonable doubt as a DNA molecule.[9] This historic experiment established for the first time that genes are made of DNA, not of protein as many had believed. It set off one of the most epic scientific races, which was won in 1953 by Watson and Crick when they deciphered the double helix.

When these investigations were performed, the mechanism of phagocytosis was not yet known. Today, thanks to electron microscopy and other sophisticated techniques, we understand the phenomenon in detail. The caught bacterium does not enter the cell through a hole, as one might suspect. It is progressively enveloped by a deepening infolding, or invagination, of the cell membrane, which remains intact. When envelopment is completed, the invagination snaps off from the inner surface without leaving any trace and turns into a closed intracellular sac, or vesicle, with the engulfed bacterium inside, entirely surrounded by the piece of membrane abstracted from the cell surface in the course of engulfment. The fluidity, flexibility, and self-sealing properties of lipid bilayers explain how such a phenomenon is physically possible.

What drives the uptake? We don't know about *Giardia,* which has not been studied in this respect, but we know of at least two mechanisms involved in other cells. One, external, relies on progressive "zipping"—better said, "Velcro-ing," if such a word existed, since we are dealing with surfaces—between the phagocyte's receptors and their complementary partners on the bacterial surface. The other mechanism is activated from inside the cell by a device that draws in membrane patches whose receptors are occupied. This mechanism allows the uptake of fluid droplets, or pinocytosis (Greek for "cell drinking"), triggered by the binding of soluble molecules to surface receptors. The general term endocytosis covers all forms of membrane-dependent engulfment, including phagocytosis and pinocytosis. The intracellular vesicles formed by endocytosis are called endosomes.

Subsequent events bring the catch into another kind of membrane-bounded intracellular vesicles called lysosomes ("digestive bodies," in Greek), in which engulfed materials are exposed to acid and to digestive enzymes, thereby suffering the same fate as food in our stomach. The acid is secreted into the lysosomes by a proton pump present in the lysosomal membrane, whereas the digestive enzymes are conveyed to lysosomes from another set of intracellular vesicles named the endoplasmic reticulum, ER for short. As digestion proceeds within lysosomes, the small nutrient molecules arising from this process pass through the lysosomal membrane into the cell proper, to participate in metabolism. In the end, the lysosomal content of enzymes and undigested residues is often unloaded into the extracellular medium by a process graphically called cellular defecation, or exocytosis in general, essentially the reverse of endocytosis. In this process, a vesicle adds its membrane by fusion to the cell membrane and discharges its contents outside.

The digestive enzymes destined to be conveyed to the lysosomes are simultaneously (cotranslationally) made and translocated into the interior of ER vesicles by ribosomes closely apposed to the vesicles' membranes. These parts of the ER, which are studded with ribosomes, are called "rough" because of the rugged appearance of the membranes in cross section. From these rough parts, the newly made enzymes are transferred to smooth (ribosome-less) parts of the ER and, subsequently, to a complex of membranous sacs known as the Golgi apparatus, system, or complex—Golgi for short—from the name of the Italian neuroanatomist Camillo Golgi, winner, with his Spanish colleague Santiago Ramon y Cajal, of the 1906

Nobel Prize in medicine. In the course of the passage of the enzymes through the smooth ER and the Golgi, their molecules undergo a number of chemical modifications generally referred to as processing or maturation.

Together, endosomes, lysosomes, rough and smooth ER, and Golgi form a complex intracellular network of sacs and vesicles, sometimes called the cytomembrane system. Consisting of up to thousands of distinct membrane-bounded compartments, this system governs a number of important cell functions grouped under the general labels of bulk import and export. Import takes place by endocytosis and usually leads to digestion in lysosomes, occasionally to storage or transcellular passage of the material taken up. Export starts in the rough ER, continues by way of the smooth ER, Golgi, and lysosomes, and ends in defecation by exocytosis. Export also follows an alternative pathway, much more important in most cells, that bypasses lysosomes and leads directly from the Golgi to exocytic discharge. This is the major pathway of secretion.

Traffic of the imported and exported materials through the many compartments of the cytomembrane system is mediated by permanent or, more often, transient connections between compartments, and directed by external railings belonging to the cytoskeleton and by internal receptors. The energy required to drive these movements comes from the splitting of ATP, with the help of special cytoskeletal-motor systems.

The cytomembrane system is a unique characteristic of all eukaryotic cells. *Giardia* tells us that this system was already laid out two billion years ago. Even the Golgi-dependent secretion machinery had been developed by that time.[10]

There are no mitochondria, chloroplasts, or other possible descendants of engulfed bacteria in *Giardia.* Although such organelles may well have been jettisoned in the course of evolution, it is of interest that the next most ancient known eukaryotes, the microsporidia, show a similar lack of endosymbiont-derived organelles.[11] These facts, together with other pieces of evidence, strongly support the view that these very ancient organisms descend from lines that branched from the eukaryotic trunk before endosymbionts were adopted. This lineage makes the organisms the closest extant relatives—although still immensely remote—of the pre-endosymbiont eukaryote.

Metabolically, *Giardia* is an obligatory anaerobe adapted to an oxygen-free environment. This adaptation could conceivably go back to when *Giardia*'s distant forebear detached from the eukaryotic line, since oxygen had not yet started to appear in the Earth's atmosphere at that time. If such is the case, *Giardia*'s entire ancestral lineage survived through the oxygen crisis and continued evolving for more than one billion years, somehow protected from oxygen until the time it found an appropriate oxygen-free environment in the gut of some animal. It is equally possible—more probable?—that the ancestral cells became adapted to oxygen and subsequently lost this property when they adjusted to anaerobic parasitic life. It is, however, noteworthy that the main electron-transfer chains of eukaryotes belong to endosymbiont-derived organelles, not to the cell membrane or the cytomembrane network. Barring an evolutionary loss, this fact suggests that the primitive eukaryote, which was probably anaerobic, like all organisms that existed more than two billion years ago, may have lacked membrane-bound respiratory chains and that its descendants managed to weather the oxygen crisis without the help of such chains until endosymbionts came to the rescue.

The genetic organization of *Giardia* appears to be of the "classical" type, as far as is known. Two points are of interest. First, the organism's ribosomes are more similar, in terms of certain molecular characteristics, to prokaryotic than to eukaryotic ribosomes. This similarity is in keeping with an early branching from the eukaryotic line, before the present kind of eukaryotic ribosomes had evolved. Next, and particularly interesting, *Giardia* has two nuclei of equal size. But before we look at this intriguing duplication, let us consider the nuclei themselves, the hallmark of eukaryotic cells.

THE EUKARYOTIC NUCLEUS

Giardia's nuclei have all the main features of eukaryotic nuclei. So named because it sits like a kernel (*nucleus* in Latin, *karyon* in Greek) in the middle of the cell, the nucleus is a voluminous, roughly spherical body entirely surrounded by a double-membranous envelope structurally and functionally related to the ER (the outer membrane is studded with ribosomes). The inner face of this envelope is bolstered by a sturdy lining made of tightly knit protein fibers. A large number of reinforced openings, or pores, inserted like portholes through the nuclear envelope, serve as regulated passageways between the nucleus and the rest of the cell, or cytoplasm.

The main residents of the nucleus are the chromosomes (Greek "for colored bodies"), so named not because they are colored but because early microscopists saw them as intensely stained bodies in cell preparations treated with certain dyes. Eukaryotic chromosomes are majestic edifices compared with their prokaryotic counterparts, which are little more than a circular stretch of naked DNA. In contrast, the eukaryotic chromosome is a highly structured entity. Imagine a miniature maypole wreathed by loops of a spirally wound garland of beads. The pole is the inner skeleton of the chromosome, constructed of protein. In the garland, the string is made of DNA and the beads consist of small protein spools around which the DNA string makes a couple of turns before moving on to the next spool. This beaded string is twisted into a thick thread, somewhat in the form of a telephone cord. The thread itself is divided into a series of ample loops anchored to helically disposed attachment points around the central skeleton. As a rule, some of the loops are uncoiled, others are packed into tight balls. When a cell starts dividing, all the uncoiled loops also become packed into balls and the chromosome assumes the shape of a thick, knobby rod. It is in such dividing cells that chromosomes were first observed as stained rods. In nondividing cells, the basic chromosome structure is hidden by the inextricable tangle of uncoiled DNA stretches. And what a tangle it is! Imagine some two miles of a very thin string looped around a two-foot rod. This is how your average chromosome would look magnified 100,000-fold.

Existence of a nucleus entails a number of fundamental consequences that make the eukaryotic organization totally different from the prokaryotic organization from which it originates. First, the nuclear envelope separates the cell into two distinct compartments that communicate with each other only by the nuclear pores. This kind of compartmentalization differs from that created by the cytomembrane network, which, with its multiple interconnected cavities lined by chemical-processing machineries and transport systems, forms a sort of halfway house between the inside of the cell and the outside world. The nuclear envelope divides actual metabolism. The principle of this division is simple. Keep in the nucleus only those functions that have a close link with DNA and leave all the rest in the cytoplasm. To maintain proper connections between the two, equip the pores with specific systems that mediate the passage of all the substances that need to go in or out, under strict control of their molecular identity.

Two functions obligatorily situated inside the nucleus are DNA replication and DNA transcription. Because of the complications resulting from chromosome structure, DNA replication requires a complex variety of disentangling systems to make the DNA accessible to the replicating enzymes. For this reason, the process is at least twenty times slower in eukaryotes than in prokaryotes. This drawback is offset by having the DNA distributed over several chromosomes—there are four in *Giardia,* forty-six in human cells—and by having multiple replication sites on each chromosome. In prokaryotes, there is a single replication site (anchored to the cell membrane) through which the entire chromosome is reeled for replication. In eukaryotes, there are a large number of such sites, so that the DNA is replicated simultaneously in many short stretches, which are subsequently joined together. Thanks to this arrangement, the entire eukaryotic genome (about six feet of DNA in a human cell) can be replicated in about one hour, little more than twice the time needed for the replication of a prokaryotic genome only one-twentieth of an inch long. When DNA replication proceeds in a eukaryotic nucleus, all the proteins needed for the construction of chromosomes are imported into the nucleus from the cytoplasm and assemble spontaneously with the newly formed DNA to form a second set of fully formed chromosomes, each joined by a bridge to its pre-existing sister, like Siamese twins.

Intranuclear DNA transcription runs into the same kind of structural problems as DNA replication, with the additional complication that the synthesized RNA products are exported out of the nucleus. Only mature RNAs are sent out to the cytoplasm. Splitting, trimming, splicing, and other RNA rearrangements are all carried out in the nucleus. A special intranuclear organelle called the nucleolus harbors the synthesis and maturation of ribosomal RNAs (the RNAs that combine with a set of proteins to form the ribosomes), which constitute by far the largest part of the RNA output of the nucleus at any given time. Other complex systems situated in the nucleus ensure the splicing of messenger RNAs, a major function in higher eukaryotes, but perhaps not in *Giardia,* which, with no split genes detected so far, may have no need for RNA splicing. Mature RNAs do not move out of the nucleus on their own, but in the company of special RNA-binding proteins, which are admitted into the nucleus unaccompanied and return to the cytoplasm with their quarry.

An important consequence of the kind of segregation just described is that the translation of genetic messages is topologically separated from their transcription. This is not so in prokaryotes, where one can often see ribosomes busily making proteins on messenger-RNA stretches that are still in the process of being transcribed from DNA. In eukaryotes, the ribosomes are all in the cytoplasm, where they function with messenger-RNA molecules that have been synthesized and properly processed in the nucleus, allowed to pass through the nuclear pores, and offered to the ribosomes in intact and accessible form. Gene expression can thus be regulated at multiple points, both in the nucleus and in the cytoplasm.

The existence of the eukaryotic nucleus creates special problems for cell division. In prokaryotes, cells divide after chromosome duplication by a simple constriction, or furrow, of the cell membrane (and of the outer wall), which takes place in such a way that each daughter cell inherits one of the duplicated chromosomes with its half of the original cell membrane. In eukaryotic cells, the cytoplasm divides by a roughly similar mechanism, but not before the nucleus has itself undergone duplication.

Nuclear division is a dramatic phenomenon, one of the few cellular events that can be observed in some detail with an ordinary light microscope. It has fascinated generations of biologists. In a "nutshell," the chromosomes are first duplicated and compacted. This is when they become visible as rods or filaments, which explains the name of mitosis given to nuclear division (*mitos* means thread in Greek). Next, the nuclear envelope is dismantled and replaced by the spindle, a complex rigging constructed from microtubules, the same structures that form the flagellar shaft. The duplicated chromosomes then congregate on the equatorial plane that divides the spindle in two halves. The rigging now comes into action, forcibly wrenching paired chromosomes away from each other and pulling one member of each pair toward one of the two poles of the spindle. Here we see the advantage of the Siamese-twins structure of duplicated chromosomes. It allows the paired chromosomes to align in a proper orientation and the rigging to work in such a way that a full identical set of chromosomes is assembled at each pole of the spindle. When this assembly process is completed, the spindle is dismantled and brand-new nuclear envelopes form around each chromosome set.

Giardia has characteristic nuclei that divide by typical mitotic division (except that, as in a number of primitive protists, the nuclear envelope does not break down).[12] We may take it, therefore, that the primitive eukaryote already possessed all the relevant structures and properties that go into the making and division of eukaryotic nuclei. But why should *Giardia* have two nuclei instead of the single nucleus commonly found in eukaryotic cells? And, as a corollary to this question, could the primitive eukaryote also have possessed two nuclei at a certain stage of evolution? We don't know the answer to these questions. But we can think of one with such tremendous implications that it deserves a separate treatment. It has to do with the origin of the most powerful force in nature: sex. I shall consider it in the next chapter.

Chapter 16

The Primitive Phagocyte

THE PICTURE IS CLEAR. When *Giardia*'s lineage branched from what was to become the main eukaryotic trunk, probably more than two billion years ago, almost all the key features of eukaryotic cells, with the exception of endosymbiont-derived organelles, had already emerged. The crucial prokaryote-eukaryote transition occurred some time during the 1.0 to 1.5 billion years following the primeval forking that led to the eukaryotic branch. During that time, a simple prokaryote developed into a primitive phagocyte, a large nucleated cell capable of capturing food and digesting it intracellularly. What pathway did this momentous transformation follow? And, especially, why was this road actually taken in reality?

Extant organisms offer a number of valuable clues to the first question, but we have only educated guesses to help us answer the second. Remember the rule: foresight excluded. There was no goal, no eukaryotic ideal beckoning from the distant future, inviting evolving cells to overcome hurdles and vanquish difficulties. Every step of this extraordinary voyage was taken in its own present context, the consequence of some chance mutation that happened to confer an immediate benefit favoring the survival and proliferation of the affected cell there and then. What hidden selective forces cut open this trail, step by step, over an immensely long period of time, to produce what was probably the most epoch-making innovation in the history of life? This question will be with us as we try to retrace the main steps of the voyage.

From what we have seen of *Giardia,* there are really two major developments to be accounted for within the context of an enlarging cell: cytomembranes and cytoskeletal elements, with a fenced-off nucleus arising through a special combination of the two. We have no clues to the origin of the cytoskeleton, which may be a true innovation. But we know the origin of eukaryotic cytomembranes. According to all available evidence, they come from the *ancestral prokaryotic cell membrane.*

SPREAD OF A NETWORK

A banal accident, which turned out to have long-term effects of enormous impact, may have initiated the whole chain of events. An ancient heterotrophic prokaryote lost the capacity to build a cell wall. This defect most often weakens survival ability but is not invariably lethal. Naturally wall-less bacteria are known, including highly thermophilic ones. In this particular case, the circumstances surrounding the accident were such that the victim not only survived but benefited from its infirmity. Perhaps the maimed organism was a resident of one of those multilayered bacterial colonies that were beginning to flourish at that time and have left their traces in the form of stromatolites. Living in the midst of bacterial mats, our denuded remote forebear was sheltered from many hazards and suffered little from its nakedness. It could continue to proliferate at the expense of its neighbors and produce similarly naked progeny. According to fossil evidence, stromatolite colonies have persisted with little apparent change from the earliest days of life to the present. If, as is possible, the prokaryote-eukaryote transition required a stable food-supplying environment for a very long time, colonies of this kind could have satisfied such a requirement.

Another event that may have happened very early is the acquisition of membrane lipids of the ester type. All eukaryotes have ester lipids. This is one of the eukaryotic features that do not fit with an archaebacterial origin. Ester lipids are characteristic of eubacteria, whereas all known archaebacteria have ether lipids. There are many possible explanations for this discrepancy, and I shall not go into them. Let us simply record that our putative wall-less ancestor probably had ester lipids, which means that it may have lived in a milder environment than did the thermophilic bacteria from which it presumably originated. In addition, the possession of ester lipids, through the increased membrane fluidity it conferred, could have been an important factor in the process whereby the loss of a cell wall turned into a benefit for heterotrophic life.

In order to appreciate this benefit, let us visualize our stripped ancestral cell. It is a shapeless, flattened blob that feeds on the remains of dead bacteria, which it digests extracellularly by means of secreted enzymes, as do all heterotrophic prokaryotes. Here is where nakedness becomes an advantage. There is no straightjacket around the cell, no barrier between cell and food supply. Helped by the flexibility of its membrane, the cell can stick intimately to the bodies on which it feeds, mold itself to their contours, and even wrap itself completely around them, helped in all these movements by surface receptors, or binding sites, capable of hooking on to certain surface components of the bacterial bodies. Thanks to these intimate contacts, the digestive enzymes discharged through the cell membrane remain trapped between cell and prey and can act optimally. In turn, the small nutrient molecules produced by digestion can enter the cell readily across the cell membrane, again without loss or delay. Our naked heterotroph is a magnificent feeder, a born winner in the struggle for food, as long as its environment is protective enough to offset the lack of a cell wall.

Second advantage: Our hero can grow bigger. The size of a cell is limited by the surface area it has available for exchanges with the outside—of nutrients inward and of waste products outward. A spherical cell surrounded by a smooth membrane cannot grow beyond a limit size because the volume increases with the third power of the radius, and the surface area only with its second power. To grow bigger, the cell must either change its shape—to a rod or filament, for example—so that it has more surface available for a given volume, or expand its membrane by infoldings (invaginations) or outfoldings (evaginations). Our naked ancestor was a champion contortionist in this respect. It could grow to any size by pleating its surface.[1]

But why should it do this? More efficient feeding is a likely answer. The more jagged a coastline, the more sheltered the inner coves within which two partners—digestive enzymes and food in the present case—can meet without disturbance. And so, natural selection would favor a larger cell with a more irregular contour. The final outcome is predictable by anybody acquainted with the self-sealing habits of lipid bilayers. As invaginations deepen, the gullets leading into them narrow until—click—there is no gullet anymore. The invagination has snapped off from the surface to form a closed vesicle inside the cell, while the scar left in the cell membrane by the amputation heals simultaneously by self-sealing. A small bubble blown into the surface of a larger bubble suddenly cuts its moorings and floats ghostlike inside—a trick some soap-bubble experts actually can produce. Inside the small inner bubble, food and digestive enzymes are now completely segregated together. From being extracellular, digestion has become intracellular.[2]

The trick, when it happened naturally, was not a trivial event. It inaugurated a crucial development in cellular evolution: the phagocytic way of life. For the first time, heterotrophs had a stomach of their own. No longer compelled to carve a stomach out of their surroundings and to reside within it, they could now afford to wander around and to survive by capturing food. This was a gigantic step on the way to cellular emancipation. From a captive held in the golden prison of a nutritive shell, like a maggot in a chunk of cheese, the cell had turned into a mighty hunter that could invade the world in search of prey.

The first cell stomach was a compound of many things. Arising from an engulfment phenomenon, it acted as a storage place for engulfed food. At the same time, the primeval stomach received digestive enzymes from ribosomes bound to its membrane. These ribosomes simply continued the job they performed on the cell membrane, with the difference that the enzymes were now collected inside the stomach, instead of being discharged into the outside medium, and acted directly on the collected food. By virtue of its origin, the stomach's membrane possessed all the transport systems present on the cell surface. These included a proton pump originally directed outward and now toward the inside of the stomach, making it acidic and thereby optimally suited to the requirements of the digestive enzymes. Like our stomach, the cell's stomach needs acid for the best performance of its enzymes. Other transport systems, which once transferred into the cell's interior the small nutrient molecules produced by extracellular digestion, now cleared the stomach of digestion products in the same way. Yet others, acting in the reverse direction, discharged into the stomach the waste products they previously excreted into the outside medium. Also present, protruding on the inner face of the stomach's membrane, were the cell's surface receptors, including those that had caused the cell to stick to some materials and engulf them. Finally, it often happened that by a reversal of the manner in which it first arose, the stomach joined back with the cell membrane, thus restoring to this membrane the patch that had been removed from it by the original internalization phenomenon, and at the same time unloading into the outside medium the stomach's contents of undigested food, waste products, and enzymes.

Put all this together and you find that the first cell stomach combined functions that, in higher organisms, would be described as ingestion, secretion, digestion, absorption, excretion, and defecation. Thanks to our foray into the future, we identify in this primeval stomach properties typical of endosomes, rough ER vesicles, and lysosomes, all in one, and we recognize as endocytosis and exocytosis the two phenomena that link the stomach reversibly to the cell membrane. Further evolution can be summed up as a progressive segregation of the various functions carried out by the primeval stomach to separate parts of an increasingly complex network of intracellular vesicles, all derived from the ancestral cell membrane. In an analogous way, but on an entirely different scale, the same functions are distributed from mouth to anus all along our own digestive tract, whereas they are carried out by a single cavity in a primitive organism such as a jellyfish.

The first functions to become dissociated were food collection and enzyme storage. This was accomplished by further membrane internalization and by migration of the ribosomes previously associated with the cell membrane to a new set of intracellular pouches. Ancestral to the rough ER, these pouches turned into receptacles of newly synthesized digestive enzymes, which were no longer discharged directly by the membrane-bound ribosomes into the outside medium or into invaginations and vesicles derived from the cell membrane. Because of this migration phenomenon, capture of food was now accomplished by ribosome-free membrane patches, and the resulting vesicles served in the temporary storage of engulfed food but not in its digestion. The initial phenomenon of haphazard membrane infolding thereby turned into what we now know as endocytosis, and the internalized vesicles became endosomes. Receptors on the cell-membrane surface made it possible for the cells to be selective and to choose their "menu" from the materials present in the surrounding medium.

The stomach proper, or lysosome, was created as a separate, acidified compartment lying between the enzyme-containing, rough-surfaced pouches and the food-containing endosomes, and connected to these two sites by a variant of the bubble trick called vesicular transport. In this process, vesicles bud off from one site, carrying with them material stored in that site, and dock at another site where they deliver their contents. In terms of the bubble analogy, a small bubble detaches from a larger one, as may happen to a soap bubble caught in an air current. The drifting, small soap bubble next bumps into another, large soap bubble and merges with it. As a result, some bubble material and a small volume of air are transferred from one soap bubble to another. In vesicular transport, membrane fabric is likewise transferred from one closed sac to another, though not with a small amount of air but with important enclosed materials. In this way, food transferred from endosomes, and enzymes transferred from rough-surfaced pouches, were led to converge into lysosomes, where the one could now be attacked by the others.

Lysosomes did not swell indefinitely as a result of this dual transfer. Small molecules arising through digestion were transferred to the cell interior by transport systems (inherited from the cell membrane) situated in the lysosomal membrane, whereas residues left at the end of digestion were unloaded from the lysosomes into the outside medium by exocytosis. As to the excess membrane material, part was removed from the lysosomal membrane by vesicles shuttling back empty to their sites of origin, and the rest was added to the cell membrane by fusion in the final phenomenon of exocytic discharge. Thanks to this continual recycling, membrane material remained stably distributed among the cell membrane and the different parts of the intracellular membrane system, in spite of the intense traffic taking place among them.

Going back to our comparison with the animal digestive tract, we have reached a stage where there is a mouth (endocytosis), a digestive cavity (the lysosome), and an anus (exocytosis), with an attached digestive gland pouring in digestive juices in the manner of our pancreas (rough ER). An important difference, besides the enormous difference in scale and, thus, in the nature of the structures involved, is that the parts are not linked by continuous channels controlled by valves, but by intermittent connections established by vesicular transport. In either case, the inside of the system is kept separated at all times from the rest of the body, except for the transfer of selected materials across the tract's lining, mediated by specific machineries.

During the subsequent evolution of this primitive intracellular digestive tract, additional intermediate stations were inserted into the main traffic lines to serve for temporary storage and specific chemical processing of the passing materials, or for their sorting and selective rerouting by means of special receptors. In this way, the smooth parts of the ER and the different components of the Golgi complex became intercalated between the rough ER and the lysosomes. The endosomes, on the other side, became subdivided into several sections, which allowed some of the materials taken up to be saved from lysosomal digestion and diverted to other directions inside or outside the cell.

A major diversion came to be inserted between the exit from the Golgi and the lysosomes, so that materials traveling along the ER-Golgi pathway were conveyed directly to the cell surface for discharge into the outside medium, without passing through the lysosomes. Eventually, this line turned into the major pathway of secretion, the process whereby cells discharge around them the components of extracellular structures and a variety of complex materials, such as enzymes, hormones, and other active agents, that they manufacture for export. These substances all consist of proteins that are made in the rough ER and are further trimmed and fitted with a variety of carbohydrate, lipid, and other components as they pass through the smooth ER and the Golgi. This new line bypassed the lysosomes, thus avoiding damage to the channeled materials. The original line leading to lysosomes was maintained, but under the control of receptors that admitted only molecules bearing a specific chemical tag common to all the digestive enzymes destined for lysosomes.

Swept up in the general process of membrane internalization was a special patch of prokaryotic cell membrane to which the chromosome was hooked by the system serving in DNA replication. Such attachment is a common feature of all prokaryotes. Internalization of this specialized membrane part moved the chromosome and its replicating system to the cell interior, where they became progressively surrounded by a sealed double-membranous envelope derived from the cell membrane and structurally and functionally connected with the rest of the intracellular membrane network. This membrane migration was a development of cardinal importance. It initiated the birth of the nucleus, the cell part to which eukaryotes owe their name.

My historical reconstitution of the development of the eukaryotic cytomembrane network is hypothetical, since descendants of intermediate forms have not been uncovered. However, the model is supported by a great deal of information written into the structures and functions of existing molecules. Scattered throughout the intracellular membrane network of eukaryotes are unmistakable molecular relatives of systems associated with the cell membrane of prokaryotes: characteristic transport systems in one part of the network, translocating ribosomes in another, lipid-synthesizing enzyme complexes in another, translocating carbohydrate-assembly systems in another, traffic-directing receptors in another, chromosome attachments in yet another, and so on. Formation of the eukaryotic membrane network by internalization and differentiation of an ancestral prokaryotic cell membrane is highly probable. Only the details are missing.

Any model of evolution must demonstrate that almost every proposed step provided a selective advantage. I have taken as the main driving force *the progressive conquest of greater heterotrophic autonomy* through the enhanced ability to find, take up, and utilize food, which, for a heterotroph, is the key condition of survival and reproductive success. This explanation is plausible and makes sense in the context of present knowledge. Furthermore, it allows for the progressive unfolding of an evolutionary process that comprised a very large number of successive steps and stretched out over a very long duration. Each little step of the proposed scenario may be seen as associated with a slight improvement in phagocytic efficiency.

THE INDISPENSABLE PROPS AND MACHINES

Membrane expansion and internalization alone could not have brought about the developments I have described. What the cells needed in addition were internal props—a cytoskeleton—that would save their growing bulk from collapse without, however, impairing their protean ability to change shape. In addition, the cells needed motor systems to perform the work involved in uptake, transport, and discharge of materials through the different cavities of the growing membrane network. An astonishing number of complex molecular systems emerged to satisfy these requirements. It is remarkable that related assemblages have so far not been detected in prokaryotes. Bacterial supporting structures and flagella are entirely different from their eukaryotic counterparts. Unlike the cytomembrane network, which clearly originates from the ancestral cell membrane, eukaryotic cytoskeletal and motor systems seem to be true innovations made during the prokaryote-eukaryote transition. This makes their emergence particularly critical to the transition. Unfortunately, no clue to the origin of these elements has yet been uncovered. They are either absent or present in their fully sophisticated form. There is no sign of the many intermediate forms that must have marked their development.

Many intracellular and extracellular structures are built of long, threadlike molecules, either proteins or carbohydrate polymers, that are intertwined into a variety of fibers, bundles, webs, sheets, plates, baskets, and other three-dimensional arrangements. As a rule, these structures are stable and static. In multicellular organisms, they give many cell types their specific shape or provide external frameworks for cells to assemble into characteristic tissues, such as those that make up skin, bones, joints, mucosae, viscera, and so on.

Structures of this kind would have been of little help to our fledgling phagocyte. They would simply have replaced an external straightjacket by an internal one. What evolution came up with, not once but at least three times, are Lego-like protein molecules capable of assembling reversibly into a rigid arrangement. Two of these molecules, actin and tubulin, are constructed according to similar principles. Imagine a set of identical building blocks that can join together by some sort of complementary peg-and-hole devices, like the pieces of construction toys. Each block has a peg at one end and a hole at the other, so that blocks can be linked together indefinitely in a linear fashion to form a rod or thread. In addition, each block also has a peg on one side and a hole on the other, allowing the threads to join laterally. In actin, this lateral link is such that two threads wind into a double-helical fiber. The link is such in tubulin that thirteen threads join spirally into a hollow, cylindrical tube, or microtubule.

Actin fibers and microtubules have in common that they can be dismantled and reassembled into different configurations with the help of ATP. They thus serve to shore up cells into a variety of temporary shapes, which sometimes even initiate movements by their changes. Rigidity is allied with plasticity. Actin fibers and microtubules are often connected with special ATP-splitting proteins that change shape when they split ATP and thereby act as converters of chemical energy into mechanical work. Both structures also participate, together with their associated molecular motors, in the construction of stable edifices of great complexity, which underlie the most elaborate forms of eukaryotic motility.

We have already encountered microtubules as important cytoskeletal elements in *Giardia.* This organism holds two magnificent specimens of the extraordinarily varied structures that can be built with microtubules and their associated motors. One, transient, is the mitotic spindle, which is assembled at each cell division and dismantled again afterwards. The other, stable, is the flagellum, a slender, cylindrical structure built from nine parallel pairs of partly fused microtubules surrounding an axial shaft made of two microtubules (the 9 + 2 structure, also characteristic of cilia). Some five hundred additional proteins complete the assemblage. Among them is a special ATP-splitting protein, named dynein, that has the remarkable property of bending forcibly when it splits ATP. The mechanical and the chemical process are obligatorily coupled. One cannot happen without the other. Thus, if the two ends of dynein are attached to separate structures, the molecule causes the structures to move closer to one another, using the energy released by the splitting of ATP to overcome an opposing resistance. This phenomenon is responsible for the wavy movement whereby flagella propel cells forward.

The presence in the most ancient known eukaryote of structures that are among the most elaborate molecular assemblages found in the whole living world is an impressive fact. It suggests strongly that the development of such structures played an essential role in the prokaryote-eukaryote transition, perhaps to the point of setting the pace of this transition, as we are obviously dealing with an extremely long succession of evolutionary steps.

To my knowledge, actin has not been detected in *Giardia,* or, for that matter, searched for in this organism. Thus, we don't know whether actin is as ancient as tubulin. This seems likely, however, since actin is found in a variety of protists, as well as in all higher eukaryotes, often in the form of variously disposed bundles aligned against the cell membrane or in the form of cables stretched across the cell like telephone wires. Bundles of actin fibers are often joined end to end by axial shafts made of molecules of an ATP-splitting motor protein called myosin. When supplied with ATP and activated by calcium ions, the myosin shaft acts like a ratchet pulling the two actin bundles toward each other. Depending on what cell parts the actin filaments are attached to, this movement may cause all kinds of intracellular displacements and cellular deformations, including a sort of crawling known as amoeboid movement from the name of a protist, the amoeba, that typically moves in this way.

The most elaborate actin-myosin arrangements exist in animal muscle cells. They consist of parallel arrays of interdigitating actin filaments and myosin fibers, held together by a number of additional proteins to form the muscle fibrils. These beautiful structures have given electron microscopists some of their most pleasurable aesthetic experiences, rivaled only by those provided by the contemplation of flagella, cilia, and other microtubule assemblages.

A third kind of protein building block capable of assembling with the help of energy provided by the splitting of ATP is clathrin, which has a peculiar three-legged shape that allows assembly of many molecules into two-dimensional hexagonal meshes of variable curvature, resembling the famed geodesic dome built by the American architect Buckminster Fuller. This structure plays a key role in receptor-mediated endocytosis and in some forms of vesicular transport. When receptors become occupied on the outer face of a cell-membrane patch, they undergo a conformational change that causes clathrin molecules to be recruited on the inner face of the patch and to assemble against it into a closely adhering mesh, with the expenditure of ATP. The mesh progressively rearranges into domes or baskets of increasing curvature, drawing with it the adhering membrane patch with its attached prey and finally cutting it off from the rest of the membrane in the form of a closed, membrane-bounded vesicle containing the catch and surrounded itself by a clathrin trellis (which soon unravels). Although not yet detected in *Giardia,* clathrin could very well also have a very ancient history, in view of its many associations with membrane movements.

Another ancient cytoskeletal structure of paramount importance is represented by the internal supporting shell and associated pores of the nuclear envelope. Made of many different proteins, this complex structure dismantles, together with its double-membranous covering, with every mitotic division, and it reforms spontaneously around each set of daughter chromosomes at the end of mitosis. This reconstruction phenomenon is one of the most remarkable known instances of the spontaneous assembly of a complex structure. The recipe is astonishingly simple. Take some juice from dividing cells, throw in any odd piece of naked DNA—even DNA that has never been near a eukaryotic cell—add a little ATP, and, lo and behold, in a matter of two to three hours, a perfectly respectable envelope assembles around the DNA, complete with double membrane, inner lining, and pores. Inside this mininucleus, the DNA even has formed a beaded string coiled into a miniature chromosome. In this process, hundreds of distinct pieces that were scattered around in the cell extract come together in seemingly miraculous fashion, summoned by no more than the added DNA and supplied with energy by ATP. We are not far from Hoyle's Boeing 747 arising ready to fly from a tornado-swept junkyard, except for a fundamental difference: There is information in all the pieces. They are not junk but pieces of a jigsaw puzzle, shaped to occupy a specific location in the overall picture. In contrast with the puzzle pieces, however, which are cut from a pre-existing picture, the nuclear building blocks are all the products of blind groping by mutations and of sifting by natural selection. The manner in which the pieces are put together is far from haphazard, however. A nuclear envelope assembles in a strictly reproducible succession of steps that are programmed by the properties of the assembling pieces and by those of the catalysts that mediate the process.

The example of the nuclear envelope can be generalized to every complex cytoskeletal structure. Cut off a flagellum, for example, and the whole edifice will grow back from its root by a defined succession of steps. In all cases, structures are the products of spontaneous self-assembly operating according to a program genetically inscribed into the properties of the assembling building blocks.

How could such molecules as actin, tubulin, clathrin, the nuclear building blocks, and the numerous other cytoskeletal proteins ever have emerged? The key word, I believe, is complementarity, which offers a clue as to what may have been the first critical mutations. Proteins were altered in a way that fitted them with complementary means of mutual attachment. The first props that allowed cells to grow bigger without collapsing arose in this manner. After that, a long succession of further mutations, each providing some evolutionary advantage, honed these proteins to their present degree of perfection and added hundreds of new proteins that joined with them to build structures of increasing complexity, often endowed with motility. Most intermediates in this evolutionary process have been wiped out by natural selection, but molecular kinships that may help in reconstructing the history of the proteins concerned are beginning to be recognized by comparative sequencing.

As with every evolutionary problem, the question arises as to what advantages drove natural selection at each tiny step of the long, drawn-out process whereby cytoskeletal proteins were developed and refined. A likely explanation is that the new proteins all played a role in helping the cells to enlarge their volume and expand their surface membrane into a network of increasingly elaborate intracellular compartments. Enhanced heterotrophic autonomy may have provided the major selective factor that drove, in mutually supporting fashion, the coevolutionary development of the eukaryotic cytomembrane network and of the cytoskeletal-motor systems. The fact that many new proteins had to be developed to make a eukaryote out of a prokaryote may well explain the exceedingly long time this transformation required.

WHY TWO NUCLEI? SEX AND THE SINGLE CELL

Giardia has two apparently identical nuclei. Same size, same shape, same four chromosomes, same genes. Or so it seems. The evidence is not all in yet, but present indications point in this direction. According to Karen Kabnick and Debra Peattie[3] from the Harvard School of Public Health, there is a good possibility that *Giardia*'s two nuclei each contain a complete copy of the same genome. In technical jargon, each nucleus is haploid (from the Greek *haplous,* single), and the cell is diploid (from the Greek *diplous,* double). These are two passwords to the whole of eukaryotic evolution—worth remembering.

It is easy to see how a binucleated cell could arise. A cell "forgot" to divide after nuclear duplication, saddling its progeny with two nuclei that were henceforth duplicated and bequeathed in pairs from generation to generation. Or, alternatively, two cells, each with a single nucleus, fused into a binucleated cell by a variant of the bubble trick involving the merger of their peripheral membranes. This phenomenon, which would have been favored by the absence of a cell wall, occurs commonly in nature and is readily provoked. It earned the Argentinian-born British scientist Cesar Milstein and the German Georges Köhler the 1984 Nobel Prize in medicine when they fused an antibody-making cell with a cancer cell. The resulting hybrid cell, which combined the property of making a given type of antibody with the cancer cell's ability to divide indefinitely, turned into a self-reproducing factory for making monoclonal antibodies on a large scale. Such factories are now at work all over the world and supply invaluable tools for research and medicine.[4]

Having a second nucleus puts an additional duplication load on a cell. This state would not have been perpetuated had it not entailed an evolutionary advantage. In fact, the benefit of diploidy is enormous and manifests itself whenever a gene suffers a mutation. Suppose the mutation is harmful. Whereas a haploid cell looses out, the diploid cell has a spare copy of the gene and survives. In the rare case the mutation is beneficial, diploidy is also advantageous. It allows the cell and its progeny to enjoy the benefit of the mutation and even to explore its further evolutionary possibilities, while the unmutated gene of the pair goes on doing its job. An initially harmful mutation may even be made beneficial in this way by one or more additional mutations of the same gene. A consequence of all this is gene diversification. The same gene undergoes different changes in different cells. Thus, many different varieties of the same gene, or alleles—another password—come to be present in the gene pool—still another password—of the species.

Diploidy typifies a new kind of evolutionary strategy characteristic of eukaryotes. Although bacteria occasionally indulge in gene duplication—and derive evolutionary benefits from it—their main strategy, helped by rapid multiplication, relies on large-scale genetic experimentation by individuals to take care of almost any contingency. Quantity is exploited, rather than quality. Eukaryotes, blessed and burdened at the same time with an increasingly complex organization and a correspondingly slower rate of proliferation, were led to evolve a strategy that allowed a similar kind of genetic experimentation while putting greater value on the individual. Diploidy was the solution.

Two additional developments turned the new strategy into a novel form of the genetic combinatorial game of immense importance. Occasionally, binucleated cells "remembered" belatedly to divide, and the resulting mononucleated cells later rejoined with different mononucleated partners to give rise to binucleated cells having two nuclei of different origin. Because of genetic diversification of individual nuclei, this reshuffling of nuclei often resulted in new combinations of genes that were put to the test of natural selection. The gene pool was stirred and the range of genetic experimentation was expanded. We don't know whether this back-and-forth movement between diploidy and haploidy ever happens in *Giardia,* but it is tempting to believe that it happened at some stage in the evolution of eukaryotes, as it provides the simplest explanation for the origin of sex.

In all forms of sexual reproduction, diploid cells give rise to haploid cells in a special kind of cell division called meiosis. Fusion of two haploid cells then generates a diploid cell with its own characteristic set of genes different from that of either of the two diploid parental cells. In the human species, for example, all the cells of the body are diploid, with the exception of the germ cells. Maturation of both sperm cells and egg cells proceeds through meiosis and produces haploid cells. At fertilization, a haploid sperm cell fuses with a haploid egg cell, giving rise to a diploid fertilized egg cell with a unique combination of genes.

In its crudest and presumably earliest manifestation, sex amounted to whole nuclei being exchanged between diploid cells. An important refinement was introduced when the two haploid nuclei of binucleated cells fused into a single diploid nucleus containing all chromosomes in pairs. This merging of the two nuclei into a single one required a major innovation of the mechanism whereby diploid cells gave rise to two haploid cells. Simple division into two mononucleated cells was no longer possible. Two mitotic divisions preceded by a single duplication of the chromosomes did the job and produced four haploid mononucleated cells from a single diploid cell. This is the mechanism of meiosis, a highly complex process that must have arisen through a long succession of steps, each driven by some selective advantage. We don't know the details of this evolution but we can guess its main advantage: genetic diversification and the consequent ability to adapt to a variety of circumstances.

In a first stage, meiosis allowed individual chromosomes to be exchanged instead of nuclei. This led to a considerable increase in possible combinations. For example, a diploid cell containing four pairs of nonidentical chromosomes can give rise to sixteen different haploid combinations of chromosomes; the number of possible combinations is about eight million with a haploid number of twenty-three, as in the human species. Then, the combinatorial range was increased almost to infinity thanks to crossing-over. In this phenomenon, homologous chromosomes (that is, bearing the same genes, often in the form of different alleles) are closely juxtaposed in a way that permits homologous stretches of DNA to "cross over" reciprocally from one chromosome to the other. The chromosomes that are exchanged after rearrangement by crossing-over are no longer the original parental chromosomes but mosaic chromosomes combining more or less randomly selected pieces of both. This virtually guarantees that each haploid cell formed by meiosis from a given type of diploid cell, and, hence, each diploid cell arising from the fusion of two such haploid cells, has a unique genetic make-up, except when genetic diversification is hampered by inbreeding.

The eukaryotic form of sex is very much superior to bacterial conjugation. It has given eukaryotes an enormously powerful means of diversification and adaptation, which accounts for much of their variety and success. It is interesting that sexual reproduction is resorted to by primitive protists only in times of crisis. This is in keeping with a general feature of evolution, which long ago invented the "if it ain't broke, don't fix it" principle—not by applying common sense, of course, but because of the simple fact that mutations are rarely beneficial when all goes well. As long as organisms are adapted to their environment, evolution is largely conservative. Cells multiply by simple division, perpetuating the same genome. But let the survival of the cells be endangered by some environmental upheaval and they suddenly go into a frenzy of sexual debauchery, which, interpreted in anthropomorphic evolutionary terms, amounts to a frantic search for a genetic combination better adapted to the new conditions. For single cells, sex is an emergency measure, not a piece of cake.

Chapter 17

The Guests That Stayed

WITH THE EMERGENCE of the primitive phagocyte, the major part of the prokaryote-eukaryote transition was accomplished. Considering the many innovations needed for this development, the uptake and adoption of endosymbionts may be seen as an almost banal event. Yet it was of paramount importance for future evolution. With rare exceptions, today's eukaryotes all belong to the post-endosymbiont era. There may be a good reason for this, possibly connected with oxygen.

AN AGE-OLD BATTLE

The possession of flagella by *Giardia* tells us that the primitive phagocyte was a motile, fully emancipated cell that had long left the shelter of the bacterial colonies within which it supposedly was born. Perhaps it tended to browse around the rich and easily accessible food supply offered by its erstwhile abode, but it could also have taken advantage of its freedom to move out into any stream, lake, sea, or ocean where bacteria were present. Quite possibly, it spread in different directions and diversified into a variety of species adapted to different environments. Products of this early diversification that have survived to this day include the diplomonads, the microsporidia, and, perhaps, other members, still awaiting discovery, of the large, incompletely inventoried group of protists.

It is likely that our distant eukaryotic ancestor had improved its chances of heterotrophic survival by acquiring some of the properties that help phagocytes today. It probably possessed chemotactic surface receptors sensitive to certain types of molecules and connected to the flagellar apparatus in a manner that made the cell move toward a potential food supply and away from noxious substances. Most likely, it also had endocytic receptors to help it catch and engulf its prey. Perhaps, like our own white blood cells, it complemented the digestive enzymes of its lysosomes with special killing agents. It may even, like some protists today, have had stinging tentacles with which to stun its victims by means of exocytized toxic chemicals.

Predictably, bacteria did not simply submit to their phagocyte-inflicted fate. Thanks to their remarkable potential for meeting all sorts of contingencies—remember the antibiotics—they no doubt evolved a number of countermeasures similar to those practiced by pathogenic bacteria today. Some, like the famous agents of bacterial pneumonia, may have evaded detection and capture by modifying their wall. Others, like the "streps" (streptococci) and "staphs" (staphylococci) that visit upon us a variety of unpleasant infections, may have responded to attack by the release of toxins that injure membranes, thereby escaping after capture and killing their catcher at the same time. Yet others, like the related agents of tuberculosis and leprosy (mycobacteria), may have evolved a strategy of survival within endosomes or lysosomes, multiplying inside these membrane-lined pockets to the point of causing their host cells to enlarge enormously and eventually disintegrate. Others may have combined the escape and survival strategies by first breaking open the surrounding membrane of their endosomal or lysosomal prison and then proliferating in the cytosol.

Occasionally, a stalemate was reached in this perpetual warfare, a sort of truce or mutual nonaggression pact, in which the captured bacteria and their captor spared each other. Such situations, when beneficial to both captor and prisoner, were favored by natural selection and evolved into lasting relationships. The captor became host and the victim guest. Many cases of endosymbiosis are known today and we may take it that many were established in those days when the first eukaryotic phagocytes started to roam the world in pursuit of bacterial prey. There is, however, a strange discrepancy.

If our timing is correct—huge uncertainties affect all such reconstructions—primitive phagocytes existed before two billion years ago, when *Giardia*'s distant ancestor branched from the main eukaryotic line, whereas lasting endosymbionts were adopted only about 1.5 billion years ago.[1] There is thus an immense gap of several hundred million years between the time cells first became capable of capturing endosymbionts and the time when lasting endosymbionts were actually adopted. Rather than dwelling on this discrepancy in unprofitable speculation, I wish simply to point to a coincidence that may or may not be significant. The first permanent endosymbionts to be adopted were oxygen-utilizing bacteria, and their adoption coincided roughly with the great oxygen crisis. Add to this the fact that the primitive phagocyte most likely was anaerobic and, perhaps, poorly equipped to deal with oxygen, and the possibility comes up that most of the primitive phagocyte's descendants, with rare exceptions, such as *Giardia*'s ancestor, fell victim to the oxygen holocaust, leaving as main survivors those that were rescued by oxygen-adapted endosymbionts. We have no proof for this, but it is an attractive hypothesis. Descendants of these life-saving guests include the mitochondria and, perhaps, the peroxisomes.

MITOCHONDRIA: THE CELL'S POWER PLANTS

Mitochondria (singular, mitochondrion, from the Greek *mitos,* thread, and *khondros,* grain) are conspicuous particulate components of the great majority of protists and of all plant, mold, and animal cells. Their shapes vary from spherical to filamentous. Their sizes, on the order of one twenty-thousandth of an inch, recall that of their bacterial ancestors. Their numbers may reach several thousand per cell. They are surrounded by two membranes, of which the inner one is pleated by ridge-like infoldings, or cristae. This inner membrane, which is derived from the cell membrane of the bacterial ancestor, is crammed with oxygen-linked respiratory chains that generate ATP by way of a protonmotive force. The inside, or matrix, of the organelle contains powerful metabolic systems that break down a great variety of substances to provide the electrons that are fed into the respiratory chains. The outer mitochondrial membrane, which is relatively porous, probably originates from the outer membrane of the (gram-negative) bacterial endosymbiont or, less likely, from the membrane of the endocytic vesicle within which the ancestral bacterium was initially caught. Mitochondria are the main sites of oxygen utilization and metabolic ATP production in all aerobic eukaryotic cells. They are the power plants of the cells.

According to sequencing results, mitochondria share a closest common ancestor with a group of present-day aerobic microbes known as purple nonsulfur bacteria. When first caught, these ancestral organisms became established in the cytoplasm of their captors, which supplied them with plenty of food in return for being kept largely oxygen-free. For this relationship to last, the proliferation of the bacterial settlers had to be adjusted to the slower rate of reproduction of their hosts. How this happened in the short run is not known. In the long run, the problem was solved by the transfer of endosymbiont genes to the host-cell nucleus. This phenomenon occurred on a remarkably massive scale, to the point that only a few of the original bacterial genes are left in mitochondria today. These remnants of a past autonomy have fortunately survived, along with appropriate replication, transcription, and translation machineries, to provide us with unmistakable proof of the bacterial origin of mitochondria. In keeping with this ancestry, the mitochondrial genome is contained in a circular, relatively unstructured chromosome of characteristic bacterial type, and the mitochondrial ribosomes also have several bacterial properties that clearly distinguish them from the ribosomes present in the surrounding cytoplasm of the same cells.

There is nothing very remarkable about the actual transfer of DNA to the nucleus. Such a phenomenon takes place routinely in transfection, a genetic manipulation in which DNA is introduced into the cytoplasm of cells by means of a microneedle or otherwise. This DNA readily becomes integrated within the nucleus in a manner such that the foreign DNA is replicated synchronously with the nuclear

DNA, as well as transcribed into messenger RNAs that are correctly translated in the cytoplasm. We may take it that the same would have happened to DNA released into the cytoplasm from endosymbionts, following an injury to the endosymbionts, for example. However, the proteins encoded by the transferred genes would then have been synthesized in the host-cell cytoplasm, where they would have been of little use. To perform their functions, these proteins needed to be translocated into the endosymbionts, a phenomenon that required some important innovations.

Today, mitochondrial proteins made by cytoplasmic ribosomes are translocated posttranslationally into the organelles by complex, energy-dependent machineries present in the mitochondrial membranes and capable of recognizing special targeting sequences in the proteins. Similar machineries exist in the bacterial cell membrane and (originally brought in by infoldings of this membrane) in certain parts of the eukaryotic cytomembrane network. The mitochondrial system presumably developed from one of these machineries, but with an adaptation to a different kind of targeting sequence.

Two facts render this development a little less improbable than it appears. First, there was no time pressure. While evolution was playing with all kinds of mutations, enough normal endosymbionts still endowed with the transferred gene remained to save the population from extinction. Next, once the right tandem of translocation machinery and targeting sequence had been established for one protein, there was need only for the same targeting sequence to appear in the other proteins, an event that could have happened by mutations or by transposition of the corresponding DNA sequence and for which, again, plenty of time was available. To paraphrase two aphorisms, there was safety in numbers and the first step was the hardest. However, there remains to be explained why gene transfer from endosymbiont to nucleus occurred on such a large scale and why endosymbionts still in possession of transferred genes were eradicated in favor of those that had lost the genes. Strong evolutionary advantages obviously drove the genetic subversion of the endosymbionts by their hosts. The most powerful such advantages probably resulted from having the genes gathered together in a central, specially equipped location, where concerted replication could take place, a variety of genetic rearrangements could occur, transcription could be regulated, and RNA products could be processed. An additional benefit could have been that the unencumbered endosymbionts were able to devote themselves more fully to their main tasks of freeing their hosts from oxygen and supplying them with ATP.

While they were playing these evolutionary games, mitochondria even indulged in the almost unique luxury of tinkering with the genetic code. These mitochondrial deviations occurred late, as they are not the same among plants, animals, and molds and even differ among some species of animals or of molds. They were probably made possible because the number of genes concerned was small enough to allow adaptation to a different genetic language.

Mitochondria have elaborate respiratory chains, adapted to retrieve a maximum of energy in usable form from the flow of electrons passing through them. So have their closest bacterial relatives and so, presumably, had the common ancestor of both. This quality fits the present role of mitochondria in eukaryotic cells, most of which have become vitally dependent on mitochondrial respiration and an adequate provision of oxygen for their energy supply. This kind of sophistication, however, has the appearance of a late product in the evolutionary adaptation to oxygen. If aerobic endosymbionts indeed rescued eukaryotes from death by oxygen, one would expect the first rescue operation to be accomplished by a more primitive form of aerobic microorganism. There is an attractive candidate for the role of descendant from such an organism. It is called microbody by cell morphologists and peroxisome by biochemists, and it is present in the vast majority of cells that contain mitochondria, whether in plants, molds, or animals.

PEROXISOMES: PROTECTORS AGAINST OXYGEN TOXICITY

Peroxisomes are particulate entities somewhat smaller than mitochondria and surrounded by a single membrane unrelated to the general cytomembrane system and possibly inherited from an endosymbiotic ancestor. They contain a variety of metabolic systems and they detoxify oxygen and some of its harmful derivatives, especially hydrogen peroxide, as their name indicates. Unlike mitochondria, they do so in a manner that is strictly unproductive in terms of energy retrieval, resembling some primitive aerobic bacteria in this respect. Their proteins are made in the cytoplasm and transferred posttranslationally to the particles by a system that relies, like that of authentic endosymbiont descendants, on the recognition of specific targeting sequences. The snag is that peroxisomes contain no trace of a genetic system. This in no way invalidates an endosymbiont origin. If mitochondria lost more than 99 percent of their genes to the nucleus, the older peroxisomes could very well have lost 100 percent. However, without the remnants of a genetic system, the case for an endosymbiont origin becomes very much weaker. In addition, sequencing results have so far provided little evidence of a kinship with bacteria. The matter is left open.

If peroxisomes preceded mitochondria as protectors against oxygen toxicity, one may wonder why they were not subsequently eradicated after the more efficient mitochondria were adopted. A likely answer to this question is that peroxisomes, in the course of evolution, came to accomplish some vital functions, unrelated to oxygen detoxification, that neither the host cells nor the mitochondrial precursors could fulfill. Human genetic pathology does, indeed, support this possibility. A number of severe inborn deficiencies of lipid metabolism, some of them resulting in the early death of afflicted infants, have been identified as peroxisomal defects. One such disease, adrenoleukodystrophy (ALD), characterized by an inability to break down some special fatty acids, has been widely publicized by the film *Lorenzo's Oil.*[2]

CHLOROPLASTS: THE EUKARYOTIC LINK TO THE SUN

After mitochondria were adopted as regular components of virtually all eukaryotic cells, a second major implantation of bacterial endosymbionts took place. More correctly, it was a wave of such implantations, since there is evidence that the same phenomenon occurred several times. In all cases, the guests were cyanobacteria, that is, representatives of the more advanced, oxygen-producing, phototrophic bacteria. The hosts were various eukaryotic cells, all well provided with peroxisomes, mitochondria, and perhaps other systems to help them resist the production of the toxic oxygen in their very cytoplasm. Without such equipment, phagocytes would have been unable to host cyanobacteria. The engulfed cyanobacteria evolved into the characteristic organelles of phototrophic eukaryotes, the chloroplasts. The protists that adopted them turned into various types of green, red, or brown unicellular algae, of which one group was to give rise later to all green plants. Lines that did not adopt chloroplasts led, besides a number of protists, to all molds and animals, which were allowed to maintain the heterotrophic way of life of their ancestors thanks to the luxuriant proliferation of their distant phototrophic relatives.

Distinctly larger than mitochondria, chloroplasts are surrounded by two membranes and filled with membranous stacks that contain the phototrophic machineries. These stacks are related to similar formations present in cyanobacteria. The matrix of the organelles, which is derived from the cytosol of the cyanobacterial ancestor, contains a number of metabolic systems, most prominently the key enzymes involved in the assimilation of carbon dioxide, the hallmark of autotrophy. Chloroplasts have the main characteristics of endosymbiont descendants. Like mitochondria, they have a rudimentary but active genetic system, only with more original genes left, in conformity with their younger age. They obey the universal genetic code. Most of their proteins are made in the cytoplasm and taken up post-translationally through the mediation of specific targeting sequences. Their kinship with cyanobacteria is supported by sequence homologies.

OTHER POSSIBLE ENDOSYMBIONTS

The possibility has been evoked that other components of eukaryotic cells besides mitochondria, peroxisomes, and chloroplasts may be derived from endosymbiotic bacteria. Such an origin has been postulated for hydrogenosomes, which are membrane-bounded, cytoplasmic organelles about the size of mitochondria, uniquely characterized by the ability to produce molecular hydrogen.[3] Discovered by Miklós Müller, from the Rockefeller University in New York, in a special group of anaerobic protists called trichomonads, which are parasites of the genital tract in humans and some animals, hydrogenosomes have been detected also in a number of other protists unrelated to trichomonads and in some fungi. Hydrogenosomes have the main properties of endosymbiont-derived organelles, except that, like peroxisomes, they lack evidence of a genetic machinery. The biological distribution of these intriguing organelles, although highly restricted, suggests that they may have originated more than once.

Lynn Margulis[4] has proposed that flagella and, therefore, the whole microtubular cytoskeletal system were brought into eukaryotic cells by flagellated bacteria belonging to the group of spirochetes (which includes the causal agent of syphilis). Some evidence, including the possible association of DNA with centrioles (eukaryotic components derived from flagellar roots), has been adduced in support of this hypothesis. Interpretation of the data remains, however, uncertain. It has been mentioned before that bacterial and eukaryotic flagella are totally unrelated chemically. No evidence to the contrary has yet been obtained.

The suggestion has also been made that the eukaryotic nucleus, and with it the whole DNA-based genetic system, may have been imported with an engulfed bacterium.[5] This hypothesis implies the existence of a primitive phagocyte using an RNA-based genetic system. It seems to me difficult to visualize how the resulting cell could have simultaneously handled an RNA and a DNA genome.

A LAST LOOK BACK

Among the profusion of prokaryotic branches that have sprung from the common ancestor to fill every niche available on our planet, the line leading to eukaryotes stands out as a towering trunk, rising high in solitary splendor before suddenly breaking out into a canopy of luxuriant ramifications that dwarf and overshadow the variegated spread below. One is left with the impression of something unique, almost uncanny, an aberrant growth that developed among millions of "normal" offshoots as a result of some extraordinary combination of circumstances or, perhaps, a unique chance event. This impression could be misleading.

The lonely eukaryotic tree started as a small bush like all its prokaryotic relatives. Its growth was far from unerringly straight. Most likely, its real shape is gnarled and twisted, knotted by the stubs of numerous abortive growths and shriveled limbs. Like all other evolutionary processes, the transition from prokaryotes to eukaryotes was groping and exploratory, with each advance selected from many attempts that have left no trace. However, as choices were made or, rather, imposed by environmental selective forces, the range open to further advances became increasingly narrow. Evolution had to tinker—to cite an expression coined by François Jacob[6]—with what it had available and it had to await a favorable mutation within the limits of previous commitments. There was no question of starting a brand-new tack.

Unfortunately, no record is left of the long pathway that converted a prokaryote into a large, mobile, nucleated, phagocytic cell. There is hope, however, that some evidence may be uncovered in the future. Strange life forms may still await isolation and characterization in the rich world of unicellular organisms.

Of all the many changes that marked the appearance of the ancestral phagocyte, those that led to the development of a cytoskeleton and related motor systems probably played the most decisive role. These parts were needed to support the formation of an intracellular cytomembrane system and demanded a large number of genetic innovations. We don't know how the new structural proteins emerged, but we may be sure that they did not do so by the stroke of a magic wand. Their birth was slow, progressive, and stepwise. A major directing factor in their evolutionary shaping was their ability to join together into higher-order structures by self-assembly, a process based itself on chemical complementarity. We have already encountered chemical complementarity as the property underlying base pairing and many other phenomena. We now find it to be the key to the self-assembly of cell structures, mostly from protein building blocks. Because of the rich array of chemical groupings offered by the twenty amino acids that compose proteins, opportunities for protein-protein associations were almost limitless and depended only on some chance mutation in order to materialize.

It is noteworthy that the two most important eukaryotic structural proteins, actin and tubulin, both display complementary regions on the same molecule, so that self-assembly can take place reversibly from a single kind of building block.[7] The polar complementary regions allow end-to-end associations of indefinite length, whereas the lateral complementary regions determine the three-dimensional organization of the resulting threads as either double-stranded filaments or hollow tubules made of thirteen threads. No evidence of sequence analogy between the two molecules has been uncovered so far. It would seem, therefore, that strong selective advantages favored the development of protein molecules with the ability to bind to their own kind. Subsequently, other molecules appeared that could attach to the first structures to shape more complex assemblages or to provide them with motility.

No reliable clue as to the origin of these two key proteins has yet been found in the prokaryotic world. Nor are bacterial proteins known that display a similar pair of complementary regions. Perhaps these are simply cases of lack of detection due to incomplete sampling. Another possibility is that the kind of mutations that give rise to such arrangements are very rare events, which happened to take place only in the eukaryotic line. However, the fact that they happened twice in this line argues against this possibility. A more likely explanation is that bacteria have no use for self-assembling proteins or may even be hampered by them, so that the relevant mutations, when they happened, were rejected by natural selection. Only in the special case of a naked, sheltered heterotroph, provided with full opportunities for cellular enlargement and membrane expansion, did the mutations find fertile ground for positive selection and further improvement along the long road that led to actin and tubulin, those masterpieces of protein engineering. This road must indeed have been long, considering the degree of perfection attained by the two proteins. It is also a typical example of progressive evolutionary narrowing. Actin and tubulin are both highly conserved proteins displaying closely similar structures throughout the eukaryotic world. This means that they were completed, with virtually no room for further improvement, two billion years ago or earlier.

Actin and tubulin are only the two most remarkable products of this extraordinary evolutionary adventure, which saw the birth of many other proteins, all unknown in the prokaryotic world, that provided the structural props and motile machineries of the first eukaryotes. Probably the main selective forces were the same in all cases and related to the improvement of the phagocytic way of life. The final prize, emancipation, was probably long in coming and achieved only thanks to the remarkable constancy, possibly extending over several hundred million years, of the physical and chemical conditions provided by the environment within which this epoch-making metamorphosis took place. For all we know, there may have been many other trials in the same direction that eventually aborted because environmental constraints did not allow them to come to fruition. Some may even have been successful but produced lines that became extinct for one reason or the other.

In the second phase of eukaryotic evolution, after the primitive phagocyte had emerged, this organism was able to diversify and invade a variety of environments, though apparently without undergoing major evolutionary changes, until the wave—possibly precipitated by the great oxygen crisis—of endosymbiotic adoptions that gave rise to modern eukaryotes. This is a typical feature of evolution. A given group may remain static for a prolonged period of time, marked only by the kind of point mutations that do not influence the performance of the affected molecules but provide valuable signposts of evolutionary distance. Then, rather suddenly, most often as a result of climatic or other environmental changes, a fairly rapid transformation occurs, giving the impression of an evolutionary jump, though only in relation to the long static period that came before. The pace of evolution is variable but not discontinuous.

After the appearance of the first endosymbiont-containing protists, evolution once again settled into a relatively static mode, engaging mostly in diversification—endless variations on the same basic themes of oxygen-producing phototrophy and aerobic heterotrophy, without a truly novel theme emerging. Then some eukaryotic cells "discovered" the advantages of getting together and pooling efforts. Why it took them so long to make this discovery is not clear. An enhanced interest in sex could be, at least, part of the answer, together with some major environmental change that made intercellular cooperation advantageous. This question will be examined in the next part.

PART V

THE AGE OF MULTICELLULAR ORGANISMS

Chapter 18

The Benefits of Cellular Collectivism

CELLS REMAINED SINGLE for about three billion years. Bacteria still are today; they do sometimes form colonies—remember the stromatolites—but not true organisms.[1] This may be related to their "selfish" mode of life, geared entirely to producing as many progeny as possible in as short a time as possible.

Even eukaryotic cells have clung to singleness for hundreds of millions of years. Cells endowed with all eukaryote attributes, including endosymbionts, have been around for well over one billion years. Yet there is no trace of multicellular life before 600 to 700 million years ago. Unicellular protists are still abundant in the present-day world.

What prompted some eukaryotic cells to join is not known, except in a general sense. We may take it that cells first got together as a result of chance mutations that favored their association, and that they stayed together because they were reproductively more successful as a group than single. Once they took hold, the advantages of collectivism were swiftly exploited further by evolution, to generate the rapidly expanding and diversifying worlds of plants and animals. Why was this discovery not made before? And why was it made when it was, and then almost simultaneously by autotrophs and heterotrophs? It is possible that some major environmental change made cooperative behavior more advantageous, perhaps by putting a premium on sexual reproduction. This possibility is supported by the behavior of slime molds, organisms that may be viewed as intermediate between unicellular and multicellular.

SLIME MOLDS: AN INSTRUCTIVE EXAMPLE

The most ancient attempt at eukaryotic, heterotrophic cooperation on record was performed by the remote ancestors of organisms named slime molds, or myxomycetes. This is a misnomer: These organisms have nothing to do with molds or mycetes. Neither are they plants or animals in the usual sense of these words. They are the survivors of an evolutionary experiment made more than one billion years ago that never really caught on. They do, however, convey an interesting message.

Slime molds consist of unicellular, heterotrophic protists similar to amoebae. Like amoebae, these organisms wander about in search of prey, which they catch by phagocytosis and digest intracellularly. Let the food supply become scarce, however, and the cells exchange a chemical signal—the agent is cyclic adenosine monophosphate (cAMP), a universal chemical transmitter derived from ATP—that causes them to aggregate into a single mass. This collective then starts crawling, leaving behind a slimy trail, and progressively builds itself into an erect structure called a fruiting body. This structure produces, sometimes through a sexual process, a special kind of protected cells, called spores, that are shed and lie dormant as long as conditions remain unfavorable. When circumstances improve, the spores mature into amoeba-like forms, which resume their unicellular mode of life.

The formation of spores is a common phenomenon in the unicellular world. Many bacteria and protists react to adverse environmental changes by encasing themselves within a protective shell and entering a state of metabolic torpor, awaiting "better days." Slime molds are the first example of cooperative sporulation, a phenomenon exhibited by many plants and fungi.

Slime molds also illustrate a mechanism of interest with respect to the emergence of animals, though in a different context. Upon exposure to cyclic AMP, the unicellular forms of the organism express new surface molecules with mutually complementary lock-and-key arrangements that keep cells stuck to each other after a chance encounter. The cells are also held together indirectly by means of surface receptors that bind them to the viscous, extracellular material they secrete. This "slime" serves as a glue, as a carpetlike substrate, and as a recognition trail. Animal cells are likewise held together by surface adhesion molecules that join the cells to each other (cell adhesion molecules, or CAMs) and to extracellular scaffoldings (substrate adhesion molecules, or SAMs).

A third lesson we learn from slime molds is the role of sexual reproduction as an emergency measure. In the history of multicellular eukaryotes, this mode of reproduction progressively became a major factor of evolutionary resilience and diversification.

THE IMPORTANCE OF SEXUAL REPRODUCTION

Bacteria engage in conjugation and genetic recombination. Real sex, however, with its systematic alternation between diploidy and haploidy, is a typical eukaryotic prerogative, which was probably first engaged in by the primitive phagocyte. The main advantages of sexual reproduction were considered in chapter 16. What was not considered in that chapter is the influence of sexual reproduction on evolutionary mechanisms.

When cells multiply by simple division, entire genomes are reproduced with, occasionally, the appearance of a mutant combination that becomes, itself, subject to reproduction. Selection operates among these different forms of the same genome. Some may continue to diverge side by side.

Things are more complex in sexual reproduction. Mutant genes become associated with different gene combinations at each generation. Their evolutionary effects must be assessed on a statistical basis, by their ability to spread into the gene pool of the population. For this reason, two evolutionary lines can separate only in reproductive isolation, that is, if they are unable to interbreed. A special discipline, population genetics, has been developed to deal with these problems. Its methodologies are too complex to be elaborated on here, but its existence deserves to be mentioned, as I shall make little reference to it in the simplified accounts given in the coming chapters.

PRINCIPLES OF CELLULAR COLLECTIVISM

A central tenet of modern Darwinian theory asserts that evolution proceeds by random mutations, the effects of which are screened by natural selection. All the findings of molecular biology support this view. This does not mean that evolution is haphazard. Running through the processes of multicellular complexification are a few unifying threads: association, differentiation, patterning, and reproduction.

Association

Every cell is born by division, next to a sister cell. If something tends to keep the two cells together, they will stay together. This will happen if sister cells stick to each other or if they remain within a shared housing. When each of the two sister cells divides, the same phenomenon keeps the resulting foursome together. Repetition of this process gives rise to a colony of increasing size.

Since plenty of mutations may occur that favor or impede the associative behavior of cells, it is left for natural selection to weigh the advantages of congregation against its disadvantages. The main drawback: Cells are likely to have less access to sources of nutrients and energy when grouped together than when isolated. On the plus side are better protection against predators and environmental injuries and, especially, the benefits of cooperativity. Colony growth cannot be indefinite, however, and must at some stage give way to colony reproduction.

Differentiation

True colonies composed of the same kind of cells are a rarity. Association gains a particularly powerful advantage from differentiation, whereby genetically identical cells become different by no longer expressing the same genes to the same extent. The seeds of differentiation lie in gene regulation, which governs many adaptive behaviors. The way in which bacteria adapt to milk sugar by turning on the genes coding for the necessary enzymes is a typical example (see chapter 14). Regulation by transcriptional control of gene expression also takes place in multicellular eukaryotes. In puberty, for example, what causes a girl to develop breasts or a boy to grow facial hair is the transcription of certain genes in the cells concerned, induced by the hormones whose secretion sets off puberty.

Transcriptional control of genes is particularly important in development. It explains why cells possessing the same set of genes may be very different—liver cells, muscle cells, nerve cells, and so on, or, in plants, root cells, bark cells, leaf cells. It all depends on which genes they express. These effects are mediated by special proteins, called transcription factors, endowed with the ability to interact with certain specific regions of DNA. The genes coding for such transcription factors are designated regulatory genes, as opposed to the genes that code for enzymes or structural proteins, which belong to the group of "housekeeping" genes.

Differentiation allows cellular specialization and, thereby, division of labor among members of a collective; it is the secret of cellular cooperativity and evolutionary complexification. Differentiation runs through all the ramifications of the tree of life. Between seaweed and magnolia, between sponge and eagle, one important difference is the number of distinct cell types that compose the organism. But this is only part of biological diversity. Another is patterning.

Patterning

The body of a human adult contains several trillion cells but only about two hundred cell types. Essentially the same cell types serve to build the bodies of a mouse or a whale, or even, with few differences, the bodies of a frog or a fish, just as the same types of bricks and planks may serve to build different dwellings, from cottages to mansions. The paramount significance of patterning is clear. If we wish to understand evolution, we must pay special attention to what the American biologist Gerald Edelman has called topobiology,[2] the study of the mechanisms that cause differentiated cells to assemble into characteristic three-dimensional patterns. Since evolution proceeds by way of genetic alterations, the changes that produced a mouse, a whale, or a human from their common mammalian ancestor, or even a fish, a frog, or a mammal from a primitive vertebrate, must largely result from mutations affecting pattern-controlling genes.

Reproduction

All multicellular organisms arise from a single cell—spore or fertilized egg—genetically programmed to enact with great precision a scenario of coordinated divisions, differentiations, and patternings resulting in organisms similar to the parent organisms and capable of perpetuating the species in a similar way. The same scenario is re-enacted at each generation. This reproductive behavior holds several implications basic to the process of evolution.

First, the target of mutation is the progenitor cell. Only mutations affecting progenitor cells are relevant to the evolutionary fate of multicellular organisms. A somatic mutation (from the Greek *sôma,* body) may have a major effect on the viability of the affected organism but, not being transmissible, cannot affect the organism's progeny.

Second, the target of selection is the organism. A progenitor cell that has undergone a mutation must produce a complete organism if the mutation's effect on viability and reproductive success is to be evaluated by natural selection, at least if selection is to be positive. Negative selection can occur any time after fertilization.

Third, the developmental blueprint of an organism is written into the genome of progenitor cells. To have an evolutionary impact, progenitor cell mutations must affect genes that control development, that is, regulatory genes.

And finally, evolution operates within the constraints of an existing developmental blueprint. The more complex this blueprint, the more severe the constraints. With only a few lines sketched out, a drawing still has the potential to become a landscape, a still life, or a nude, depending on the whim of the artist. As more details are put in, the commitment becomes increasingly stringent. This rule, all-important to our understanding of multicellular evolution, explains why only a small number of distinct body plans, all dating back to early stages of evolution, underlie the profusion of different organisms that have arisen.

Chapter 19

The Greening of the Earth

ONE BILLION YEARS AGO, the continents were barren expanses of rock and lava, deserts baking in the sun by day and freezing by night, rarely refreshed by rainfall and unable to retain moisture for lack of topsoil.[1] In contrast, the oceans were filled with all kinds of unicellular life. Bacteria were abundant. So were unicellular eukaryotes, which already had diversified into a variety of phototrophic and heterotrophic species, many of which had developed a sexual mode of reproduction as an alternative, under special conditions, to their usual vegetative way of multiplying by division. In this watery laboratory, protists formed all sorts of associations, most of which failed to survive. A few of these associations turned out to be advantageous and developed further.

Multicellular eukaryotic forms of life probably arose initially from small clones of cells that remained associated after their production, by successive divisions, from a single parent cell. The cells were held together either by intercellular connections or by a shared external wall or shell. Roughly speaking, the former mechanism led to animals and the latter to plants and fungi. This division reflects key differences in lifestyle. The heterotrophic animals had to maintain freedom of movement in order to catch prey, even if this freedom meant greater fragility. The phototrophic plants needed only to catch sunlight (and dissolved mineral nutrients) and could afford to remain immobile, even derived an advantage from being immobilized in a favorable location. Fungi, which developed a scavenging form of heterotrophy based on the breakdown of dead organisms by means of secreted digestive enzymes, were able to forsake mobility for the advantages of a protective coating. Because of such fundamental differences, these three kingdoms followed very different evolutionary pathways.

Most easily reconstructed is the early history of plants,[2] because species that may be representative of successive evolutionary stages still exist today. There is danger in this extrapolation from present to past. Extant alleged "missing links" all evolved over long periods and may in no way resemble their distant ancestors. One might even say that they could not possibly resemble their ancestors. Otherwise, why were they not wiped out by natural selection? This is a problem, though not an insurmountable one. Evolutionists no longer tend to view change as a necessary concomitant of evolution, but as something that happens only if enforced by circumstances, most often an environmental change. If a form of life is well adapted to its surroundings, it may persist unchanged as long as its niche remains unaltered. Even a poorly adapted form may survive indefinitely if competition is weak. The prudent recourse to sexual reproduction on the part of many protists illustrates nature's inherent resistance to change.

ALGAE AND SEAWEEDS

Mementoes of early multicellular plant life are found today in the variegated world of algae and seaweeds, from the tiny organisms responsible for the emerald tinge of many a pond to the thick, brown kelps that cover coastal rocks with glistening manes, undulating with the ebb and flow of assaulting waves, or that used, so the legend goes, to ensnare the imprudent navigators who ventured into the Sargasso Sea. At least three distinct evolutionary lines of algae exist, each related to the endosymbiotic adoption of a different kind of phototrophic cyanobacterium. They are, in order of decreasing ancientness, the red, brown, and green algae. With rare exceptions due to secondary regressions to parasitic life, all are phototrophic and produce molecular oxygen. Their chloroplasts all contain green chlorophylls, but with various amounts of accessory pigments of different colors.

There is great diversity of size, shape, chemical composition, metabolism, developmental pattern, and reproductive behavior within each of the three groups. A common feature is the construction of external walls made of carbohydrate polymers, among them cellulose, a glucose polymer that plays a dominant structural role throughout the whole plant world, and a variety of viscous or gummy substances, several of which are used industrially. Whenever you savor an ice cream, there is a good chance that the smoothness caressing your palate is due to alginic acid, a carbohydrate polymer extracted from certain kelps.

The morphological organization of multicellular algae is usually simple, consisting most often of branched filaments, sometimes of flat leaflike sheets, unconnected by a vascular system. Their most prominent specializations include an anchoring structure, called a holdfast, whereby many seaweeds are attached to solid surfaces; bladders or air sacs, which serve as floats; and primitive sex organs. The reproductive behavior of algae varies along a scale of complexity that is often depicted as a recapitulation of the evolutionary history of reproductive function.

All algae have a sexual mode of reproduction, involving the fusion of two haploid gametes (from the Greek *gamos,* marriage) into a diploid zygote (from the Greek *zygos,* yoke). Haploid cells have a single set of chromosomes; diploid cells, two. In the simplest and, probably, most primitive form of sexual reproduction, the two gametes look identical. They may be mobile and depend on flagellar motion to find each other. Or they may lack mobility and be brought together passively by appropriate adaptations of the enclosing walls. At the other end of the spectrum, gametes show extensive sexual dimorphism. One is small and flagellated, like male spermatozoa; the other is large, immobile, and stocked with reserve substances, as are female egg cells. Both kinds of gametes are usually produced by the same plants. These are hermaphroditic, joining the attributes of the Greek god Hermes with those of the goddess Aphrodite.

In algae, meiosis, the kind of cell division whereby the number of chromosomes is halved and haploid cells are formed from diploid cells, rarely produces gametes directly. The first haploid cells to be formed, called spores, may go through more or less complex phases of multiplication and development before giving rise to the gametes. In the extreme form of this growth pattern, the organism is haploid in all its stages, with the exception of the zygote, which goes through meiosis immediately after being formed. The other extreme, in which the organism is entirely diploid except for the gametes, is also known. In many cases, the situation is intermediate between these two extremes. A spore develops into a haploid organism, which produces gametes, which fuse into a zygote, which develops into a diploid organism, which produces haploid spores, which start a new cycle. The haploid and diploid organisms often have similar shapes. Known as the alternation of generations, this pattern is characteristic of many algae and has become, under innumerable variations, a leitmotiv of plant life.

MOSSES INVADE THE LANDS

Simple though they are, algae are perfectly adapted to their aqueous milieu and have thrived in it ever since their emergence. What caused some to move out of their balmy abode to confront the rigors of land? Overcrowding, exclusion by more successful species, excessive grazing by animals are possible explanations, though not very convincing ones. The odds to be overcome were of such magnitude that little short of a life-and-death situation could account for the transition. Most likely, certain bodies of water became cut off from the oceans and slowly dried out, leaving to survive only those forms of life that succeeded in adapting to increasing dryness.

Adaptation was progressive, following the gradient of decreasing wetness along the coastal edges. The least adapted forms remained nearest to the water; the best adapted survived farthest from it. At first, the plants were still intermittently provided with essential water and minerals by tides and waves, so the first hurdle was to avoid desiccation between wet episodes. Plants that acquired an impermeable waxy covering, or cuticle, gained a selective advantage. However, this advantage was curtailed by the demands of nutrition. Modifications that allowed the plants to imbibe mineral-laden moisture from the soil were favored by natural selection. So were openings in the cuticle, the precursors of today's stomata, that made it easier for the phototrophic cells to absorb atmospheric carbon dioxide and get rid of oxygen. Also useful were any surface projections that anchored the plants to the ground and prevented them from being blown away from vital moisture by the wind. Some of these appendages doubled as imbibing structures, or rhizoids, prefiguring roots.

One last development was needed for plants to become fully established on land. Their reproduction had to be ensured without the participation of aquatic progenitor cells. Alternation of generations provided evolution with the appropriate mechanism. The haploid spores developed protective coverings and served as vehicles for aerial dissemination. In the soil, the protected spores could remain dormant until enough moisture was present to trigger germination. The haploid plants arising from germinating spores produced motile male gametes and immobile female eggs in neighboring structures that were kept sufficiently humid to allow the male gametes to swim toward the eggs and fertilize them. The resulting diploid zygotes, after going through an abbreviated developmental phase, then gave rise to haploid spores. Thanks to these adaptations, primitive mosses began to cover the shores with furry, green carpets, extending farther inland as their rhizoids penetrated deeper in the soil to catch water and minerals.

Apparently, only the green variety of algae succeeded in colonizing dry land. They did so by way of a smoothly progressive adaptation of their ancestral algal blueprint. We can account for the entire evolutionary sequence by small additions or changes to this blueprint, favored at each step by improved fitness to survive and reproduce on land. The continually receding water line exerted considerable selective pressure in favor of these modifications, which would have been of little value in a watery milieu. These facts illustrate the power of environmental factors to influence the direction of evolution and the inherent constraints that force evolution to proceed within the framework of an established body plan.

Once a successful strategy for survival has been developed, environmental pressures relax while intrinsic constraints become more stringent. What then follows is mostly secondary radiation, invasion of an increasing number of ecological niches through an increasing diversification of details. This is why mosses still thrive today, divided into some 15,000 distinct species adapted to a wide variety of climates, from tropical to arctic, and clinging to a great diversity of supporting structures, from waterlogged bark in a rain forest to the barest of rocks.

VASCULARIZATION, A CRITICAL ACQUISITION

Early land plants were mostly confined to humid coastal fringes, leaving large expanses of dry land still barren and open to invasion. Conquest of the ancient deserts was achieved inch by inch by mutant plants that progressively acquired a root system capable of penetrating deeper into the soil and absorbing water and mineral nutrients more efficiently. Several other changes accompanied this development. The body of these plants became polarized into two distinct growth zones: the colorless, subterranean root tips and the aerial, green buds, separated by a system of connecting stems. At the same time, the plants became sensitive to the Earth's gravitational field (geotropism) and tended to adopt an upright position. Finally, and most importantly, they developed conducting channels that allowed water and minerals absorbed from the soil to flow up from the roots to the other parts of the plant, and the organic, photosynthetic products made in the green parts to flow down to the roots and other colorless parts. Thanks to this vascularization, the plants were able to grow bigger while expanding their light-catching, photosynthetic parts into a ramified system of flattened leaves. A major step in evolution was accomplished.

Reproductively, the first vascular plants, like their predecessors, went through alternating haploid and diploid generations and used haploid spores as the means of dissemination, but with a marked shift in emphasis from the haploid to the diploid stage. Whereas the dominant form of mosses was the haploid, gamete-producing form, as in many algae, that of the early vascular plants became the diploid, spore-producing form. Shed spores, after germinating in the soil, gave rise to mature gametes at the end of an inconspicuous, often subterranean developmental phase. The gametes then fused into the diploid zygote from which the main plant grew, often attaining considerable size.

With these developments, the stage was set for one of the most fateful events in the saga of life. About 400 million years ago, the green armies marshaled from the oceans began invading land on a massive scale, helped by climatic and geographic changes that occurred at that time and to which they themselves contributed with the water they pumped out of the soil. The atmosphere became more humid, the rains more abundant, and the soil better able to retain moisture. Bacteria of all kinds accompanied the invaders and were soon followed by the first land-based fungi and animals, further enriching the biotope. The plants grew bigger and developed a tough, polymeric substance called lignin that allowed the building of solid trunks. Trees appeared, up to forty feet high and three feet in diameter. Much of the land turned into enormous tropical swamps, harboring rich vegetation that grew much faster than the heterotrophic organisms that fed on them. The dead remains of these plants were left to accumulate and fossilize, creating the huge deposits of carbon-rich material now mined as coal. Hence the name Carboniferous given to the geologic era, between about 360 and 286 million years ago, when those swamps flourished.

Most of the pioneers in the conquest of land are long extinct. Their closest extant relatives, according to the fossil record, include the horsetails (*Equisetum*), the lycopods, or club mosses, whose flammable, powdery spores were used in my youth to envelop stage monsters in fiery veils, and, especially, the ferns, of which some nine thousand species are known. These plants are mere shadows of their ancestral glory, surviving relics of a bygone era. What caused their downfall? As in most evolutionary upheavals, changes in geographic and climatic conditions were responsible.

THE PERMIAN CRISIS AND THE FORMATION OF SEEDS

After 50 million years of spectacularly successful development, the great Carboniferous swamps began to dry out and their forests slowly withered away. Not just the land plants but much of marine life was wiped out as well, in what may well have been the most dramatic mass extinction in the history of life on Earth, the great Permian crisis (the Permian is the geologic period extending between 286 and 250 million years ago).[3] Probably the main cause of this catastrophe was the drifting together of all the lands of the Earth into a single continent, Pangaea. Much of the interior of this mass turned into an immense land-locked desert, like the Gobi Desert today. In addition, the climate became much colder, due, perhaps, to catastrophic volcanic eruptions in what is now Siberia, which obscured the skies and blotted out the sun. A good part of Pangaea was situated over the South Pole and was covered by a thick ice sheet. Glaciers lined its coasts with massive frozen cliffs, which crumbled periodically into huge icebergs that were carried away by the currents to cool the seas right up to the tropics. The water level dropped, and much of the sunlight that fell on the Earth's surface was lost by reflection. The Earth had entered the harshest ice age in its history.

Plants reacted against this cataclysmic situation by replacing spores with seeds as a means of dissemination. Or, more likely, some seed-producing species already existed but did not flourish until circumstances turned this property into a vitally important asset, allowing survival where the spore-bearing plants could not hold out.

The transition from spore to seed signaled female emancipation. The first step, already accomplished in certain seedless plants, such as lycopods, was a separation of the sexes at the spore level. Instead of a single kind of spore giving rise to a hermaphroditic haploid organism that produced both kinds of gametes, two kinds of spores germinated into two different gamete-producing entities. Large macrospores developed into female organisms, small microspores into male organisms. It was then necessary for the male organism's sperm cells to seek a female organism to find egg cells to fertilize. Subsequent events took place as with hermaphroditic organisms. The fertilized egg cell developed into an early embryo, which eventually became implanted in the soil.

This separation favored outbreeding over inbreeding, the main disadvantage of hermaphroditism, and allowed experimentation with all kinds of different diploid genomic formulas. But it decreased the chances of fertilization. The next evolutionary step obviated this drawback by shifting the site of fertilization from the soil to the plant itself. Macrospores no longer were shed to germinate in the soil. They completed their maturation on the plant, in special organs called ovules in which the eggs arose within a cocoon of protective and nutritive structures.

Male spores continued to be dispersed as wind-borne pollen grains, which, however, were now programmed to pursue their maturation only in a compatible ovule. The haphazard character of this mode of dissemination was obviated by the vast number of pollen grains produced. A pollen grain landing on an ovule matured into a sperm cell, which entered the ovule to fertilize an egg cell. The ensuing zygote developed into an early embryo, up to a stage where further development was arrested and the ovule closed around its occupant. This protected, dormant embryo became the seed, which was shed.

After seeds were dispersed, their covering protected the embryos against cold and dryness, awaiting favorable circumstances that induced the embryos to resume development and break out of their protective shells. Reserve substances included within the seeds provided the nutrients necessary to sustain the embryos during the time needed for the first rootlets and leaflets to become functional and support autonomous growth. Such an adaptation would be of little use to a plant living in a swamp but would save the species if geographic and climatic conditions became harsher. Seeds are sturdier vehicles of dissemination than spores; they are capable of resisting extreme physical conditions for months, if not years or even centuries—the record, held by some lotus seeds found in a peat deposit in Manchuria, exceeds one thousand years[4]—until a favorable moment, be it fleeting, arrives to permit germination and implantation.

After the first wave of spore-bearing plants was decimated by drought and cold, a second wave, equipped with hardy seeds in lieu of spores, invaded the inhospitable lands of Pangaea. The extant descendants of this second army include the seed ferns, the palm treelike cycads, the ginkgoes, and, especially, the pines, spruces, cypresses, redwoods, and other conifers. Together, these groups form the superfamily of gymnosperms (from the Greek *gymnos,* naked, and *sperma,* seed). Actually, their seeds are hardly naked; they are described as such in contradistinction to the angiosperms (from the Greek *aggeion,* envelope), whose seeds are contained within fruits.

FLOWERS AND FRUITS: THE CROWNING ACHIEVEMENT

Angiosperms are the most advanced and abundant forms of plant life on Earth. They started spreading across the continents about 100 million years ago. By that time, Pangaea had moved northward and broken up into continental masses that were drifting apart toward their present-day positions. Nobody knows how the new plants came about, but we may imagine. One day in those remote times, a seed plant suffered a mutation that caused the leaves around the sex organs to lack chlorophyll and turn white or, perhaps, yellow or pink if they retained some accessory pigments. The uniformly green landscape of the primeval fields and forests thus became dappled for the first time with bright patches, which acted as beacons for insects that happened to be genetically programmed to move toward light. Because of this fortuitous circumstance, the genetic accident suffered by the plant turned into a benefit. In the course of their visitations, the attracted insects collected pollen grains on their bodies from the male organs of the plant and dropped some grains again on female organs. The pollination record of the mutated plant increased, and so did its reproductive success. The insects also profited from the plant's mutation, which guided them to nutritious nectar; they proliferated in large amounts. Once initiated, the new evolutionary process moved on ineluctably. Further plant mutations created new shapes and colors and a variety of scents, attracting all sorts of pollinating insects and also other animals, such as birds and bats. Propelled by the most far-reaching instance of mutually advantageous relationships ever established between plants and animals, a revolution was launched that sprinkled the green expanses of the Earth with countless dabs of color. Flowers were born.

The key property of flowers is to develop into fruits. This term encompasses not only the oranges, grapes, apples, plums, berries, and other fruits we call by that name; it also includes nuts, ears of corn, pea-filled pods, and the many winged or fluffy seed-containing gondolas that weeds and trees send out floating in the wind on a summer day. A fruit may be defined as one or more seeds enveloped in a hull—although unfertilized flowers can be coaxed to yield pipless oranges and seedless grapes for the fastidious. The hull, derived from the female part of the flower, distinguishes angiosperms from gymnosperms. It consists of a protective covering and nutritive tissues. It owes its origin to a special process called double fertilization, which is unique to flowering plants. While one male sperm fuses with an egg cell to form the zygote out of which the embryo is to grow, a second sperm cell fuses with a diploid cell in the female part of the flower. The triploid cell arising from this second fertilization develops into the fruit's hull. Such is the basic theme on which evolution has composed an extraordinary number of variations, to the delight of our senses.

In this new evolutionary phase, plants with male and female organs grouped in a single flower gained a selective advantage, although this never became the rule. Plants with separated sex organs or even plants with two distinct forms, each bearing a single kind of sex organ, also exist. Many different kinds of insects and other animals participate in the great pollination game, with flowers displaying an astonishing array of specialized lures and traps selected to ensure that the pollen reaches its target.

Summarizing the evolutionary history of plants on Earth, figure 19.1 shows in

FIGURE 19.1
A Bird's-Eye View of Plant Evolution

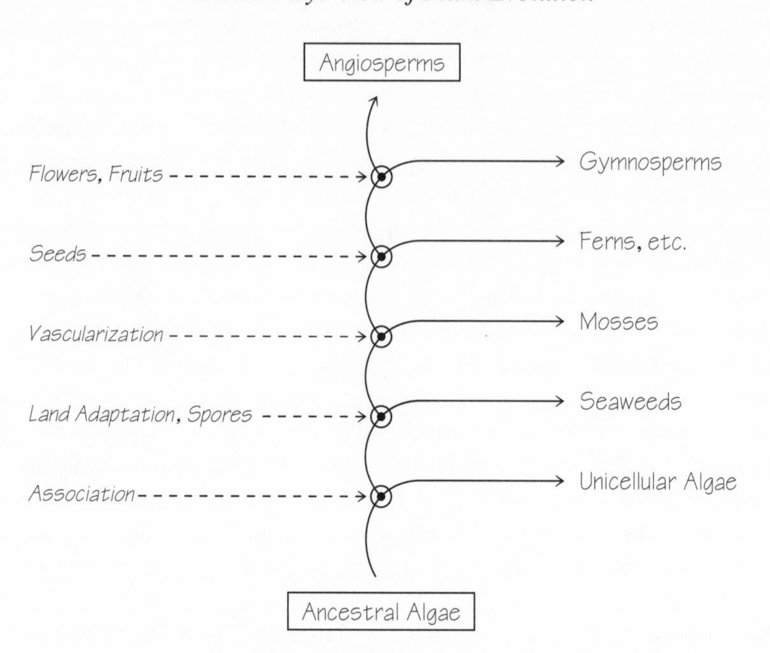

This figure depicts the main steps in the rise of plants in the direction of greater complexity. At each fork, a mutant evolutionary line that underwent the genetic change in body plan indicated on the left diverges upward from the unmutated line—represented by the arrow curving to the right—leading to existing phyla.

highly schematic form how key mutations of "fork organisms" led to significant evolutionary advances, leaving the descendants of unmutated organisms to provide us with some information on the properties of the fork organism, the last ancestor they have in common with the more evolved species.

UNDERGROUND INFILTRATORS

When plants started invading land, they were soon followed by a horde of scavengers. Related to the unicellular yeasts, these opportunistic fellow conquerors resembled primitive plants in forming branched, tubular structures, or hyphae, encased by tough carbohydrate polymers, and in reproducing by means of spores; but they were colorless and strictly heterotrophic. Unlike other heterotrophic organisms, they were immobile and unable to catch prey, relying entirely on a primeval form of extracellular digestion for their subsistence. Clinging tightly to their companion plants or to the remains of dead plants, sometimes strengthening their hold by means of tiny rootlets, they attacked these substrates with powerful digestive enzymes, which they secreted from their cells, and they absorbed the soluble products of this digestion by permeation through their surface.

This primitive, apparently precarious way of life has turned into a remarkably successful formula, the prerogative of more than 200,000 species forming the vast group of mycetes (from the Greek *mykês,* mold), which includes yeasts, molds, rusts, smuts, mushrooms, toadstools, puffballs, and sundry other fungi. Long classified as members of the plant kingdom and viewed as originating from degenerate plants that had lost their chloroplasts, mycetes are now considered a separate kingdom, distinct from the plant and animal kingdoms. Contrary to former belief, the mycetes have been found by molecular sequencing to be more closely related to animals than to plants.[5]

Mycetes are prime scavengers and play an important role in the recycling of bioelements. A number are parasitic, causing a variety of diseases in plants and, less frequently, in animals. Others have long been used for their ability to catalyze various fermentations useful in the preparation of breads, cheeses, and alcoholic beverages. The misadventures, sometimes deadly, of unwary mushroom eaters and the opposite experiences of millions of patients saved by penicillin and other fungal antibiotics are witness to the chemical versatility of mycetes. Many fungi live essentially underground and make their presence known only when they suddenly sprout some reproductive structure that emerges from the soil to disperse its spores. Some have established lasting symbiotic associations with green algae, forming the lichens, one of the hardiest forms of life.

The body plan of mycetes has remained simple, made mostly of a network of interconnected hyphae called a mycelium. This network may extend over huge areas of up to several square miles. The nuclei of fungal mycelia are always haploid, but the cells themselves are often binucleate, as are those of *Giardia,* the most ancient of presently known eukaryotes. This is due to a delay, often of considerable duration and including a number of cell divisions and other developmental events, separating nuclear fusion from cell fusion in sexual reproduction. In plants and animals, the two phenomena follow each other rapidly in the course of fertilization. Nuclear fusion does eventually take place in mycetes, and the resulting diploid nucleus then almost immediately goes into meiosis and forms haploid nuclei. These emerge with a reshuffled genetic content thanks to meiotic chromosome rearrangements, the main evolutionary benefit of sexual reproduction. The uninucleate, haploid cells produced by meiosis give rise to spores, which are shed to disseminate the species.

Sporulation is the major event in the life of most mycetes. It is accompanied by the development of special structures, of which mushrooms are the most spectacular examples. In some cases, these organs are built to forcibly eject and spread the spores around. The most famous of fungal spores was released by a common mold, *Penicillium notatum,* on a September morning in 1928, and landed on a bacterial culture in the laboratory of a Scottish microbiologist, Alexander Fleming, at St. Mary's Hospital in London. The spore developed into a fluffy, greenish colony that killed all the microbes around it, creating a clear circular zone that, fortunately, was noticed by Fleming. The outcome, fifteen years later, was the miracle drug penicillin, thanks to the persevering efforts of an Australian pathologist, Howard Florey, and an émigré German chemist, later naturalized British, Ernst Chain; thanks also to the special circumstances created by World War II, which justified an extraordinary outlay of energy and money that might perhaps never have been expended under peaceful conditions.[6]

Chapter 20

The First Animals

AT THE TIME unicellular phototrophic algae started to assemble into the first primitive seaweeds, heterotrophic protists were also led by the vagaries of mutation to experience the virtues and drawbacks of multicellular association, leaving it to natural selection to pronounce the final verdict. Because of the overwhelming need for food that dominates heterotrophic life, the selective advantages that drove animal evolution were different from those that propelled the evolution of plants, depending mainly on improved feeding and reproduction through cooperative association among cells.

The outcome is an amazing diversity of life forms, which the combined efforts of taxonomists, comparative anatomists and physiologists, paleontologists, and, more recently, biochemists and molecular biologists have ordered into a majestic genealogical tree depicting the evolutionary history of extant and extinct animals.[1]

PHYLOGENY AND ONTOGENY

The first elaborate animal tree was drawn by the nineteenth-century German naturalist and philosopher Ernst Haeckel, an early and enthusiastic disciple of Darwin, as well as an imaginative master in the art of weaving sparse facts into daring, persuasive generalizations. The most famous of these is summed up by the aphorism "ontogeny recapitulates phylogeny," by which is meant that animals, in the course of their embryological development (ontogeny), go through successive stages that recall the stages of their evolutionary history (phylogeny). Known as the recapitulation law,[2] this statement, though not to be taken literally, expresses a profound truth. Recent acquisitions of molecular biology have shown that development is the main key to animal evolution, which has proceeded largely by way of genetic changes affecting body plans.

A possible misapprehension must be corrected. Many of us, when looking at a representation of the animal evolutionary tree, tend to visualize our lineage as passing through successive stages in the form of sponges, jellyfishes, worms, mollusks, and so on. This view is false. The animals we are familiar with are terminal twigs on the tree of life, final products of long evolutionary histories. *Our early ancestors make up the trunk of the evolutionary tree.* In order to reconstruct them, we must backtrack mentally from the tips of twigs, through branches of increasing importance, down to a major fork where a principal limb separated from the trunk. What we find there are forms considerably less specialized than those that occupy the twigs. By definition, these "fork organisms" make up key ancestral populations that were split by a mutational event into two groups that parted company and started to evolve in different directions. As a rule, these directions were related to two distinct habitats, of which each gave one of the two groups a reproductive edge over the other. One direction branched out into a complex system of ramifications that produced an extant major group of animals. The other direction prolonged the trunk of the tree, up to the next fork from which a new principal limb separated. It is the succession of these fork organisms that makes up our ancestry and that we must reconstruct. It is an uncertain exercise, since the tree of life is known to us only by the terminal twigs that are still alive and by sparse fossil vestiges whose position on the tree is often difficult to ascertain. However, thanks to comparative sequencing, we are now in a position to evaluate—with still limited but ever-increasing confidence—the distance that separates two twigs from their last common fork organism.

THE AWAKENING OF ANIMAL LIFE

The first successful association experiment on record involved ancient representatives of the family choanoflagellates, which are monoflagellated, heterotrophic, aerobic protists so named because their flagellum emerges from the bottom of a food-collecting funnel (*khoanê* in Greek). Such cells may have joined initially into hollow, spherical arrangements combining cooperative propulsion with cooperative feeding.[3]

With time, further mutations flattened the sphere into a miniature, double-walled pancake, with a back and a belly made of different kinds of cells. The thick, ventral (belly) cell layer served for crawling and food gathering; the thinner, dorsal (back) layer for protection and swimming (see figure 20.1). The animal sometimes raised its central part above the sea floor, creating a space that served as a primitive alimentary cavity. Like the parent protist, the organism could reproduce sexually under certain conditions, such as excessive crowding. It did so by way of large, nutrient-laden egg cells, which were released after fertilization and developed into copies of the parental organism.

FIGURE 20.1

Some Key Steps in Early Animal Evolution

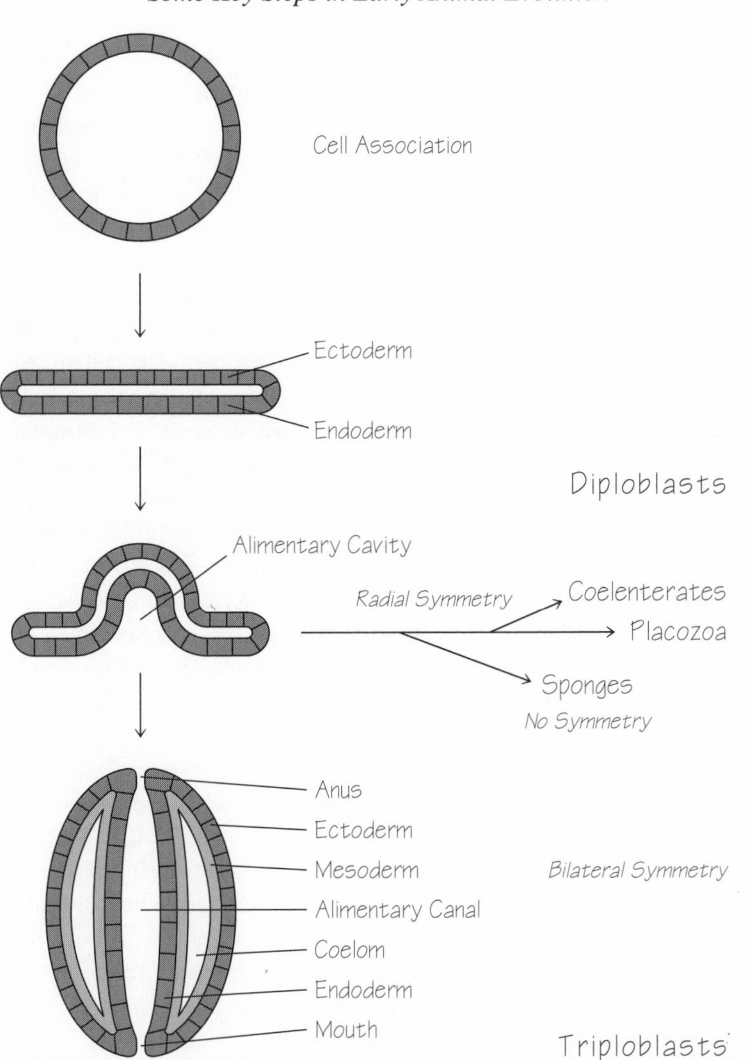

On top is depicted the evolutionary conversion of a spherical cellular monolayer into a primitive placozoan, ancestral to all diploblasts, by way of a flattened pouch—with a differentiated ventral endoderm and dorsal ectoderm—that subsequently bulges in the middle to form an alimentary cavity lined by endodermal cells. Below are shown some early steps in the formation of triploblasts from diploblasts: (1) development of a third cell layer, the mesoderm, lining an internal cavity, or coelom; (2) body elongation and acquisition of bilateral symmetry; and (3) conversion of the alimentary cavity into an alimentary canal, which eventually becomes open at both ends and unidirectional, with a mouth and an anus.

This account tells how the basic features of cellular collectivism—association, differentiation, patterning, and a genomically inscribed body plan—could have been first realized in the animal kingdom. The account is imaginary, but its outcome is not, being based on a description of *Trichoplax adhaerens,* a member of the small phylum placozoa, itself a member of the group of diploblasts, which includes all the most ancient known forms of animal life. The term diploblast refers to the two-layered body plan, with one layer, termed ectoderm, derived from the dorsal layer of the ancestral organism and eventually forming the skin of the animals, and the other, named endoderm, derived from the ventral layer of the ancestral organism and evolving into a mucosal, digestive lining.

Two major lines diverged from the primitive, ancestral diploblast. While one continued to exploit the original body plan, another began to rearrange the two-layered pancake into a network of interconnected channels. The cells lining these channels kept water flowing by their beating flagella; they fished out bacteria and smaller food particles from the flowing water, which also provided them with mineral salts and oxygen and carried away waste products. In the meagerly supplied habitat these organisms occupied, this modification proved advantageous. Instead of relying on sluggish creeping to find food, the organisms did better by filtering large amounts of water through their channels. No longer in need of mobility, they became fixed, developed an extensive supporting skeleton, and went on exploiting the advantages of their new body plan by creating an ever more complex labyrinth of cavities and channels. Their present-day descendants are the sponges, whose protein skeleton, cleaned and processed, strokes our skin with a softness no plastic material has yet matched.

In the other diploblast line, the tendency of *Trichoplax* to raise its middle part into a space in which food was retained and digested was accentuated and further exploited. Eventually, the organisms took the shape of miniature double-walled pouches that opened to the outside by a narrow orifice. Thanks to this transformation, the organisms gained a segregated digestive cavity, a marked advantage provided enough food entered the cavity. Mutations endowing the rim of the pouch with food-catching appendages took care of this requirement. The outcome was a tiny, primitive medusa, the common ancestor of hydrae, polyps, sea anemones, jellyfishes, and other related organisms known jointly as coelenterates.

The kinship among these animals—even their animal nature—may not be obvious to the naked eye. But look closely and you will see, projecting from myriad tiny chambers in a coral reef or from the bole of a sea anemone, entities built according to the same general plan as a Portuguese man-of-war. A number of species exist that alternate between a fixed polyp form and a free-swimming medusa form.

The bodies of all these organisms are constructed in radially symmetrical fashion around a central alimentary cavity communicating with the outside by a single opening, which serves as both mouth and anus. A variety of tentacles armed with stinging, sometimes deadly poisons, are arranged around the opening, serving to capture prey and draw it into the cavity where the catch is digested and assimilated. Residues are discharged through the same opening. These animals are built of a number of differentiated cell types, sometimes including muscle and nerve cells. They are equipped to float and drift with the currents. Some move actively by a kind of jet propulsion induced by contraction of the central cavity. They reproduce sexually by means of typical sperm and egg cells. Most are hermaphroditic, but some have distinct male and female forms.

Sponges, coelenterates, and a few related animals represent today's outcome of the diploblast experiment. Their ancestors shared the seas only with algae and seaweeds—and with a multitude of microbes—sometime between 600 and 700 million years ago. If their variety at that time was anything comparable to what it is today, they would have been enough to delight a scuba diver. Visit some tropical, underwater "animal garden," ignore the fish, the crabs and other crustacea, the worms, the octopuses, the hard-shelled mollusks, and you are left with a view of what the seascape might have been in those early days of animal evolution—and might perhaps still be today had not some mutant initiated a new kind of body plan.

THE WORM'S FINEST HOUR

Two major changes characterized the new body plan (see figure 20.1). First, the symmetry, from radial, became bilateral; the body shape, from circular, became elongated; the alimentary pouch became a canal, first blind-ended and later open at both ends with a mouth-to-anus polarization allowing the directional transit and graded digestion of food. With these changes came the emergence of a head, in which nerve cells began to congregate around the mouth to create the first rudiments of a brain, and the appearance of well-developed excretory and reproductive organs.

Preceding, accompanying, or following these developments—no information is available on the time sequence—a third cell layer known as the mesoderm arose from the ventral, or endodermal, layer of the ancestral diploblast, giving rise to the three-layered body plan characteristic of triploblasts, which include the major part of the animal world.[4]

Still entirely soft-bodied, these newcomers have left no fossil remains. However, petrified mud dating back more than 600 million years has kept traces of their trails, tracks, and burrows, revealing signs of their erstwhile importance and diversity. Their most primitive present-day descendants are the flatworms. Several of these have become adapted to parasitic life and have an atrophic alimentary system, among them the tapeworms, which inhabit the mammalian digestive tract, and the schistosomes, or flukes, the agents of several grave tropical diseases. Next are the nemertine worms, which are the most ancient animals having a one-way alimentary canal, and a number of other primitive, wormlike animals, including the roundworms, or nematodes, which are found everywhere and are said to be the most abundant kind of animal life in the world. Nematodes comprise several parasitic forms that found their niche in mammals, including the more than one-foot-long ascaris worms that thrive in the intestine of horses; the common pinworm familiar to many parents of young children; and the much more dangerous agents of such dreaded diseases as trichinosis, ankylostomiasis (hookworm disease), and filariasis (elephantiasis).

The body plan of these lowly worms is understandably described as primitive by zoologists, who have seen so much greater sophistication in insects, fish, birds, and mammals. Yet it would have struck an observer who happened to explore the oceans in those remote days as a marvel of exquisite intricacy. We can get a glimpse of this remarkable complexity by looking at the small nematode *Caenorhabditis elegans,* currently the most completely known of all animals.[5] It consists of exactly 959 cells, each of which has been located with great precision and traced back to its origin from the egg cell through between eight and seventeen successive rounds of mitotic division. Many studies have thrown light on how this development unfolds from the program inscribed in the egg cell's genome. All sorts of mutations have been induced, single cells have been killed with miniature laser guns, delicate surgical procedures have been practiced, all helping to unravel the network of genetic switches that command the unfolding of the program and to elucidate the mysterious signals that allow each of the 959 cells to differentiate correctly and find its proper place. The emerging picture is mind-boggling in its complex precision and will require many more years of research before it is completely understood.

THE "*MILIEU INTÉRIEUR*" AND THE OXYGEN CONNECTION

Roundworms inaugurated, in still imperfect fashion, an evolutionary development of major importance, namely, the opening of an internal cavity, or coelom (from the Greek *koilos,* hollow). The alimentary cavity, whether with one or two openings, is not a true internal cavity. It communicates with the outside. The coelom does not. In its simplest form, it is a hollow, double-walled sheath completely lined by mesodermal cells (the third cell layer that distinguishes triploblasts from diploblasts) insulating, so to speak, the endodermal alimentary canal from the ectodermal skin (see figure 20.1). In the human body, the coelom is represented mainly by the abdominal and thoracic cavities; its mesodermal lining forms the peritoneum, which surrounds the abdominal viscera, and the pleura, which envelops the lungs.

The coelom and the one-way alimentary canal running from mouth to anus were crucial additions to the basic triploblastic body plan. They gave evolution an enormously expanded range of potentialities to exploit, resulting in a profusion of new marine animals, some of which, for the first time in history, possessed hard parts. Marked by a sudden abundance and rich variety of fossil remains, many belonging to bizarre, long-extinct animal species, this period is described by paleontologists as the great Cambrian explosion, or radiation.[6] (The Cambrian is the geologic period between 600 and 520 million years ago.)

What could have triggered this explosion and how does it fit within the general framework of animal evolution? A likely factor, though perhaps not the only one, according to the Harvard scientist Andrew H. Knoll,[7] was a rise in the oxygen content of the atmosphere. The emergence of cyanobacteria equipped with the oxygen-generating photosystem II led, after mineral oxygen sinks became saturated, to a steady rise in atmospheric oxygen content, which, in turn, provoked a major crisis in the prokaryotic world. After oxygen-adapted microbes started appearing, the oxygen produced began to be consumed in increasing amounts, until a steady state was reached in which consumption equaled production and a stable level of atmospheric oxygen was established. According to Knoll, this level was substantially lower than the present level of 21 percent of atmospheric pressure, and a second important rise occurred in Precambrian times, perhaps as a result of the abundant proliferation of eukaryotic algae. The Cambrian explosion allegedly coincided with this second rise and was made possible by it. Isotopic analysis of Precambrian carbon deposits[8] does indeed indicate an excess of photosynthetic activity—which is known to select the lighter carbon isotope ^{12}C against the heavier ^{13}C—over the capacity of aerobic organisms to oxidize the organic matter made. This disparity would have left a net increase in atmospheric oxygen.

As Knoll cautions, this explanation remains hypothetical. However, a relation between the Cambrian explosion and oxygen seems likely. But the cause of the sudden expansion in diversity could have been the increased ability of animals to consume oxygen rather than the increased availability of oxygen, or it could have been a combination of both factors. All animals have an absolute need for oxygen. This requirement puts severe constraints on marine animals, which must obtain their oxygen from surrounding water, which itself receives it from the atmosphere. Because of the low solubility of oxygen in water, aquatic animals need a large supply of freshly oxygenated water and an efficient way of removing oxygen from water. Diploblasts and primitive triploblasts meet these requirements by maintaining swift water currents along or through their bodies and by having virtually every cell in direct contact with the circulating water. Higher marine animals, however, could not survive without a mechanism for extracting oxygen from the surrounding water and distributing the vital gas to all parts of the body. Evolution had to await the development of such a mechanism before more complex organisms could arise. Once an effective mechanism was in place, further evolution could have been very rapid, thereby producing the Cambrian explosion.

Evolution's key solution to the oxygenation problem was the creation of what the great nineteenth-century French physiologist Claude Bernard[9] has called the *milieu intérieur,* an internal fluid of specific composition bathing all the cells of the body. Thanks to this fluid, oxygen taken up by cells in direct contact with sea water could be transferred to more deeply situated cells. Three acquisitions increased the efficiency of this transfer: (1) the formation of gills, specialized surface exchange organs made of thin, highly expanded skin folds allowing a rapid flux of oxygen from surrounding water into the internal fluid; (2) the addition to the internal fluid of special oxygen-carrying molecules—either the red hemoglobin, an iron-containing hemoprotein, or the blue hemocyanin, a copper-containing protein—that greatly increased the oxygen capacity of the fluid; and (3) the development of pumps (hearts) to move the fluid and facilitate the transport of oxygen. In their earliest forms, hearts were no more than contractile thickenings of a tube communicating directly with the main body cavity (open circulation). Eventually, tubes supporting the hearts joined into a closed network (closed circulation) and the *milieu intérieur* became subdivided into two compartments: the circulating blood, present inside the tubes, and the stationary lymph, or intercellular fluid proper. This division made it necessary for the circuit of blood vessels to form two intercalated, highly ramified networks of small, thinly sheathed conduits (capillaries), each network providing a total surface area large enough for oxygen exchanges to proceed at a sufficient rate. One of these networks traversed the gills and served to collect oxygen from the surrounding water into the blood. The other, traversing the tissues, allowed an efficient delivery of oxygen from the blood to the tissues.

Two important fringe benefits accompanied these developments. Nutrients arising in the alimentary canal from digested foodstuffs could be collected—this necessitated a new capillary network around the canal—and distributed throughout the body. Conversely, waste products produced in the body could be discharged into the blood and delivered—through yet another capillary network—to special collector cells assembled into excretory organs, the nephridia, or primitive kidneys.

The establishment of rudimentary exchange mechanisms for oxygen, nutrients, and waste products by way of a (circulating) *milieu intérieur* marks a turning point in animal evolution. Henceforth, organisms could become more than a few cells deep, and a variety of organs could develop. The cells lining the alimentary canal began to differentiate into several types, some of which formed separate organs (glands) that manufactured digestive enzymes and discharged them into the canal through ducts. Secretion was born. Mobile cells, specialized in various forms to defend against pathogenic microorganisms, started patrolling the body by way of the internal fluid and, later, by way of the blood. Immunity was initiated. In response to increasing size and bulkiness of the organism and its organs, certain cells turned into builder cells that constructed extracellular supporting frameworks made of proteins and carbohydrate polymers. Contractile cells joined into muscles, appropriately disposed to accomplish coordinated movements. Special reproductive organs, often voluminous, also developed for the formation and maturation of gametes and for ensuring fertilization by a variety of mechanisms, including direct copulation. Finally, to meet the growing need for regulation and coordination, nerve cells wove increasingly complex networks, while modified gland cells making chemical transmitters (hormones) discharged their products into the internal milieu instead of into the alimentary canal.

Thus was perfected the basic animal blueprint, with its key functions of food capture, digestion, and absorption, oxygen uptake, waste elimination, locomotion, and reproduction, linked by circulation and coordinated by neural networks and chemical transmitters. To an investigative scuba diver, at that time, the world of worms might have appeared as essentially completed, with only details to be added here and there by evolution. What our explorer could not have foreseen is the creative power of duplication.

Chapter 21

Animals Fill the Oceans

MUSICIANS have discovered that if they have a good theme, they can create a rich work by repeating the theme many times in different variations. Composers of serial music have exploited this formula to the utmost, by moving from one variation to another in almost imperceptible steps. Something similar happened in animal evolution. After the basic blueprint was completed, further progress was made by duplication and variation of the blueprint.

BODY DUPLICATION: THE ROAD TO INNOVATION

The first step in the new direction was accomplished by a remarkable genetic modification that led to the formation of a multisegmented animal looking for all the world like a string of primitive worms joined end to end by minimal connections. Each segment of this strange creature was, in itself, an essentially complete organism, with an alimentary canal, two nephridia, a circulatory network linked to a pair of gills situated in lateral outgrowths of the body, male and female sex organs, a rudimentary innervation radiating from a centralized group of nerve cells (a ganglion), a set of circular and longitudinal muscles, and a surrounding reinforced skin, or cuticle. Separated by incomplete partitions, the segments were linked to each other mainly by the skin; by the alimentary canal, which was continuous; by two large blood vessels, one running along the back, the other along the belly of the organism; and by a nerve cord joining the ganglia. In this early form of the organism, the head and tail were constructed like the other segments.

This organism clearly originated from a number of copies of the same individual, linked end to end. Yet a molecular biologist sent to Earth at that time would have found that most genes were present in the organism's genome in only single copies. Only a small number of genes, all grouped together, were present in as many copies as there were segments, arranged along the chromosome in the same order as the segments. The protein products resulting from the translation of these genes were neither enzymes nor structural proteins; they were proteins that reacted with other genes or sets of genes to turn them on or off, indicating that the duplicated genes belonged to the superfamily of regulatory genes. In the unsegmented ancestral organism, these genes commanded central switches in the realization of the body plan. Duplication of these genetic switches led to repeats of the instructions, with formation of repeats of the organism. At first, all segments were identical. But soon the duplicated genes underwent different mutations and the segments became different. The first repeated genes to be affected by mutations in this evolution were those situated first and last on the chromosome, leading to the development of specialized head and tail parts.

Several of the regulatory genes involved in these phenomena have been identified and sequenced. They share a highly conserved sequence of 180 base pairs, called the homeobox (from the Greek *homos,* same). The stretch of sixty amino acids encoded by this box in the corresponding proteins is characteristically shaped to bind to DNA, as befits proteins influencing the transcription of certain genes (transcription factors). Homeotic genes are very ancient. They have been recognized throughout the animal world and even in plants and fungi.[1]

Segmentation represents a major mechanism of evolutionary diversification, perhaps the most important one in the history of life. It initiated an extraordinary combinatorial game involving complete, originally viable modules that could be mutated, fused, reduplicated, deleted, and otherwise reshuffled, all by the magic stroke of single or sparse genetic modifications, to offer natural selection a large variety of body plans to test. Investigators working on the fruit fly *Drosophila,* the central object of classical genetic research, have discovered the amazing creative flexibility of this game. They have been able to produce, by single homeotic gene mutations, headless, two-tailed flies; animals with an extra pair of legs or wings; or strange monsters sporting legs in front of their heads in lieu of antennae. Using the means of modern molecular biology, these investigators have begun to unravel the remarkable molecular mechanisms whereby homeotic genes control development.

INVERTEBRATES GALORE

The first products of this new game were most similar to present-day annelids, so named because their bodies appear like a series of rings (*anulus* means ring in Latin). The land-adapted, common earthworm is a familiar member of this phylum, which also includes a variety of sea worms, among them a number of organisms (fan worms, peacock worms) that live in solid tubes of their own making and catch prey by means of beautifully colored appendices that they wave from the orifice of their abode. Leeches form another class of annelids.

Evolution did not stop at the still fairly repetitious body plans of annelids. After considerable further variations, of which few intermediates are known, it ended up producing the most extensive group of animals living on Earth today: arthropods, or animals with articulated limbs (*arthron* means joint in Greek and *pod* comes from the Greek word for foot), represented in water by the crustacea (shrimps, crabs, and the like) and the chelicerates (sea spiders and horseshoe crabs) and on land by spiders, scorpions, ticks, centipedes, millipedes, and, foremost, the immense group of insects.

Look at a lobster or shrimp. You have no difficulty recognizing the segmented body plan inherited from the ancestral organism. There are one pair of gills and one pair of legs per segment, but with considerable variation in the specialization of each segment. Several front segments are fused together to form the head, and their legs are converted to a variety of antennae, claws, masticating devices, and other appendages serving as sensory organs or feeding aids. The limbs consist of several articulated parts. The main body and tail segments are largely muscular, while the viscera are grouped in the front part of the body. Circulation is open and the oxygen carrier is hemocyanin. This is in contrast with the more ancient nemertine worms and annelids, whose present-day representatives have a closed circulatory system and use hemoglobin as the oxygen carrier. The arthropod body is entirely covered by a tough carapace made of chitin, a highly resistant, celluloselike carbohydrate polymer. This shell is shed at intervals to permit growth. The molting animal is temporarily vulnerable until it has built a new shell. Amateurs of soft-shelled crabs particularly appreciate this tender state.

Somewhere along the annelid-arthropod line, a major branch detached that led to the large phylum of mollusks, which range from the many conches, clams, oysters, mussels, and sundry other hard-shelled animals we tend to identify with this name to the very different-looking squids and octopuses.

The triggering event that started some segmented worm on the way to becoming a mollusk could have been a mutation that endowed a protein constituent of scaly structures on the back of the animal with the ability to seed the formation of calcium carbonate crystals. The horny scales became hard, mineralized plates, which gave the animal additional protection and thus provided it and its descendants with a selective advantage. Remnants of this ancestral structure are still visible in chitons, primitive mollusks that have an elongated structure with bilateral symmetry, an open alimentary canal with a mouth in front and an anus in the rear, two lateral rows of gills, and a series of protective dorsal plates hardened by calcium carbonate deposits.

In the further evolution of the ancestral mollusk, the dorsal plates fused into a single shell, segmentation was largely lost, and the body folded and coiled in such a manner that mouth, anus, gills, and excretory and genital outlets all came to be grouped in the front part of the animal, between a head, containing the rudiments of a brain and primitive sense organs sensitive to light (eyes), touch (antennae), and gravitation (otocysts), and a muscular, ventral foot serving for locomotion. Later evolution played largely with the shapes of the shells, to the gratification of shell collectors and fossil hunters alike, who both profit from the durability of the mineral deposits.

Among the many shapes adopted by mollusks, some may have carried a selective advantage, but most variations on the shell theme were probably the result of evolutionary quirks that did not greatly affect the reproductive potential of the animals. This fact illustrates an important aspect of evolution: A change does not have to be advantageous to be retained by natural selection, especially under weak competitive conditions; the change need only not be adverse enough to cause eradication. Major developments in the history of mollusks that were presumably the outcome of positive selection were the duplication of the shell, leading to the bivalve mollusks, and its atrophy to a mere internal plate in squids, and complete loss in octopuses.

This sketchy survey of the vast world of mollusks—the second largest phylum in the animal kingdom, with more than 50,000 living species and almost as many extinct ones—would have completed our description of the animal tree of life—or rather, there would have been no description, for want of a describer—were it not for an astonishing head-to-tail conversion that happened to some ancestral annelid, or, more correctly stated, to its developing embryo.

A FATEFUL FLIP-FLOP: FROM MOUTH FIRST TO MOUTH SECOND

In order to understand this new, particularly far-reaching forking of the tree of life, we must take a brief look at embryological development. With Haeckel, who first pointed out the similarity, we note that the developing embryos of animals indeed appear to recapitulate their evolutionary history.[2] The cells arising from the early divisions of the fertilized egg first form a sphere, the blastula, which turns into a double-walled pouch, the gastrula, with a single opening, the blastopore. The gastrular cavity later becomes the digestive tract and in all but the most primitive animals acquires a second opening and turns into a canal. Here is where development was modified by a mutation to initiate a new evolutionary line. In the animal groups considered so far, the blastopore becomes the mouth and the new opening the anus. They are called protostomes (mouth first) for this reason. The historic flip that started the new line made the blastopore the anus and the new opening the mouth, thereby initiating the deuterostomes (mouth second), the group out of which all vertebrates would someday arise. It is possible that, without this fateful switch, there would be no fish, no amphibians, no reptiles, no birds, no mammals, no humans.

Many a biologist in the past has pondered and wondered about the possible mechanism of the astonishingly abrupt change in body plan that caused the deuterostome limb to branch from the protostome trunk. Our wonderment is no less today, but our knowledge of homeotic genes offers a glimpse of a possible explanation. If a single homeotic gene mutation can replace the head of a fly by a second rear, perhaps one or more such mutations could have interchanged mouth and anus in some ancestral annelid. This is pure speculation, but it is difficult to imagine such a radical change in developmental program occurring otherwise than by some major upheaval of the kind homeotic gene mutations are known to bring about.

The first consequence of the upheaval detectable in extant organisms (acorn worms) was an anatomical modification that drew together the structures responsible for food and oxygen uptake. The front part of the alimentary canal, or pharynx, was converted into a sort of bilateral straining device formed by two opposing rows of narrow slits lined by gills. These branchial slits (*branchia* means gill in Latin) corresponded to successive segments in the body plan. Animals equipped with this new machinery took in large amounts of water through their mouth and chased the water out through their branchial slits. Oxygen was absorbed from the passing water by the gills, while food particles were retained by the slits, which served as filters, often helped to this effect by fine comblike surface structures. Food collected by the slits was subsequently sent farther down the alimentary canal. This combined mode of food and oxygen gathering by filtration recalls to some extent the primeval mechanism of maintaining water currents used by sponges, polyps, and jellyfishes.

THE BIRTH OF VERTEBRATES

Further key events in the evolution of deuterostomes led to the development of a segmented, hollow structure running along the back of the animal and containing the main parts of the nervous system. If Haeckel's law is to be believed, an early event in this development was the centralization of nerve cords in a dorsal neural tube, developmentally derived from an infolding of the ectoderm called the neural crest. Then, underneath the neural tube, there was formed a tough, resilient rod, the notochord, the hallmark, if not always in the adult at least at some embryonic stage, of the whole chordate phylum, of which the earliest representatives are the lancelets. Finally, about 500 million years ago, the neural tube and the notochord became surrounded together by segmented cartilaginous structures, the first vertebrae.

The most important advantage associated with this development was protection of the fragile neural tube within a solid sheath. Think of the hazards that would beset our highly vulnerable spinal cord were it not protected within our backbone. Segmentation turned out to be extremely useful in the building of this protective sheath, as the vertebrae could be made of hard, unyielding material without the drawback of an overly rigid body. The segmented spinal column maintained enough flexibility to allow all the backbone movements needed for locomotion. The continuous notochord, which had provided a valuable scaffolding for the construction of the spinal column, later became more of a hindrance than an asset; it eventually came to play only a transient role in early embryological development, subsequently to be broken down. There was a price to pay for the advantage of a segmented backbone, as is well known by many a sufferer of a slipped disk and, more severely, by the paralyzed victims of spinal injuries. However, this price did not burden evolution, as it was exacted only half a billion years later, after some primate adopted an upright posture.

The first vertebrates had cartilaginous bones and resembled worms more than fishes, having no jaw and only rudimentary fins. According to the fossil record, some were bizarre, ferocious-looking animals covered with armored plates. Their closest present-day descendants are the lampreys and hagfishes, which are very different from their remote relatives but share some primitive features with them.

The next major development was the formation of a hinged jaw, probably from cartilaginous arches supporting the anterior gill slits. At the same time, the body acquired a variety of fins supported by cartilaginous bones and moved by muscles. The animals turned into powerful swimmers and dangerous predators. Cartilaginous fish, which include sharks, rays, and skates, are their nearest present-day relatives.

As in the development of mollusks, the last major change was the acquisition of a structural protein capable of seeding the formation of mineral crystals. In the present case, the crystals consisted of mixed calcium phosphate and carbonate, as found in the mineral hydroxyapatite. Resilient cartilaginous structures turned into solid bones. Most present-day fish are descendants of these first bony fish.

THE ECHINODERMS: AN EVOLUTIONARY QUIRK

Before the developments that led to the first vertebrates, a bizarre forking event took place in the deuterostome limb shortly after its split from the protostome trunk. As a result, perhaps, of some homeotic gene upheaval, the elongate, bilaterally symmetrical body plan of the ancestral deuterostome, still retained at the larval stage, gave place in the adult to a fivefold symmetrical blueprint in which a greatly compressed and coiled alimentary canal was surrounded by five virtually identical segments. The resulting monster somehow found a favorable niche and thrived, to give rise to the sea urchins, the starfishes, the sand dollars, the sea cucumbers, and other animals characterized by a fivefold radial symmetry. They are grouped under the name echinoderms (*echinos* means hedgehog in Greek), even though not all possess the typical spikes of sea urchins.

The evolution of invertebrates, from their unicellular ancestor to the first vertebrates, is summarized schematically in figure 21.1. As in figure 19.1, the graph shows how key modifications of the body plan of "fork organisms" led to significant evolutionary advances, while the descendants of unmutated organisms give some idea of the body plan of the fork organism, the last ancestor they have in common with the more evolved species.

FIGURE 21.1
A Bird's-Eye View of Invertebrate Evolution

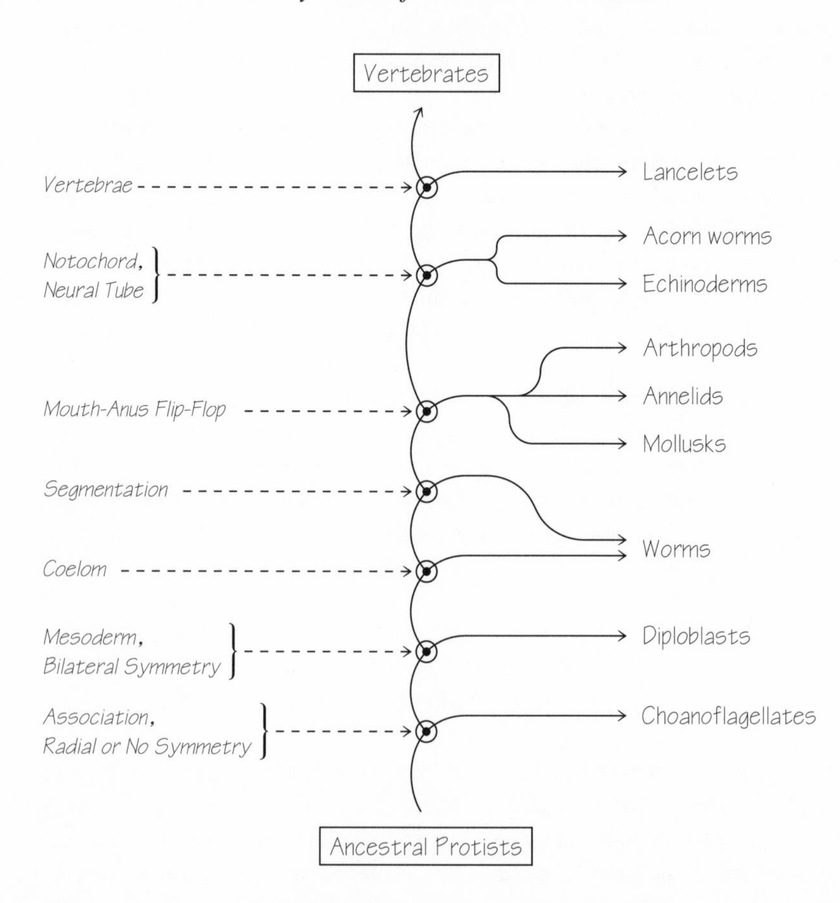

This figure, similar to figure 19.1, depicts the main steps in the rise of animals in the direction of greater complexity, from the ancestral choanoflagellates to the first vertebrates. At each fork, a mutant evolutionary line that underwent the genetic change in body plan indicated on the left diverges upward from the unmutated line—represented by the arrow curving to the right—leading to existing phyla.

Chapter 22

Animals Move Out of the Sea

ONCE PLANTS AND FUNGI started invading the lands some 400 million years ago, new pastures became available for animals to exploit. These new opportunities did not long stay neglected. Modifications that had been of no use to aquatic animals in earlier days now became advantageous in the changed surroundings. What water-adapted animals mainly needed in order to take advantage of the rich new sources of food offered by land plants was to be able to resist loss of water, to utilize atmospheric oxygen (respiration), to move on land (for those that lacked such means), and to reproduce away from water. These adaptations were gradual and occurred first on coastal fringes and in marshy areas still exposed to intermittent flooding. Most aquatic animals, except for the lower invertebrates, developed solutions of one sort or another to the problems of life on land. I shall consider only two types of animals, the arthropods and the vertebrates, which together account for the major part of the terrestrial fauna.

INSECTS AND THEIR RELATIVES: THE GREAT LAND CONQUERORS

Arthropods had it easiest, being already shielded by a waterproof covering and equipped with functional legs. However, their fragile gills could not long have resisted desiccation. What helped arthropods to utilize oxygen was the formation of thin, tubular invaginations of their carapace. These tenuous air ducts, or tracheae, progressively developed into a highly ramified network of passages that penetrated all parts of the body, allowing them to be in close contact with outside air. The thin walls of the passages allowed oxygen to diffuse into the tissues and carbon dioxide to diffuse out of them. Body movements served to move air in and out of the ducts, thus restoring the depleted oxygen and removing the accumulating carbon dioxide.

All kinds of arthropods developed the same form of respiration by tracheae, including the wormlike, multisegmented centipedes and millipedes; the small creatures known as wood lice or pill bugs, which are among the rare terrestrial crustacea; the spiders and scorpions, which, with other chelicerates, are related to horseshoe crabs; and the innumerable species of insects, almost all of which are terrestrial. These diverse animals did not all inherit tracheae from a common ancestor. What they inherited was a body plan that admitted only one solution to the problem of respiration or, perhaps, favored this solution over others because of some property of the chitinous cuticle that all arthropods have in common. This is a typical instance of convergence.

Many marine arthropods reproduce by copulation. Thus, the ancestors of terrestrial arthropods did not need water for spermatozoa to find egg cells to fertilize. Their principal requirement in order to reproduce on land was protection and nutrition of the fertilized egg and the embryo. At first, they simply used water as a medium for larval development, as did their marine forebears and as mosquitoes and many other insects still do today. Then, in the course of time, an amazing number of different housings, usually kept moist one way or another, were either adopted or constructed for the successful development of the young away from water.

Thanks to popularizers such as Britain's David Attenborough, television has vividly brought to everyone's attention the extraordinary operations carried out for the sake of their progeny by dung beetles, termites, bees, wasps, and many other insects. The remarkable coordination and apparent purposefulness of these rituals, faithfully accomplished generation after generation by tiny creatures equipped with a brain no bigger than the head of a pin, have struck many an observer as feats of almost miraculous organization, not readily compatible with a materialistic, Darwinian view of life and evolution. Yet these complex behaviors are child's play in comparison with the stupendous molecular and cellular events that govern the development of the same animals from fertilized egg cells. Should our eyes be able to follow what happens inside a larva deposited in a beehive alveolus, we would not pay another second's attention to the construction of the housing itself.

Many insects even go through two entirely distinct consecutive developmental programs. From caterpillar to butterfly, from silkworm to moth, from maggot to fly, the animal veritably dies and decomposes within a self-built tomb—cocoon or other pupal covering—leaving alive only some embryonic remnants (imaginal disks). Out of these—the tomb turning into womb—a brand-new organism then arises according to a completely different blueprint. Next to such architectural wizardry, inscribed into a couple of feet of DNA, what are a few additional stereotyped gestures serving to build some primitive dwelling? It is like admiring the builders of the Taj Mahal for their ability to make a hut out of straw and mud.

AMPHIBIANS: THE FIRST FISH OUT OF WATER

Fish also moved out of water, but they had greater obstacles to overcome. They took their time and went through an intermediate, half-aquatic, half-terrestrial stage stable enough to give rise to an important extant class of vertebrates, the amphibians. We don't know how the transition occurred but we can hazard some guesses.

A key evolutionary event may have been the development, in some fish, of an air-filled pouch communicating with the pharynx. The air in the pouch came from the gills by way of the blood, which circulated through an increasingly rich network of capillaries surrounding the pouch. An advantage the fish derived from such a pouch was adjustable buoyancy, the main function of what is now the swim bladder. Another advantage was that the fish, in the manner of a scuba diver, carried a reserve of oxygen it could use in case of emergency, when its blood oxygen fell to a dangerously low level. In such an event, oxygen would diffuse in the reverse direction, from the pouch into the blood. This adaptation opened the way to breathing, the pouch acting as a primitive lung. We can watch this in our fishbowl when a goldfish surfaces to take a breath of fresh air and, more dramatically, during the dry season in many a tropical lake of Africa, South America, and Australia, where lungfish survive for months in the drying mud, awaiting the next rainy season. This, most likely, is how amphibians "learned" to breathe, while retaining the ability to use dissolved oxygen.

Stranded fish capable of breathing would no doubt wriggle their bodies and move their fins in efforts to find shade, moisture, and food. The nimblest at this exercise were animals with two pairs of fleshy, lobed, ventral fins that could help them crawl as well as swim. Fish of this sort were abundant 100 million years ago according to the fossil record. They were believed to be long extinct until a day in December 1938, when one landed in the nets of a fishing trawler off the east coast of South Africa.[1] The unusual catch was brought to the attention of the curator of the New London Museum, Marjorie Courtenay-Latimer, who described it to a local ichthyologist, James Leonard Briefly Smith, who recognized it for what it was: a living fossil, described by paleontologists under the name coelacanth. It took fourteen years of adventurous episodes, including the posting of rewards in many remote fishing villages along the Indian Ocean and the provision of a special plane by the president of South Africa, Daniel F. Malan, before a second specimen of the rare fish, caught off the Comoro Islands, became available for thorough examination. Coelacanths are deep-sea fish and do not use their fins for walking. But they share ancestors with ancient, lobe-finned, freshwater lungfish that invaded swampy lands soon after plants did, some 400 million years ago, thanks to a succession of chance mutations that turned the fins into articulated legs. Such changes would probably not have been retained by natural selection in a watery habitat. On land, they became valuable acquisitions.

Before these conquerors could settle definitively in their new surroundings, they had to solve the problem of reproducing on land. Most did not do so and retained the customs of their aquatic ancestors. They spawned in water and their eggs developed first into swimming larvae. This was so because water was available everywhere and there was a lack of selective pressure in favor of true terrestrial reproduction. Evolution rarely moves without some selective inducement. What drove the animals to perfect their breathing and walking machineries was not scarcity of water but abundance of food. This was the time when the great Carboniferous forests began to flourish and to build the surfeit of organic matter that now fuels many of our stoves and furnaces. In addition to plants, plenty of insects, snails, and worms also had become available on land to satisfy animals with more carnivorous tastes.

Amphibians thrived in those days but many were later eradicated by the great Permian crisis. Among the survivors, some, such as newts and salamanders, retained the tail of their marine ancestors. Others, among them frogs and toads, kept this appendage only in the free-swimming, fishlike larval stage. The subsequent transformation of a tadpole into an adult frog represents another striking instance of metamorphosis, less dramatic than the total reincarnation exhibited by some insects, but impressive nevertheless. Atrophy of the tail and the sprouting of four legs are among the more conspicuous changes that accompany this transformation.

These events are triggered by the secretion of thyroxin, an iodine-containing hormone essential to growth in all higher vertebrates. In humans, lack of thyroxin in early developmental stages causes dwarfism and mental retardation. When this substance enters cells, it binds to an intracellular protein receptor, which is thereby turned into an activator of a number of genes. This is an interesting variation on the theme of regulatory supergenes that, like homeotic genes, control by their products the transcription of a number of other genes. There is a twist in this case. The gene product—the thyroxin receptor—is active as a transcription factor only when it has bound the hormone. Other examples of hormones acting in this fashion are ecdysone, which affects molting, pupation, and metamorphosis in insects, and the steroid sex hormones, which control many aspects of the sexual activity of mammals, including the onset of puberty, the menstrual cycle, and pregnancy.

These various phenomena illustrate another fact of general importance: the role of programmed cell death in developmental processes. The decomposition of the caterpillar body and the melting of the tadpole tail are spectacular examples, but there are many others. In the conversion of lobed fins to articulated legs that occurred in the course of the fish-amphibian transformation, the tips of the legs became divided into five fingers, not as a result of budding but through the selective death of intervening tissues. This sculpting is still "recapitulated" in embryonic development. The limbs of a fetus grow first as rounded buds, which are later cut into fingers by selective, programmed cell death.

REPTILES "INVENT" THE AMNIOTIC EGG

Millennia of sustained drought and wintry cold ushered in the great Permian crisis. The luxuriant Carboniferous forests of ferns and lycopods withered away. The swamps dried out. Marine animals, accustomed to the balmy environment of tropical lakes and seas, became extinct in catastrophic numbers. Amphibia also took a heavy toll. But for the fossils unearthed by geologic upheavals, the chance findings of observant wanderers, and the painstaking searches of paleontologists, we would not have the slightest inkling of all this past splendor or of the planetary cataclysm to which it fell victim.

As happened many times, life rallied; evolution responded to ecological challenges by appropriate adaptions. It even turned disaster into success, driven by the great Permian crisis to accomplish one of its most decisive advances. While seed plants took over the cold, dry swamps left barren by the decimation of sporulating plants, some obscure amphibian suddenly soared into prominence by developing the animal equivalent of the seed: the fluid-filled egg.

Instead of delivering fertilized egg cells for development in some body of water—the normal amphibian mode—the female of this key transition species enclosed its fertilized egg cells in a fluid-filled sac, the amnion, within which the embryo could pursue its normal aquatic development. After Claude Bernard's *milieu intérieur* to bathe all cells and tissues, here was a re-created *milieu extérieur* to shelter the developing embryo. A hard, porous shell protected this substitute marine incubator, while a highly vascularized membrane, the allantois, produced by the embryo and lining the inner face of the shell, served in gas exchanges and waste disposal. Another sac, filled with a richly nutritious yolk, provided the embryo with necessary foodstuffs. Thus, the complete development of the organism up to a stage where it could survive on land took place within the protective, well-stocked, and appropriately renewed environment of the amniotic fluid. True terrestrial reproduction was initiated. The first reptile was born.

This creature enjoyed modest success until the great Permian crisis. After that, reptiles developed and radiated tremendously, to the point of even producing species that lost their legs, while remaining on land, or that returned to an aqueous habitat for living but, paradoxically, moved out of the water, as do sea turtles today, to lay their eggs on land, at great peril to their young. Lizards, snakes, and turtles are the main extant reptiles, but the most spectacular representatives of this group are the dinosaurs, the most celebrated of all fossils. I shall not dwell on the saga of these extraordinary beasts, some of which reached enormous sizes and assumed the most bizarre and, to us at least, terrifying shapes. For a vivid representation of the age of dinosaurs, visit the Peabody Museum at Yale University. A mural, 110 feet by 16 feet, painted by Rudolph Zallinger between the years 1943 and 1947, depicts in arresting colors, amid the contemporary vegetation, the whole history of the dinosaurs, from the emergence of the first amphibia some 400 million years ago down to the great dinosaur extinction, more than 300 million years later.[2] We can now see the dinosaurs in full action, thanks to Steven Spielberg's blockbuster movie *Jurassic Park.*

The disappearance of the dinosaurs 65 million years ago has become one of the most gripping scientific whodunits of all time. Dinosaurs were not the only victims of this mass destruction; they are only the most conspicuous to strike our imaginations. Many other animals were eradicated at the same time, for example, the beautiful, spiral-shelled ammonite mollusks. Flowering plants also were decimated, to be replaced for a while by ferns. Many explanations of this mysterious holocaust were proposed, until, in 1978, the American physicist and Nobelist Luis Alvarez, together with his son Walter and other coworkers, made a remarkable observation. They found that a thin layer of sedimentary rocks deposited at the time of the extinction was twenty times richer in the rare element iridium than the adjoining layers.[3] Iridium is more abundant in cosmic material than on Earth, and the workers had measured it in order to time the rate at which the material that witnessed the great extinction was deposited at the bottom of the ancient seas. If sedimentation had been fast, the material would have included less cosmic dust, that is, less iridium. The opposite would be true if sedimentation had been slow. The investigators were looking for modest changes; they had not bargained for the huge increase they found. Here is one more example of serendipity, the magic mother of many a scientific discovery, a fairy that cannot be courted but sometimes gratuitously favors those who search for truth, even if they do so with the wrong idea in mind. But one must be able to recognize such a blessing. As the great Louis Pasteur once said, chance favors only the prepared mind.[4]

In the present case, the gift from chance could hardly be missed. The scientists could think of only one explanation for the iridium anomaly: A huge asteroid, six or more miles in diameter, fell on our planet 65 million years ago. First received with considerable skepticism, this suggestion is now widely accepted. Corroborative evidence has been found in many parts of the world, and the probable impact area, almost two hundred miles wide, has been located at a site, Chicxulub, on the north coast of the Yucatan Peninsula in Mexico.

How could such an event, a mere prick on the skin of the Earth, cause such a worldwide catastrophe? By sheer brute force. It is estimated that the impact released the equivalent of 100 million megatons of energy, or as much as 10,000 times the energy that would be released by all the atomic bombs of the world exploding at the same time! Clouds of dust, smoke, and soot obscured the sun for years. Raging fires destroyed plant and animal life over large parts of the continents. A period of piercing cold (impact winter) was followed by intense warming due to the greenhouse effect of released gases. Acid rain poisoned the waters. Against this doomsday scene, the biblical picture pales to insignificance and the warnings of ecologists become derisory. Once again, however, the irresistible force of evolution came to the rescue and turned disaster into blessing. And what a blessing, at least from our selfish, anthropocentric point of view, since we might well not be here had an asteroid not hit the Earth 65 million years ago and wiped out the dinosaurs.

THE MAMMALIAN WOMB: THE ULTIMATE GENERATION MACHINE

Were the dinosaurs cold-blooded, like all extant reptiles? Or were they warm-blooded? This question is the object of a lively debate. Even though we don't know the answer, we may state that at least one dinosaur branch had or acquired the ability to control body temperature around 100°F. These animals remained active in the cold—in contrast to the other, more sluggish reptiles, which attained such a temperature only when basking in the sun—but they had to pay for this advantage by needing more food. They used their agility to satisfy this requirement and became carnivorous hunters. A thick fur pelt came to cover their bodies, protecting them against heat loss and allowing them to thrive in cold areas where the run-of-the-mill reptile could not survive. Finally, the females adopted the habit, advantageous to the survival and propagation of the species, of covering their eggs until hatched and subsequently shielding their young in a warm clasp. The hungry young, in turn, came to lick the fatty material secreted by skin glands on their mother's chest. One thing leading to another, in evolution's usual way of combining chance mutations with natural selection, the secretion turned into milk, and the skin glands into specialized, hormonally controlled feeding organs, the mammary glands.

Mammals led a modest and inconspicuous existence for nearly 200 million years. They rarely exceeded the size of a rabbit and kept out of the way of the increasingly voracious and ferocious dinosaurs. But when the big test came, the monstrous beasts succumbed, whereas the little, furry animals survived. The rest, as the saying goes, is history. Except that one more development of major importance needs to be mentioned. At some time, a female mammal stopped laying eggs and kept them to incubate and hatch inside her body instead. At first, the young were delivered in a very immature state of development, so fragile as to require immediate transfer to a protective ventral skin fold, or marsupium, within which they had access to the mammary glands for feeding. Later, embryos became able to prolong their stay in the mother's womb and to achieve a much higher degree of development. They succeeded in doing so by drawing nutrients and oxygen from the maternal blood by means of rootlike extensions, the chorionic villosities, inserted into the wall of the womb, which underwent corresponding adaptations. This intimate fetus-womb connection became the placenta.

Today, placentals rule the world, which they have filled with a wide variety of species adapted to every possible kind of environment, including the oceans. Extant egg-laying mammals (monotremes) are rare, for example, the platypus. Marsupials are largely confined to the Australian continent, where they became geographically isolated and never suffered competition from placentals until recent times, when these were brought in by European settlers. If we allow natural selection to have its way, the Australian marsupials will soon be outcompeted by the placentals, as they were in other parts of the world.

Among the many branches that grew from the mammalian limb, the tree-dwelling primates deserve a special mention. This branch, which detached from the main limb tens of millions of years ago, when dinosaurs still roamed the Earth, went through a long succession of evolutionary forkings and adaptations out of which there emerged, a mere six million years ago, somewhere in East Africa, the twig—indistinguishable from the others at first—through which life entered the Age of the Mind. This will be the subject of the next part of this book.

The later steps of animal evolution, from the first vertebrates to the human species, are depicted in figure 22.1, which, like figure 21.1, is drawn so as to highlight key genetic modifications of the body plan of "fork organisms."

THE CONQUEST OF THE SKIES

Unlike humans, no animal ever flew because it wanted to fly. The conquest of the skies was entirely a matter of accidental opportunism. The simplest such accident was any anatomical modification that helped an animal to extend the range of a jump by gliding. Flying fish and flying squirrels are examples of animals that are helped in this way by membranous expansions of their fins or limbs. If gliding is useful, natural selection takes care to perpetuate the expansions. A more advanced form consists in the flapping of such expansions as a means of sustaining and propelling the body through the air. Flying dinosaurs called pterosaurs are believed to have done just that, with a span that sometimes exceeded thirty feet, and flying mammals, or bats, do it today, using a remarkable sonar device to direct themselves in the dark and locate the insects on which they feed. The most extraordinary and mysterious conquerors of the skies are the insects and the birds.

Nobody knows how dragonflies, butterflies, bees, mosquitoes, and other flying insects won their wings. It is not even known whether they inherited their wings from a common ancestor or achieved flying separately by convergent evolution. Unlike the wings of other flying animals, those of insects are not modified limbs. They are formed by flattened outfoldings of the chitinous covering of the animal's back, which are moved by muscles of extraordinary performance efficiency. How such an amazing arrangement ever came into being is anybody's guess.

The last major bequest of the dinosaurs before they disappeared were the birds. These landed on the world some 150 million years ago, as revealed by the famed *Archaeopteryx,* a fossil discovered in 1864 in a schist quarry in Eichstätt in Bavaria.

FIGURE 22.1

A Bird's-Eye View of Vertebrate Evolution

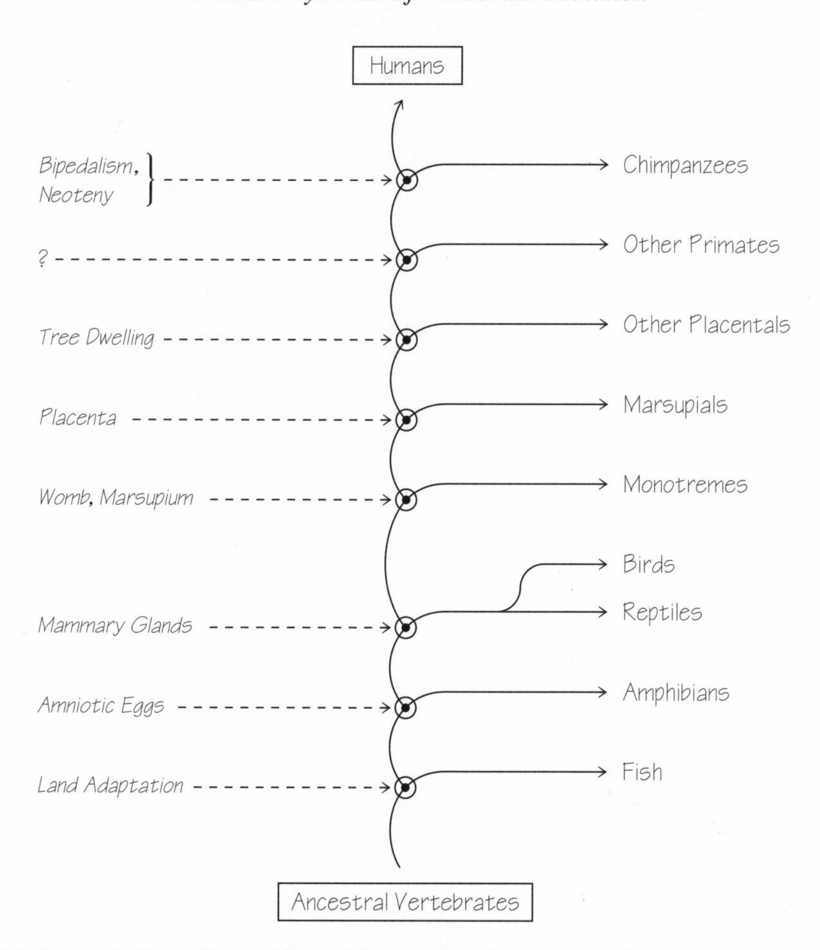

This figure, similar to figures 19.1 and 21.1, depicts the main steps in the rise of animals in the direction of greater complexity, from the first vertebrates to humans. At each fork, a mutant evolutionary line that underwent the genetic change in body plan indicated on the left diverges upward from the unmutated line—represented by the arrow curving to the right—leading to existing phyla.

This weird animal would have passed for a small dinosaur by any test were it not for the imprint of feathers miraculously preserved in the soft stone. Feathers, indeed, turned a reptile into a bird. These remarkable appendages are related to hairs, horns, nails, and scales, and are likewise constructed from a special, tough structural protein, keratin. Feathers obviously did not come into being in one shot, suddenly converting their fortunate owners into flying machines. It must have taken many successive steps to transform a pelt into such a beautiful arrangement of quills and barbs. Flying was out of the question during all that time. It came later, as a fringe benefit, so to speak, albeit one of tremendous value. Some other evolutionary advantage must have driven natural selection. Many have pondered on the possible nature of this benefit. Better thermal regulation is the explanation considered most likely at present. Whether that or another explanation is correct, the phenomenon itself is a remarkable illustration of the devious ways evolution sometimes follows to achieve results that have nothing to do with the primary driving force of the process. Once the developing feathers began to allow even the most primitive form of flying, greater proficiency in this extraordinarily advantageous form of transportation became a powerful driving force for the evolutionary improvement of the machinery. Birds, like mammals, have now invaded every possible ecological niche and adapted their feeding habits accordingly. Some have even given up their main evolutionary asset and returned to a walking life.

According to palynologists, those scientific sleuths who reconstruct the history of the world by looking at fossil pollen, flowering plants enjoyed a remarkable diversification some 50 million years ago. This success is attributed to invasion of the skies by pollen-carrying insects and birds.

THE DRIVING FORCE OF EVOLUTION

The history of plants and animals on Earth highlights the groping, unpredictable ways of evolution in its progression toward complexity, mediated at each step by a long-extinct fork organism that offered chance an opening for progress. The two evolutionary lines also illustrate the constraints existing body plans impose upon further advance. Animals have been more "inventive" than plants in this respect, having come up with such spectacular changes as body duplication and the protostome-deuterostome inversion. The evolution of plants has been more conservative, proceeding along the single theme of growth by branching.

Each ruled by different selective criteria—plants need light, animals food—the two lines shared a number of problems, to which they evolved comparable solutions. Increasing body size and complexity was one common problem, which, in both cases, was solved by vascularization. Also shared were the problems posed by the invasion of land, which imposed strict, water-saving measures on both evolutionary lines. Most important of all common problems was the need for successful reproduction, the selective criterion *par excellence.*

It is significant that both lines adopted sexual reproduction right from the start. This emergency measure of unicellular protists became an essential means of genetic diversification in the evolution of multicellular organisms. Further progress was linked in both lines with the development of more efficient fertilization mechanisms, better protection of the fertilized egg, and improved fostering of the growing embryo. From aquatic fertilization and development to spores, seeds, and, finally, flowers and fruits in the plant line (see figure 19.1), and to copulation, the amniotic egg, and the mammalian womb in the animal line (see figure 22.1), the trend is unmistakable. Also impressive is the division of reproductive functions between the sexes. In both plants and animals, feeding and sheltering the developing embryo is a female prerogative. The male role is largely restricted to fertilization, compensating for a lack of elaborate specializations by an extravagant production of pollen or sperm.

The role played by natural catastrophes in the evolution of plants and animals is noteworthy. Evolution is punctuated by massive extinctions, sometimes of cataclysmic proportions. Almost invariably, life's response has been remarkably innovative. Apparently, when evolution becomes sluggish, it is not so much for want of an appropriate chance mutation as for the lack of a worthy environmental challenge.

Chapter 23

The Web of Life

IN RETRACING THE HISTORY of life on Earth (see figure 23.1), we have looked mostly at the core structure of the tree, the line traced by the successive appearance of living forms of rising complexity. But each major step in this progression has also produced side branches that extend their ramifications to the present day. The history of life is not just vertical growth in the direction of complexity; it is also horizontal expansion in the direction of diversity. Each cross section of the tree becomes more varied with advancing time, recapitulating the tree's previous history by means of what were the terminal twigs of the branches at the time considered. A cross section at three billion years ago would show two sturdy, moderately diversified clusters bearing the archaebacteria and eubacteria existing at that time, almost hiding a tiny, isolated bud, which no observer could have suspected would one day turn into the massive eukaryotic trunk. A cross section at 400 million years ago would show a diversity of bacteria of both types, many kinds of protists, an abundance of algae, some primitive mosses and fungi, a variety of sponges, coelenterates, worms, mollusks, arthropods, and echinoderms, many of them long since extinct, and, sprouting from what we now know to be the main trunk, a number of primitive fish. The display would be richer than the earlier one, yet would include no trees, no flowers, no insects, no amphibians, no reptiles, no birds, no mammals. In such reconstructions, we identify the trunk in retrospect, as the branch that was to lead to the most important innovations in the future. This identification often would not have been evident to contemporaries and is liable to change with time. Today, we place our own species on top of the tree. At least, most of us do so. Ten million years from now, however, we could be on a side branch, or nowhere at all. The new trunk could prolong what looks today as a side branch; it could bear a form of life more complex than the human and beyond the power of our imagination.

Cross sections through the tree of life do not just consist of separate dots, as do cross sections through a real tree. In the tree of life, the dots are interconnected by an intricate network of relationships; they form a web. As the dot pattern increased in complexity, so did the web. In this chapter, we shall look at some critical aspects of the development of this web.

THE PRIMORDIAL LINK

Most of us think of life as a creative process, generating form through the synthesis of proteins, nucleic acids, and other specialized molecules. The plastic age has drawn our attention to the importance of biodegradation. Proteins are not intrinsically more fragile than many artificial polymers. What renders them fragile is their

FIGURE 23.1
A Summarized History of Life on Earth

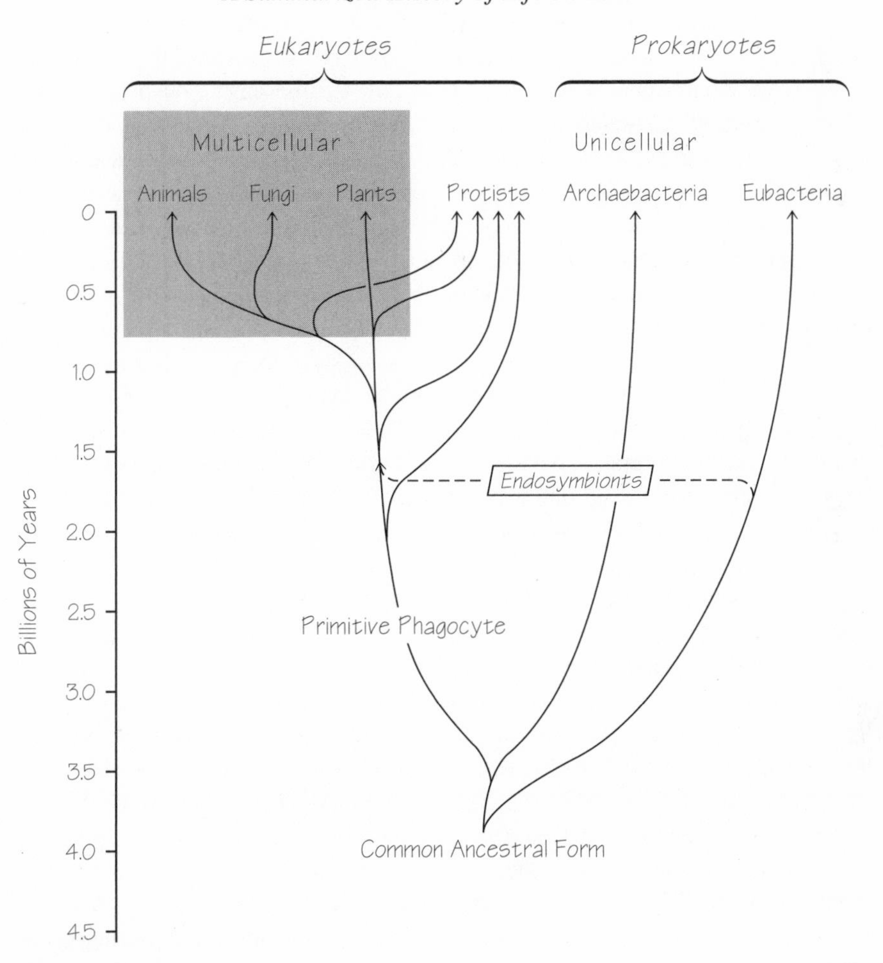

This figure summarizes, in highly schematic form, the history of life on Earth. Notable is the late appearance of multicellular organisms after three billion years of unicellular life. Any horizontal section through this "tree" shows the forms of life present on Earth at the time indicated.

biodegradability. Had life not developed the means to break down the products of its own industry, there would be no biosphere, only an inert shell of biopolymers, a "plastosphere" of the kind human ingenuity is beginning to create.

The link between biosynthesis and biodegradation is the primordial link in the web of life. Most likely, it existed already in the first common ancestor of all life. Even if this organism was autotrophic, it must have had the ability to dismantle biopolymers; it must have possessed digestive enzymes. These are the simplest biological catalysts and can hardly have failed to arise early. Furthermore, they would have been needed to allow the organism to survive in the absence of an energy supply—a phototroph in the dark, for example—by consuming the remains of dead cells or part of its own substance. All autotrophs do this today.

For such processes to function, the cells needed safeguards against suicidal self-digestion. One such safeguard in present-day bacteria is extracellular discharge of the enzymes concomitantly with their synthesis by ribosomes bound to the cell membrane. Other safety measures involve a variety of chemical controls that keep digestive enzymes inactive inside the cells and unleash them only when and where needed. The main digestive enzymes we secrete into our stomach and intestine are made in this way, as inactive "zymogens," which are activated only when exposed to the gastric or intestinal milieu. If the enzymes are prematurely activated, as in pancreatitis, deadly damage to the tissues may ensue. A third kind of protection depends on confinement of the enzymes within membrane-bounded digestive pockets (lysosomes), lined by an enzyme-resistant inner layer. Injuries to this lining may also result in widespread tissue damage, as seen in many pathological conditions.

In the beginning, biosynthesis exceeded biodegradation and bacterial life progressively covered large surfaces of the Earth with thriving, self-sustaining colonies. These soon turned into feeding layers for mutant forms that had lost the capacity for autotrophic growth and turned into obligatory heterotrophs. As revealed by stromatolites, multilayered associations of autotrophic and heterotrophic bacteria had developed in several parts of the world by 3.5 billion years ago, perhaps earlier. These mats became organized in a manner that reflected the requirements and tolerances of their constituent organisms.

The surface was occupied by phototrophs that needed the maximum intensity of light. Below them were phototrophs capable of using the light filtered by the top layers. Then came a number of heterotrophs each adapted to their immediate surroundings. Individual layers in this colony were connected with those above and below by mutual relationships that tended to stabilize the structure of the colony in a steady state. By necessity, phototrophs adjusted their proliferation to the consumption ability of the underlying heterotrophs, while the heterotrophs limited their voracity to a level compatible with the maintenance of the food-supplying autotrophs.

Thus, stromatolite-generating colonies may be viewed as pseudo-organisms, made of several kinds of cell types and held in dynamic equilibrium by a number of self-regulatory circuits. This organization arose spontaneously and was maintained by automatic mechanisms, guided by no more than the blind screening effect of natural selection, which eliminated the forms that did not fit within the colony's economy and favored those that did. Comparison of stromatolites of different ages with similar extant colonies suggests that the basic organization of such formations may not have changed substantially over 3.5 billion years.

As the tree of life developed, so did the bonds between autotrophs and heterotrophs. These bonds still rule the major equilibria of the biosphere. They have become more complex through the emergence of heterotroph-eating heterotrophs, organisms that feed on autotrophs indirectly, sometimes by way of a long food chain. A squid may owe its sustenance to the biosynthetic activity of phototrophic marine microorganisms (phytoplankton) by feasting on a crab that had eaten the remains of a fish that had enjoyed a meal of phytoplankton-fed shrimp. We may ourselves derive energy from the sun by eating meat that was built in a bull thanks to the presence in the animal's stomachs of microorganisms able to break down grass into usable nutrients.

Dynamic equilibria such as exist within a living stromatolite also stabilize the different parts of the biosphere. A population of foxes cannot outgrow the rabbits on which they feed. In turn, the rabbits are limited in their expansion by the proliferation of the foxes that kill them. Hence, the numbers of foxes and rabbits go through cyclic oscillations in which the rise of one coincides with the decline of the other. This classic example is known as the Lotka-Volterra cycle, from the names of the scientists who studied it theoretically in the 1920s.[1] However, such simple predator-prey interactions hardly depict the complex relationships that link together the components of real ecosystems. Even the simplest of fields or ponds are multifactorial systems stitched into intricate networks by dynamic interactions among the plants, animals, fungi, and microorganisms they contain. Such systems, in turn, join to create larger, more complex fabrics, eventually closing into a single, gigantic web of formidable complexity that envelops the entire Earth: the biosphere.

Understanding this web has become a major aim of ecological research. In spite of extensive field studies and increasingly powerful computer simulations, this area of investigation is still in its infancy, so complex and, to some extent, unpredictable are the interconnections it tries to unravel. A rare insect may control the balance of an entire rain forest because of its role in the pollination of certain essential plants. Some key regulatory principles of general significance remain applicable.

BIOSPHERIC METABOLISM

Basically, the biosphere and its constituent subsystems continue to be ruled by the primordial link between autotrophs and heterotrophs, acting as recycling converters of matter and energy. The green mantle traps energy from sunlight, carbon from carbon dioxide, nitrogen from nitrate or atmospheric nitrogen, sulfur, phosphorus, sodium, potassium, calcium, magnesium, iron, and other needed elements from dissolved mineral salts, and water from whatever source is available. These materials are converted into biological constituents, with oxygen as the major byproduct.

Part of the store of bio-organic products laid up by the phototrophs enters the food chain and serves, directly or indirectly, to nourish animals and other heterotrophs. These organisms utilize their foodstuffs to build their own constituent molecules and to fuel their energy requirements. In doing so, they use oxygen and break down much of their foodstuffs into carbon dioxide and other waste products. The job is completed by worms, fungi, and bacteria, which decompose dead plants and animals, as well as incompletely metabolized waste products such as uric acid, urea, and ammonia. Thus, the oxygen produced by the phototrophs is consumed, and the carbon dioxide, nitrogen, nitrate, and other mineral constituents they use up are regenerated. Parts of these cycles occur anaerobically, depending on fermentation processes or the intervention of electron acceptors other than oxygen. Some autotrophic activities take place without the help of light, fueled by the oxidation of mineral electron donors.

By and large, these phenomena are governed by the same kind of self-regulating mechanisms that maintained the primitive bacterial colonies in dynamic equilibrium. Synthesis and breakdown tend to balance each other so that the biosphere is held in a steady state and the main biogenic elements are recycled. But there are local variations and, sometimes, large-scale ones. Peat, coal, and oil deposits remind us of a time when biosynthesis greatly exceeded biodegradation. The fossil record holds repeated examples of major upheavals in the composition of the biosphere. Not least impressive, as well as worrying because we are responsible, are the increasing threats inflicted upon natural balances by human interventions. In order to understand these problems, we must pay attention not only to the interactions among living members of ecosystems but also to the interactions between living organisms and their environment.

THE ENVIRONMENT

The web of life is intimately connected with the environment by a dense network of mutual interactions. That life depends on the environment is obvious. Temperature, sunlight, rainfall, availability of essential nutrients, and other environmental factors delimit the ability of certain plants to grow in a given area and thereby define the possibility for various animals, fungi, and microorganisms to develop in the area. Less evident, perhaps, are the influences exerted by living organisms upon their environment. Yet these influences were, and still are, of fundamental importance. Without life, our planet would be entirely different from what it is.

Life has entirely altered the oxidation-reduction balance of the Earth. The large amounts of ferrous iron that filled the Archaean oceans have become locked up, partly in the ferric form, in minerals such as magnetite and pyrite. Hydrogen sulfide has been mostly oxidized or trapped in minerals and is now found only in certain volcanic areas, where it emerges from cracks in the Earth and is quickly oxidized by atmospheric oxygen. The most dramatic change is the rise in atmospheric oxygen itself, which is essentially due to the activity of phototrophs. This change, in turn, has modified the composition of many rocks and given rise to the ozone layer (the ozone molecule, O_3, consists of three oxygen atoms), which now shields the Earth and its inhabitants against excessive ultraviolet radiation from the sun.

Another major effect of life is the abundant presence of water on our planet. But for living organisms, the Earth would, like Mars, be almost entirely desiccated today. Its water would have been progressively split by UV irradiation. The resulting hydrogen would have escaped into outer space, and the oxygen would have been trapped by mineral "sinks." Life splits water also, but in a manner that saves the hydrogen and allows the water to be restored. Living organisms also play an important role in keeping the soil humid and in generating the atmospheric currents that bring rain to the continents. Without life, the land masses would have remained arid and dry, as they largely were until living organisms started migrating out of the oceans.

Also largely influenced by life is the manner in which carbon dioxide and its salts, the carbonates, are distributed on the Earth. At present, carbon dioxide constitutes only 0.0315 percent by volume of atmospheric gases and is dissolved in the oceans at a correspondingly low concentration. Its prebiotic levels may have been as much as one hundred times higher. A predictable consequence of such a high atmospheric content of carbon dioxide would have been a considerable heat retention due to what is known as the greenhouse effect (see chapter 30). It so happens, however, that the young sun was cooler in those days and delivered about 25 percent less heat to the Earth than it does now. The high carbon dioxide content of the atmosphere provided a compensatory blanket, so that the surface temperature of our planet was kept around 20° to 25°C (68° to 77°F), according to the best estimates by experts.[2] As life developed and started using up carbon dioxide, more heat escaped from the Earth, but more also was received from the warming sun, so that the two effects largely canceled each other.

The carbon abstracted by life from prebiotic carbon dioxide was woven into the organic fabric of the biosphere and, through Carboniferous luxuriance, partly stored in vast underground deposits from which it is now being returned to the carbon dioxide pool by fossil fuel combustion. A good part of the prebiotic carbon dioxide also became immobilized, mostly as calcium carbonate, in the shells and other structures built by marine organisms. Sedimentation, metamorphism, and resurfacing by tectonic movements have produced the fossil-studded limestones, marbles, and other calcareous rocks that now emerge in many parts of the world. The white cliffs of Dover, the majestic natural cathedrals carved by hidden rivers running deep below our feet, the gleaming Carrara stones out of which so many masterpieces were born would not exist had not life arisen and developed on our planet.

The biosphere is thus not just a pellicle of living matter that envelops the Earth like a coat. It is intimately linked to the Earth by myriad reciprocal connections, a giant, sun-powered surface processor that both draws from and acts on the crust, oceans, and atmosphere around it, continually remodeling them and being remodeled by them. Life and Earth are so undissociable that some view them as joined into a sort of planetary superorganism, made of interconnected living and nonliving parts held together by a network of cybernetic relationships. This view has been popularized under the name of Gaia.

GAIA

She was the Mother Earth goddess of the ancient Greeks. Long forgotten, except in such words as geology, geography, and geometry, she has recently been revived by James Lovelock,[3] a distinguished English scientist, an F.R.S. (Fellow of the Royal Society), the most coveted set of initials to follow a British scientist's name. A physicist by training, Lovelock is a successful inventor of scientific instruments who enjoys a comfortable income from the patents he earned when he was young. He now lives as the proverbial "gentleman of independent means" in a converted eighteenth-century water mill near the border between Devon and Cornwall, the most southwesterly English counties. Coombe Mill (Experimental Station) is both Lovelock's home and the computer-crammed laboratory in which he simulates the whims and vagaries of Gaia.

What distinguishes Gaia from other global concepts is homeostasis, self-regulation. In Gaia, life and Earth do not simply interact haphazardly. They do so in a manner that tends to correct the imbalances they inflict on each other. An example investigated theoretically by Lovelock is "Daisyworld." A planet is seeded with a mixture of dark and light daisies that have the same growth requirements but differ by the proportion of incoming light they absorb and reflect. Dark daisies absorb more light and reflect less than do light daisies. On a cold planet, dark daisies fare better than light ones because they retain more heat. They thus spread progressively from the tropical regions, where they started, to the cooler ones, which they help warm up. If, however, the amount of light falling on the planet increases, as happened to the young Earth, the planet may become too hot for daisies to grow. The light daisies, which reflect more light and tend to cool their surroundings, are now favored and outgrow the dark ones. Dark and light daisies thus act as a thermostat. They react to changes in temperature in a manner that opposes the changes; they tend to keep the environmental temperature constant. This simple model recalls the Lotka-Volterra predator-prey model. The difference is that the Daisyworld model involves life and an environmental factor, not two forms of life.

Lovelock has progressively refined Daisyworld by introducing daisies of up to twenty different shades and even adding rabbits and foxes. The result is always the same: self-correction, even in the face of deliberately introduced disturbances. With the help of these experimental simulations, Lovelock has built what he considers an increasingly strong case in support of what he acknowledges was originally just an intuitive hypothesis. According to the Gaia theory, the Earth is a living organism that automatically regulates its environment so as to make it optimal for life.

Gaia has been greeted with mixed reactions by the scientific establishment. The concept has been enthusiastically endorsed by Lynn Margulis,[4] who has become one of its most ardent proponents. According to the late Lewis Thomas, Lovelock's observations "may, one day, be recognized as one of the major discontinuities in human thought."[5] The cosmologist Freeman Dyson has adopted the Gaia concept and writes, "Respect for Gaia is the beginning of wisdom."[6]

Others, however, have been disturbed by the seemingly teleological character of the concept and by the almost mystical language in which it was worded in the beginning. Lovelock agrees that the style of his early writings may have been misleading, but vigorously protests that Gaia is a bona fide scientific theory, open to testing by observation and experiment.

Ecologists tend to distrust the Gaia concept for a different reason. It depicts the living Earth as a robust organism capable of resisting many insults, not as the fragile structure they see as threatened on all sides by human activities. However, Lovelock hardly deserves to be charged with insensitivity toward environmental causes. He has been critical of what he considers misguided emphasis on certain threats, such as weak carcinogens or nuclear power. But at the same time he has spoken out eloquently against what he calls the three deadly Cs,[7] cars, cattle, and chain saws, which he blames for destroying the English countryside.

The history of life on Earth offers some support for Lovelock's general point of view. This history has gone through repeated catastrophes, due to such causes as tectonic movements, volcanic eruptions, climatic changes, and asteroid impacts, that wiped out much of the existing flora and fauna. Each time, life not only rebounded but came up with some decisive innovations. However, it took millions of years for this to happen. We can hardly rely on Gaia's natural resilience if we wish to save the Earth for our children and grandchildren.

Chapter 24

The Virtues of Junk DNA

A MAJOR DIFFERENCE between eukaryotes, especially the higher plants and animals, and their distant prokaryotic relatives concerns what may be called their DNA thriftiness. Prokaryotes practice the strictest possible economy in DNA content. Their genome contains hardly a single nucleotide that is not involved in coding or control. In the words of Harvard chemist and Nobelist Walter Gilbert, the bacterial genome is "streamlined," probably as a result of strong evolutionary pressure favoring fast proliferation.[1]

In striking contrast, the eukaryotic genome is made mostly of noncoding DNA without obvious function, sometimes called "junk" or "ballast" DNA. Less than 5 percent of the human DNA has a coding function. Salamanders do much better—or worse, depending on one's point of view.[2] Some of these animals have twenty times more DNA than we have, with those in the west of the United States beating those in the east by severalfold. Fortunately for our self-esteem, DNA quantity is not by itself a measure of overall quality. Western salamanders are not obviously cleverer than their eastern congeners. Having more DNA does not automatically make salamanders superior to us.

SELFISH DNA

The amount of apparently useless DNA in the genome of higher plants and animals requires an explanation. According to Britain's ethologist Richard Dawkins, the explanation lies in the "selfishness" of DNA.[3] The unit of selection is DNA, not the body. The body is no more than a means of replicating DNA, just as a chicken has been said to be an egg's way of making another egg. To quote Dawkins: "The true 'purpose' of DNA is to survive, no more and no less. The simplest way to explain the surplus DNA is to suppose that it is a parasite, or at best a harmless but useless passenger, hitching a ride in the survival machines created by the other DNA."[4] This imaginative concept does not explain the striking difference in DNA economy practiced by prokaryotes and eukaryotes, nor the fact that, albeit with wide variations (remember the salamanders), the proportion of "junk" DNA in the eukaryotic genome tends to increase with increasing evolutionary complexification.

Part of the eukaryotic DNA seems to play no evident role. There are "dead" genes, copies of functioning genes that have become useless as a result of some crippling mutation. There are also long linking stretches between genes and large stacks of multiple repeats of the same sequence that have no obvious function. In contrast, the bacterial genome contains no dead genes, no unnecessarily long linkers, no stacks of apparently useless repeats. If, as seems likely, evolutionary "streamlining" is responsible for this continual pursuit of genomic leanness, it appears that eukaryotes were not under the same kind of pressure.

Indeed, eukaryotic cells are not continually multiplying as quickly as they can. Their DNA replication is a leisurely affair, which takes only part of the time required by cell division and can be adjusted to any length of DNA simply by replicating more stretches of DNA simultaneously. This faculty is missing in bacteria, which are limited to a single replication origin. It is thus possible that eukaryotes carry "selfish" DNA from generation to generation because the advantage of getting rid of it is not sufficient to drive natural selection. On the other hand, the possibility that this DNA plays a role, for example, in chromosomal structure, or in some other unknown way, cannot be excluded.

SPLIT GENES

This is not, by far, the whole story of eukaryotic "junk" DNA, nor even its most intriguing chapter. Noncoding DNA is present not only between genes but also within them. Many genes of eukaryotic organisms consist of discrete segments, numbering from two to more than one hundred. Called exons (because they are expressed), these segments are separated by intervening sequences, or introns, that in most cases are not expressed into anything useful. Exons are short and of relatively uniform length, more than two-thirds being between 50 and 200 nucleotides long. In contrast, the length of introns is much more variable, ranging from less than 10 to more than 50,000 nucleotides. Split genes are transcribed in toto, exons and introns alike, into correspondingly segmented RNAs. These subsequently undergo an intricate processing such that the introns are excised, generally to be broken down, and the exons spliced into mature RNAs.

Imagine interspersing a text with gibberish, printing the whole crazy hodgepodge, and then carefully cutting out all the gibberish and pasting together the pieces that make sense. No sane person would willingly add so many apparently unnecessary risks of error to the processing of information. Scrambling genetic texts in this way seems particularly absurd because the gibberish adds greatly to the burden of transcription and replication. Furthermore, not a single letter may be missed or misplaced in splicing, lest the whole message become gibberish itself. Finally, the energy cost of the whole, apparently futile, exercise is far from negligible. For all these reasons, no scientist before 1977 would possibly have imagined split genes. "Colinearity" was a dogma. It thus came as an utter surprise, rarely equaled by a scientific discovery, when, in 1977, two molecular biologists, Phillip Sharp, from Boston's Massachusetts Institute of Technology (MIT), and Britain's Richard Roberts, working at Long Island's Cold Spring Harbor Laboratory, independently found unmistakable evidence that a gene was divided into several segments that were cut out and spliced together in the RNA transcript.[5] This discovery earned its authors a 1993 Nobel Prize.

Evolution selected gene splicing and honed it to a remarkable degree of precision, which means that split genes conferred substantial advantages commensurate with the risks they entailed. According to Gilbert, the most likely advantage cells gained from split genes was "exon shuffling,"[6] the ability to make various combinations of the same DNA modules and put them to the test of natural selection, the way RNA genes are believed to have been first assembled (see chapter 7). Replacing the earlier, shorter RNA modules, exons were reshuffled within the genome into a wide variety of different "mosaic" genes. Opportunities for diversity were greatly increased. Cells enjoying this flexibility were saved from becoming progressively constrained within a genomic straightjacket. They kept their options open and retained the capacity to innovate.

The same exons have, indeed, been used as building blocks for different genes, thereby allowing certain key peptide motifs to be used repeatedly in different contexts,[7] just as the same switches, microchips, and other spare parts are assembled in different ways to make different machines. Exon shuffling is re-enacted in a particularly remarkable fashion in each individual in the course of maturation of the immune system (see chapter 14).

THE ORIGIN OF INTRONS

At what time in the history of life did introns appear in DNA? This is a hotly debated question. According to the evolutionary record, introns came late and spread slowly through the sole eukaryotic line. Almost nonexistent in prokaryotes, they are rare in lower eukaryotes and tend to increase in number with increasing evolutionary progression. This fact suggests that introns entered the genome during or after the prokaryote-eukaryote transition and subsequently spread like some sort of virus to occupy more and more sites within genes (and also between them, thus accounting for some of the junk DNA). A correlate of this view, though not a necessary one, is that this invasion of the genome by wandering bits of DNA played a significant role in eukaryotic evolution, by multiplying the number and variety of genetic rearrangements that were offered for natural selection to screen.

Surprisingly, a case can also be made for the "antiquity of introns," as eloquently argued by Gilbert.[8] What is known of the modular construction of genes suggests that something akin to exon shuffling, albeit with smaller RNA modules, already played a major role in early protocells. So did RNA splicing, which served an essential function in the early combinatorial game with RNA minigenes, reemerging later in another key capacity to convert RNAs transcribed from split genes into mature RNAs. If the first DNA genes were split by exons, an uninterrupted line of descent might exist between the early form of RNA splicing and its present use in eukaryotes. On the other hand, if exons are a late evolutionary innovation, one has to explain the revival of RNA splicing after more than two billion years of eclipse.

The theory that the first genes were split by introns implies that intron loss acted as a brake to evolutionary progress. Bacteria lost virtually all their ancestral introns and remained prokaryotes to the present day. Lower eukaryotes, such as yeasts, conserved a few introns and evolved further. And so on, up to the highest plants and animals, which have retained the largest number of introns. The idea of innovation depending on prolonging a flexible, unformed state where much is still possible holds an undeniable appeal. It fits with the view that important steps in the further evolutionary modeling of body plans were accomplished by putting off the moment of some definitive developmental commitment.

The issue of evolution by gain or loss of introns will not be settled on the basis of theoretical arguments. Facts will decide. In support of the acquisition theory, a number of introns have been found to be derived from wandering pieces of DNA, or transposable elements.[9] The existence of such elements was discovered in the mid-1940s by an obscure American plant geneticist, Barbara McClintock, who concluded from the distribution of variegation patches in corn cobs that these patches must have resulted from the transfer of certain DNA stretches from one daughter cell to another in the course of meiosis. Long ignored, this revolutionary concept was eventually recognized as fundamentally important, earning its modest and retiring author belated world fame and a 1983 Nobel Prize.[10] It is now known that certain segments of DNA in prokaryotes and eukaryotes are equipped with end sequences that allow the segments to be excised from their location and inserted into another site, not only within the same genome but also between cells, between organisms of the same or different species, and even across the prokaryote-eukaryote barrier. Several such intruders have been caught in the act of landing in the midst of a gene and causing a genetic deficiency by inactivating the gene. A number of introns have been identified as transposable elements of clearly recent origin. These findings have not closed the debate, but presently available evidence seems to favor the theory that split genes are a late acquisition.[11]

THE UNIVERSE OF EXONS

A growing number of exons present in different genes have been found to be related descendants of common ancestral DNA stretches, suggesting that all the proteins found in nature may have arisen from the combination of a limited number of genetic modules. According to an estimate by the Gilbert school, no more than about 7,000 exons—with a range of 950 to 56,000—could have served in the assembly of all known eukaryotic genes.[12] Although this estimate of the "universe of exons" is far from unanimously accepted, the very fact that the problem proved approachable and open to an acceptable solution indicates that the number of different exons with which genes were constructed must have been an extremely minute proportion of the unmanageably immense number allowed by simple statistical estimation—4^{50}, or one thousand billion billion billion different possible sequences for a stretch of only fifty nucleotides. This implies that extensive exploration of the combinatorial exon "space" may have been possible even at late stages of plant and animal evolution. The significance of this point was emphasized in chapter 7.

PART VI

THE AGE OF THE MIND

Chapter 25

The Step to Human

SEVENTY FOOTPRINTS in volcanic ash—two individuals of unequal size walking side by side and a third following in the bigger one's steps—were left to petrify for 3.5 million years in what is now the arid Laetoli area of northern Tanzania until they were uncovered in 1977 by Mary Leakey, of the famous Kenyan family of fossil hunters.[1] These ancient traces bear witness to the existence, in those remote times and in that part of the world, of creatures that walked erect on feet resembling ours. Such creatures must have wandered over much of East Africa at that time. The most famous is a young female named Lucy—after the Beatles song "Lucy in the Sky with Diamonds"—who made the headlines in 1974 when her amazingly complete remains—almost half a skeleton—were found in the Afar region of Ethiopia by Donald Johanson, the founder of the Institute of Human Origins in Berkeley, California.[2] Lucy is about the same age as the Laetoli walkers. Her pelvic anatomy indicates that she too walked on two legs. So did the owner of a knee joint, dated 3.9 million years, likewise found by Johanson in the Afar region. These early hominids (prehumans) are now known under the name *Australopithecus afarensis.*

The name *Australopithecus,* meaning southern (*australis* in Latin) ape (*pithêkos* in Greek), was originally coined, with the epithet *africanus,* by Australian-born Raymond Dart, after his 1924 discovery in a cave at Taung, South Africa, of a fossil skull belonging to an immature apelike primate, now universally known as the "Taung child." On the strength of the position of the hole in the skull through which the spinal cord connects with the brain, Dart claimed that the Taung child walked erect and was an intermediate between ape and human. This claim was strongly resisted by the paleoanthropological establishment at the time. It is no longer in dispute now that it is known that the Taung child, which is "only" two million years old, was preceded by erect-walking apes that lived almost two million years earlier.

According to comparative molecular sequencing, our closest extant relatives are the chimpanzees of Central Africa, and the last ancestor we have in common with them lived about six million years ago. It is thus not surprising that the first traces of the chimp-to-human trail should be found in Africa. The fact that these early creatures walked erect supports the hypothesis that bipedalism (walking on two feet) played an important role in the initiation of the human line.

Clues of the kind I have just described, patiently gathered in Africa and other parts of the world, have allowed specialists to piece together a patchy account of the advent of humankind,[3] within the wider context of primate evolution revealed by the fossil record and by comparative molecular sequencing. I shall tell only the broad lines of the story, without detailing the many controversies that still divide experts concerning the age, identity, genealogy, and kinship of the long-deceased owner of a tooth, jawbone, skull fragment, or other old piece of bone. But I shall look more closely at those characteristics that make us specifically human and at the possible mechanisms of their evolutionary emergence.

UP IN THE TREES AND DOWN AGAIN

Many millions of years ago, when dinosaurs were still around, a small, rodentlike mammal took up residence in trees, where it found food, shelter, and safety. New habitats call for new adaptations. By natural selection, this strange newcomer to a world hitherto occupied mostly by birds and insects evolved to acquire long arms, strong gripping fingers on all four limbs, a prehensile tail, and frontally facing eyes, giving it improved stereoscopic vision. Possible intermediates in this fateful transformation are represented in today's world by the lemurs of Madagascar and the wide-eyed tarsiers of Malaysia, followed by the hordes of monkeys that shatter the calm of tropical forests all over the world with their shrieks, aerial antics, and uninhibited displays of sexuality.

The big primate adventure might have ended there had not a group of monkeys found themselves isolated, some 30 million years ago, in an inhospitable African jungle where a stronger constitution, new skills, and greater cunning were required for survival. The brawnier and brainier *Proconsul,* the ancestor of the hominoid apes, was evolution's answer to the challenge. Up to several feet tall, *Proconsul*'s sturdy descendants invaded many parts of the Old World, giving rise eventually to the gibbons and orangutans of Southeast Asia and, later, to the gorillas and chimpanzees of Central Africa.

Finally, about six million years ago, one more decisive event took place. Some tree dwellers directly related to the ancestors of today's chimpanzees left their arboreal abode for the savannah, where they started the evolutionary line from which humankind emerged. This epoch-making event occurred somewhere in East Africa, where it may have been prompted by the geographic and climatic upheavals that created the Great Rift Valley. The forest receded and became increasingly overcrowded, with less food to offer and more dangers lurking in the shadows. On the other hand, the open savannahs, which were replacing the woodlands, held rich bounties for creatures endowed with the appropriate adaptations. Bipedalism, which the chimplike ancestors had already toyed with, did the trick. It allowed the animals to remain erect for better sighting of prey and enemy through the tall grass; especially, it freed the hands.

Watch chimpanzees today and you can see all that the ancestors we share with them may have succeeded in doing with their hands: carrying their young, grooming, clearing their way through thickets, picking up edibles, plucking berries and other tidbits, peeling bananas, bringing food to their mouth, clasping sex partners, fighting foes and rivals, hunting prey, gesturing, signaling, even using stones and sticks as weapons or as means of obtaining food. Watch the intent eyes, the frowning brows, the pursed lips, while the hands carry out some delicate, obviously purposeful operation, and you can, with a little imagination, visualize the wheels of a mind turning under that slanting forehead. What happened somewhere in the African jungle six million years ago played a vital role in the emergence of the human mind. From brain to hand and back from hand to brain, a self-reinforcing shuttle of impulses was initiated that was going to change the world.

For some reason, the chimpanzees did not much capitalize on the advances of their forebears. Perhaps the forests they inhabited did not strain their survival and reproductive success sufficiently to favor a change. Evolution tends to stagnate if not prodded. The East African crisis that drove a group of apes down onto the savannah provided the prod. It put a premium on bipedalism and dexterity and on the pooling of efforts. The animals formed small bands that wandered together, hunted and fought cooperatively, shared food and shelter, bred mostly within the group, kept contact by means of sound signals, and even learned to run personal risks for the sake of the common good. These new qualities were not the product of conscious planning. They were acquired the hard way through the elimination by natural selection of offspring lacking the qualities. In the meantime, new self-reinforcing shuttles between the brain and other parts of the body were established, leading to rudiments of verbal communication and social behavior. The more successful bands increasingly kept to themselves and away from the more brutish families, eventually giving rise to *Australopithecus,* followed later by *Homo* (Latin for man).

Australopithecus afarensis, africanus, robustus, boisei; Homo habilis, ergaster, erectus, rudolfensis, neanderthalensis, sapiens—we shall leave the names to the experts and let them thrash out among themselves who came first and where, and who was related to whom. All these distant relatives of ours subsisted on hunting and gathering, moving from place to place according to what their experience had taught them of the seasons when certain fruits were most abundant or the young of their prey most vulnerable. Some found shelter in caves and settled temporarily until scarcity of food drove them away. Many sharpened stones to carve up the products of their hunting, helped by anatomical changes of the hands—most prominently, the opposable thumb—that improved adroitness. Several harnessed the fire that fell from the skies and learned how to light it again with the spark of a flint when it had died out or had been doused by rain. The males fought each other fiercely, though in some measure ritually, for possession of the females. Rival bands mostly kept their distance and covered separate territories, clashing only when both set their sights on the same choice terrain or suffered a paucity of females. They fed, protected, and tended their young communally. They emitted different kinds of grunts, each with a special meaning, and were able to communicate at a distance or in the dark. They rarely lived longer than needed to produce enough progeny. A few lasted past reproductive age and may have been taken care of by the group for the services they rendered. Natural selection favored the groups that cared for their reproductively useless elderly over those that abandoned or killed off their senior citizens, because the overall reproductive success of the altruistic groups was enhanced. For example, the old could tend the young while the more youthful adults devoted their strengths and skills to other useful purposes. In hazardous circumstances, the experience of the old could also have helped the group survive.

Were these creatures apes or humans? The boundary is fuzzy, not easily drawn. All that can be said is that they became a little less ape and a little more human over about six thousand millennia—a length of time amazingly short or amazingly long, depending on how you look at it: less than two-thousandths of the age of life on Earth, yet three thousand times as long as has elapsed since the birth of Christ, more than five hundred times the whole of recorded human history. On our time scale, the advent of humankind was a slow, imperceptible process, hardly noticeable from one generation to another, even though the leap seems huge to us today. Between the nearest ancestor we share with chimpanzees and us, more than 300,000 generations have come and gone.

MITOCHONDRIAL EVE

Landmarks in this history are few and far between, scattered over several million years and over wide areas, not only in Africa but also in Europe and Asia. Then, roughly 200,000 years ago, the trail apparently focuses back to a single point in Africa, where there lived a woman believed to be the mother of us all. This was the surprising announcement made in 1987 by the late Alan Wilson of the University of California at Berkeley and his collaborators, Rebecca Cann and Mark Stoneking. These researchers collected 157 human samples from Africa, Asia, Europe, Australia, and New Guinea, and extracted from them and analyzed the small amount of DNA present in mitochondria, the respiratory cell organelles derived from bacterial endosymbionts, of which they have retained a few genes. There were two reasons for choosing mitochondrial DNA for these studies. First, sperm cells do not contribute mitochondria upon fertilization, so that mitochondria are inherited exclusively by way of the female egg cells. This simplifies genetic analysis. In addition, mitochondrial DNA mutates much faster than nuclear DNA, so that many significant variations could be expected to appear in only a few hundred thousand years.

The recorded "fingerprints" were, indeed, diverse, but they clearly appeared all to be derived from a single ancestral molecule, which, according to the reconstructed tree, belonged to a female who lived in Africa about 200,000 years ago. It took little time for the media to publicize this story under the inevitable title of "African Eve," adorning it with a variety of comments and conclusions, most of them wrong. Scientists also pounced on the story, soon to find it full of holes. The investigators came back with new results supporting their original conclusions, which, in turn, were faulted by the contradictors.[4] The debate is not over yet, but has been muted considerably since the unfortunate death of Alan Wilson, the main defender of the African Eve hypothesis. What are we to think?

One point is clear. The method used to construct the original tree was flawed. Other trees rooted at different times and in different geographical locations can be constructed from the same data. It is unlikely, however, that "Eve" could be much older than 500,000 years (200,000 years not being excluded), and it is probable, for other reasons, that she lived in Africa. The main fact that we all go back to a single female remains unchallenged. But it may not mean much. It certainly does not mean that the whole of humankind is derived from a single couple or even that there may have been something special about Eve.

Consider an ancestral population containing a certain number of females, each of whom can start a line along which a given mitochondrial genome is transmitted from daughter to daughter. Inevitably, as time goes on and generation follows upon generation, these lines are bound to die out one after the other for lack of females left to transmit the gene. Eventually, only one line will be left. Theory shows that the time needed for this to happen in a stable population is equal to the generation time multiplied by twice the population size. Thus, with a generation time of twenty years, the time needed for "coalescence" to a single Eve is forty times the population size. In other words, if Eve lived 200,000 years ago, she may have had 4,999 female congeners, in which case the survival of her line was due to chance and has no special meaning.

The possibility does remain, however, that Eve had a trait, transmitted by the female line, presumably by way of mitochondrial DNA, that contributed to the success of her progeny in crowding out all other hominids. That a mitochondrial gene could have had such a far-reaching effect is not what most geneticists would expect. Perhaps a more probable alternative is that the present human race originates from a highly inbred population. In which case, incidentally, the decisive mutation could just as well have occurred in Adam's nuclear DNA.

Another puzzling aspect of mitochondrial Eve is that she rules out any recent admixture of "foreign" females (Neanderthal, for example) to our ancestry, which hardly fits with the usual behavior of conquering males. Unless a particular trait made the vanquished females repulsive to their new masters—their inability to speak has been evoked as a possible deterrent, but conquering males are not reputed for seeking the pleasures of conversation—it would appear that the hybrid progeny had poor survival value or, perhaps, were infertile, which would imply that Eve's descendants had diverged from other hominids to the point of forming a distinct species. This point is relevant to a standing debate pitting the proponents of a monophyletic origin of the human species against those of a polyphyletic origin. Clearly, mitochondrial Eve still has much to tell us.

ADAM'S APPLE

Whether mitochondrial Eve or some other ancestor, one of our distant forebears must have acquired a trait that gave its progeny a decisive evolutionary advantage. What was this trait? The answer to this tantalizing question could come from those linguists who have studied the origin and evolution of language. It is possible that this ancestor was born with a genetic "defect" that moved the larynx deeper down the neck. As pointed out by the American linguist Philip Lieberman,[5] from Brown University, this anatomical feature is unique to modern humans and appears late in development. In newborn babies, as well as chimpanzees and all other animals, the larynx is much closer to the mouth. Even the Neanderthals, who became extinct about 35,000 years ago, showed this disposition, according to Lieberman, although there is some disagreement on this point.[6] It is our lower larynx that gives us the ability to emit a much wider variety of sounds than any other animal. What started the modern human line may have been the ability to speak and, with it, the power to communicate in an increasingly refined way and thus to conquer the world.

The new group did not achieve preeminence rapidly, probably because many evolutionary steps were needed for true language to emerge. At each step, however, communication among individuals became richer, social bonding stronger, and the pooling of skills and efforts more efficient. At some stage, another important change occurred. It affected female sexual physiology so that, unlike the females of any other animal species, women became continually receptive. The male-female bond and kin cohesion were greatly strengthened by this unique modification, which the American physiologist Jared Diamond considers a key event in the ascent of humanity.[7]

About 50,000 years ago, the evolution of our ancestors started, rather precipitously, to bear an abundance of fruits, creating a host of new inventions in a relatively short time. The humans of that time made increasingly refined tools and weapons, built shelters next to the caves of their forebears, equipped their dwellings with fireplaces and grease-burning stone lamps, sewed pelts together into protective clothing, manufactured watercrafts sufficiently seaworthy to bring them to distant lands, and achieved remarkable skills in hunting and fishing. They inaugurated travel and trade, spread over Europe and Asia, crossed the sea to Australia, moved to Siberia and Mongolia, and trekked over the desolate ices of the Arctic to invade the Americas. They established a string of colonies, each creating its own subculture adapted to the local terrain, climate, and biotope. They began to look at nature in awe and wonderment and to worship the hidden powers behind it. They expressed their feelings in paintings and carvings, nursed their sick, buried their dead, and hung trinkets of beads, shells, and crafted bones around the neck and wrists of their women. Culture was born and, with it, the possibility of a new form of hereditary transmission that completely altered the rules of evolution.

Pottery, agriculture, animal domestication, food processing, metal smelting, and the wheeled wagon followed, arriving finally at the written word. It took humans only a few more millennia to achieve the means to walk on the Moon, engineer life, and kill tens of millions of their kind in one shot.

Chapter 26

The Brain

OF THE MANY CHANGES that mark the conversion of ape to human, the most conspicuous and far-reaching is the increase in brain size, which has tripled in a few million years. This increase is responsible for our greatly enhanced intellectual abilities.[1] It is time to take a look at these most remarkable of all eukaryotic cells, the neurons, and at the rules that govern their assembly.

THE MAGIC OF NEURONS

A neuron is essentially a miniature receiver-transmitter device. It has a cell body, which possesses a nucleus, a cytomembrane network, cytoskeletal and cytomotor elements, organelles, and all the other characteristic attributes of animal cells. The cell body takes care of all the housekeeping functions necessary for cellular life; it is the combined power, maintenance, and repair unit. Thin filamentous extensions form the receiver and transmitter parts. These extensions can be exceedingly long, up to three feet in humans, thirty feet in a whale. The transmitter consists of a single fibril, the axon, which usually branches into terminal ramifications only in the neighborhood of its target. The receiver part is most commonly made of bushy arborescences called dendrites (from the Greek *dendron,* tree). A neuron acts as a one-way relay, from dendrite to axon. If a dendrite is disturbed, physically or chemically, the axon fires, in most instances by discharging some specific chemical, called a neurotransmitter. This chemical, in turn, sets off a response in any cell displaying the appropriate receptors and connected to the discharging axon tip by a special junction called a synapse. Depending on the nature of the target cell, the response may be contraction (muscle cell), secretion (gland cell), or stimulation or inhibition of firing (another neuron).

Sending out extensions is a general property of eukaryotic cells. Such protrusions, or pseudopodia (pseudofeet in Greek), which serve in sensing, food capture, or locomotion, usually have a transient existence and are withdrawn soon after they are emitted. Assembly and disassembly of microtubules play an important role in these reversible phenomena. We may take it that a neuron first arose when such extensions became stabilized—the evanescent microtubules turned into stable neurotubules—and polarized into one-way receivers and transmitters. Plant cells, because of their encasement within solid walls, never developed into neurons. In contrast, animal cells did so very early. Except for sponges, all animals have neurons. In many respects, the whole saga of animal evolution may be viewed as written by these remarkable cells, which, thanks to their unique combinatorial properties, were able to weave the increasingly intricate regulatory networks that supported rising complexification.

The first neurons probably connected skin cells directly to muscle cells in such a way that, if the skin cell was perturbed, contraction of the muscle cell ensued. If established between suitably situated cells, such connections could have been useful, for example, by inducing avoidance or pursuit of the disturbing object. Natural selection preserved genetic body plans with the most advantageous neuronal connections.

A major advance was accomplished when neurons began to establish connections with other neurons. A chain of two or more neurons could thereby link skin cells to muscle cells. Especially important, neurons in such chains could be bridged by cross-linking neurons that triggered or blocked firing in one neuron according to what happened in the other. This was a crucial development, bound to be retained by natural selection, as it allowed the connected neurons to be "informed" of each other's activities and to "program" their own activities accordingly. The circular string of neurons found in some jellyfishes represents one of the simplest and, perhaps, earliest such arrangements. It allows the animal to contract its body in a coordinate fashion and thereby to propel itself.

Once neuron-to-neuron connections became possible, direct skin-to-muscle connections were soon replaced by indirect connections mediated by neurons. These cells became subdivided into sensory neurons, transmitting impulses from the skin to a neuron; motor neurons, transmitting impulses from a neuron to a muscle; and intermediary neurons, transmitting impulses from a neuron to a neuron. In addition, axons and dendrites became increasingly ramified, so that eventually a single neuron could simultaneously send out impulses to thousands of neurons and, conversely, receive impulses from an equally large number. Further diversification was achieved by the multiplication of neurotransmitters and their cognate receptors, to form a variety of chemically distinct synapses. Thus, a combinatorial game with almost infinite possibilities was initiated. Its most complex achievement to date is the human brain, with more than 100 billion neurons interconnected by an average of 10,000 junctions per neuron, involving at least fifty different kinds of synapses. All the world's computers together could not form such a rich information processor.

THE EQUIPMENT OF A HEAD

In the evolution of the nervous system, cell bodies progressively congregated into collectives, or ganglia, and their sensory and motor extensions joined into bundles, or nerves. In the first segmented animals, these arrangements became duplicated many times. In earthworms, each segment has a typical pair of ganglia connected to the skin and muscles by sensory and motor nerves. Cords linking the ganglia serve to coordinate the activities of the segments. Along with mouth-to-anus polarization, the front body segment progressively acquired an enriched nervous system, while a larger number and greater variety of sensory cells developed in this part of the organism, where they were best placed to warn the animal of an approaching boon or danger. Brain formation thus followed head formation. An important additional development was the replacement, in chordates, of the lateral neural cords (except for the vegetative nervous system, which controls the viscera) by a central tube (the cerebrospinal axis) capable of increasing in size by surface expansion and folding, as illustrated most vividly by the convolutions of the human brain.

The main function of the primitive brain was to collect information from the environment and from the body itself and to reprocess this information into appropriate motor responses. One readily sees how any change in the "wiring" of the system leading to a more adequate response, such as faster flight from a predator or quicker snapping up of prey, would confer a significant evolutionary advantage. As soon as neurons appeared, a relentless selective pressure drove the nervous system toward ever increasing complexity.

By necessity, this progress depended on improved information collection. Hence the appearance of specialized cells sensitive to a variety of physical and chemical signals, including pressure, stretch, sound, heat, cold, light, electricity, injury, and a host of chemical compounds. Some of these cells came to be grouped into organs of stupendous complexity and sensitivity. The object of the greatest admiration is usually the eye. But we should not forget the bat's ear, which provides split-second reconstructions of a continually changing environment from the reflections of the ultrasonic waves the animal itself emits, nor the exquisite sense of smell of the bloodhound and, especially, of those male insects that can detect attractant pheromones emitted by females hundreds of yards away. At that distance, the compound reaches the animal's olfactory cells literally molecule by molecule.

The perfection of sense organs is one of the most striking attributes of animals, often brandished by adversaries of Darwinism as evidence that evolution cannot possibly have proceeded without guidance. This argument is epitomized in the famous watchmaker allegory by the English theologian William Paley.[2] In his *Natural Theology—or Evidences of the Existence and Attributes of the Deity Collected from the Appearances of Nature,* published in 1802, this worthy cleric pointed out that if you "found a watch upon the ground," you would perforce conclude that "the watch must have a maker." He used this parable as a proof that living beings, like the watch, revealed the hand of a Creator by their organization.

Paley's book came out more than half a century before Darwin's *Origin of Species.* So he can be excused for having gone astray in his reasoning. We have no such excuse. As explained by the British ethologist Richard Dawkins in *The Blind Watchmaker,*[3] natural selection can account for the emergence of even something as complex as the eye. All that is needed, starting from some simple light-sensitive cell, of which there have been plenty since the earliest days of life, is a succession of small blueprint changes, each of which was associated with some selective advantage. The argument that 5 percent of an eye could have been no good is fallacious. Five percent of an eye—not a piece of cornea or retina, of course, but a primitive structure that, when illuminated, causes a neuron to fire—is obviously better than no eye at all. An additional fraction of a percent in the efficacy of this structure might then be sufficient to confer a significant evolutionary advantage. Multiply the number of such tiny steps over the hundreds of millions of years that were available and you might well end up with an eye, or even, as happens to be the case, with several kinds of eyes independently arrived at by convergent evolution.[4] Dawkins's conclusion cannot be faulted: The Watchmaker is blind. But does this mean, as Dawkins claims, that there is no Watchmaker? Not all orthodox evolutionists, including Darwin himself,[5] have seen this as a logically compelling inference.

At first, the brain mediated simple motor responses to sensory inputs. I use the word *simple* in a relative sense. What happens in the brain of an octopus when the animal suddenly pounces on the crab its eyes have sighted is no less than "mind-boggling" in its complexity. Yet it is almost ludicrously elementary in comparison with what goes on in your brain while you read this sentence. In lower vertebrates, the processing of sensory information still occupies a good part of the brain's functions. In a fish, for example, much of the brain mass is taken by centers for smell and sight and by their connections with motor systems. What changes as we move up the scale is the growing importance of associative structures, in particular the cortex. This is a sheet of neural tissue about one-tenth of an inch thick, consisting of six distinct cell layers, that envelops the more ancient brain parts. The size of this mantle, corrected for body weight, increases sixtyfold from the lowest mammals to chimpanzees, and another threefold from chimpanzees to humans. In humans, the cortex forms the highly folded brain convolutions, which cover a total surface area of about two square feet. This part of the brain, roughly one pound of neural tissue, is the seat of the highest mental processes.

THE WIRING OF A BRAIN

How is brain structure prescribed by the genome and how was this blueprint modified in the course of evolution? At first, every detail of the neuronal wiring was written into the genetic body plan. For example, in the roundworm *Caenorhabditis elegans,* each of the 302 neurons—almost one-third of the total cells of the animal—occupies a specific site.[6] Soon, however, the number of neurons and of alternative connections among them became such that the complete network information could not possibly be accommodated within the genome. Only the gene combinations for making the various kinds of differentiated neurons remained rigidly encoded, together with a small number of chemical guidelines that direct the assembly of the cells in time and space, mostly by way of cell adhesion molecules (CAMs) and substrate adhesion molecules (SAMs), and of specific secreted growth factors. The rest of the blueprint is filled in epigenetically, that is, in the course of development. The brain is no different from the stomach or liver in this respect. Each cell in such organs obviously does not occupy a genetically determined position either. What makes the brain unique is the intricate network of connections the cells establish with each other. What turns brain development into a problem orders of magnitude more complicated than the development of any other organ is the necessity for millions of billions of axons and dendrites to seek and find each other and become correctly linked by a functional synapse.

The answer, dictated by both logic and experiment, to the problem of brain development is that only the general lines of the wiring are engraved into the genetic blueprint. Details vary. No two individuals, even identical twins, have the same neuronal connections, and these connections change in the course of embryonic and, especially, postnatal development. The human brain has completed all the neurons it will ever make some five months before birth. Contrary to what happens to other cell types, multiplication of neurons ceases after that. Henceforth, neurons only die, starting in utero, to the tune of hundreds of thousands per day. I have lost several billion neurons since I was born. Between starting and finishing this sentence, I have lost about one hundred more. The thought is unsettling, but I take comfort in assurances from my neurobiologist friends who tell me that many brain connections are superfluous and redundant and that, even though I cannot replace my neurons, I can still rewire some connections if I keep sufficiently busy.

The residual plasticity of my brain is very limited, however, as compared with that of a newborn. A baby comes into the world with all the neurons it will ever have, but with relatively few connections among them, just enough to sustain essential body functions, as well as some elementary motor activities such as sucking, crying, symmetrical limb movements, and, very soon, smiling, this unique and uniquely endearing attribute of the human young. At this stage, the cortex is still a relatively sparse forest of thinly connected axons and dendrites. But it is the site of feverish activity, which continues for many years, through infancy, childhood, adolescence, and early adulthood, though at a gradually decreasing pace as more and more connections are sealed.

An observer watching this activity would be struck by its mixture of apparent incoherence and purposefulness. Billions of axons and dendrites sprout and retract buds in every direction, as though sniffing their way along invisible trails, until some ramification finds what appears to be the right signal and moves farther on to build progressively spreading arbors. Many contacts are made in the course of these wanderings, some of them transient, others turning into synapses and forming a slowly thickening network of interconnections. The great Spanish neuroanatomist Santiago Ramon y Cajal (1852–1934) has left arresting drawings of the growth of this luxurious jungle, which may well have served as inspiration to the Catalonian painter Joan Miró.

The development of neuronal arborizations is a remarkable mixture of deterministic and stochastic events. The deterministic aspect is revealed by the fact that different body parts are projected on the sensory and motor areas of the cortex in the form of "maps" that have the same disposition in all human beings and are obviously genetically determined. The visual area, for example, is divided into alternating stripes—slabs of cortical tissue, to be more precise—receiving impulses from a single eye. Imagine a zebra pattern, with all the black stripes connected to the right eye and the white stripes to the left eye. This much we all have in common. The exact delimitation of the stripes, however, is variable—just as no two zebras have exactly the same skin pattern—and it is influenced by circumstances that have nothing to do with genes. For example, if the lids of one eye are sewn together at birth, the stripes of the covered eye remain atrophic and those of the functioning eye are correspondingly widened, as was shown in the 1970s in experiments on cats by the American Nobelists David Hubel (born in Canada) and Torsten Wiesel (born in Sweden).[7]

This is the open-ended part of brain development. Neurons must fire regularly to establish synaptic connections with other neurons or, rather, to maintain such connections. The remarkable feature of neuronal development is that neurons start by making an extravagant number of loose connections and then progressively strengthen the connections that are used and disjoin those that are not. This all-important phenomenon has been called selective stabilization by the French biologist Jean-Pierre Changeux,[8] neural Darwinism by the American biochemist-turned-neuroscientist Gerald Edelman,[9] and, with his special brand of irony, neural Edelmanism by Francis Crick,[10] who has also shifted his interest to the neurosciences in recent years. Changeux also refers to Darwin in this sentence: "The Darwinism of the synapses replaces the Darwinism of the genes."[11]

To understand this reference, consider the mechanism of natural selection according to Darwin. It proceeds in three steps: variation, screening, and amplification. First, a large number of variants are produced by random mutations. Next, the variants are screened on the basis of their fitness in a given environment. Finally, the screened variants are amplified thanks to their higher proliferative rate. Edelman (who first gained fame—and a Nobel Prize—for unraveling the complete structure of an antibody molecule) cites as another example of such a mechanism clonal selection, the process whereby competent antibody-making cells develop in response to a microbial or viral aggression (see chapter 14). In the immune system, variation is created by the genetic rearrangements that give rise to a large repertoire of cells programmed to make different antibodies. Antigens then do the screening by binding to the cells that display the corresponding antibody on their surface. This binding triggers cellular multiplication, thus providing for amplification. A similar triad operates in brain wiring, according to Edelman. Variation is introduced in the form of a large number of more or less random neuronal connections. Screening is done by usage, which leads to amplification by reinforcement of often-used connections into permanent synapses. The analogy begs a little in this third step, which does not involve selective proliferation. The message is, however, the same with all three mechanisms, which have in common the achievement of an adaptive reaction without design, instruction, or intentionality.

The knowledge that has been gained recently on the development of the brain is of paramount importance and should be drummed into the ears of every prospective parent. The way you treat your babies literally fashions their brains. If you want them to develop a rich neuronal network, the prerequisite of a rich personality, you must speak to them from the day they are born, sing to them, cuddle them, attract their visual attention, give them colored shapes to play with; in short, you must provide them with a wealth of sensory inputs so as to help them build the innumerable alternative neuronal circuits that support unfolding mental life. A child that has been deprived of such stimulations will remain forever stunted in psychic development, as many case studies attest. On the other hand, the remarkable life history of Helen Keller,[12] who became blind and deaf at the age of two—she already had an uncommonly developed personality by then—and was taught language by an extraordinarily devoted and persistent governess who had access to the child's brain exclusively by touch, demonstrates the fantastic plasticity of the developing brain, its ability to wire itself with the help of whatever impulses it has available. Nevertheless, despite tremendous efforts by dedicated investigators, no baby chimp has ever been taught to "speak" anything beyond the most rudimentary sign language.

What we *can* do is in our genes. What we *do* with our potentialities depends on our environment, especially during the crucial years of early childhood. Such is the lesson of modern neurobiology concerning the famed nature/nurture controversy.

ON THE IMPORTANCE OF BEING RETARDED

In everyday language, the word *retarded* means "backward." Yet it may well be the secret of our success. Popular books make much of the fact that we share more than 98 percent of our genes with chimpanzees. This is actually an underestimate. The correct figure is closer to 99.9 percent. The techniques whereby the figure of 98 percent was arrived at measure all the molecular differences between human and chimp DNA. Most of these differences are present in noncoding regions of the DNA or affect coding regions in a manner that does not significantly alter the properties of the gene product. In terms of actually expressed DNA, we are virtually identical with chimpanzees. Not quite, of course. But the genetic difference is very small. It could not be otherwise in view of the very short time it took our last simian ancestor to become human. What kind of slight change in genotype could have exerted such a profound influence on the product of its expression, the phenotype? The answer, most likely, is retardation, neoteny in scientific terms.[13]

A striking difference between humans and the higher primates is in the timing of developmental events. Every stage in life, from a baby's first steps to old age and death, is reached later in humans than in hominoid apes. The only human event that is only slightly retarded is birth. The gestation period is forty weeks in humans, as compared to thirty-four in chimpanzees, thirty-seven in gorillas, and thirty-nine in orangutans. The reason for this exception is presumably that birth at a later stage would have been anatomically impossible. The human newborn is really a premature baby, forced to be delivered early because of the size of its head. Later delivery would have required major changes in the female skeletal anatomy, which would have been impossible in such a short time. The hazards of premature birth were assessed by natural selection as a risk worth running for the benefits of a larger brain. No newborn animal is as helpless as the human newborn. Just watch a young colt. It stands on its legs, gawkily but successfully, almost immediately after birth. It takes a baby eight to nine months to begin creeping on all fours and almost twice as long to walk unaided.

What progressively converted a chimplike hominoid into a human over a period of some six million years was a gradual slowing down of the developmental clock that allowed—this is the essential element—the brain to pursue its development further. This increase in brain size, at an average rate of about 160 grams (5.7 ounces) per million years, is clearly evident from the various hominid skulls that have been found. Simply to have a larger brain is not enough, however. What counts is how the brain is used, or can be used.

Our Neanderthal cousins had brains the same size as ours, perhaps even bigger. Yet they achieved only a rudimentary culture. Even our direct ancestors did not start developing a refined culture until some 40,000 years ago, the time that Diamond has named the Great Leap Forward. The size of the human brain has hardly changed during that short period. What made the difference, according to most authors, was speech. If Lieberman's theory, referred to in the preceding chapter, is correct, the Neanderthals lacked the anatomical requisites of diversified discourse and communicated only by grunts and gestures. On the other hand, we have seen that the key anatomical change needed for speech—the descent of the larynx—may have occurred in our ancestors as early as 200,000 years ago. But this does not mean that Eve's progeny turned into a Caruso or Callas overnight. It may have taken all of 160,000 years for the vocal tract to reach its present conformation and, especially, for the speech centers to undergo the appropriate organization in the brain. Only then could a true language develop and, with it, the means of creating culture and constructing civilization. Diamond sums it up as follows: "Until the

Great Leap Forward, human culture had developed at a snail's pace over millions of years. That pace was dictated by the slow rate of genetic change. After the Leap, cultural development no longer depended on genetic change. Despite negligible changes in our anatomy, there has been far more cultural evolution in the past 40,000 years than in the millions of years before."[14] For speech to be possible, however, consciousness, the most mysterious of human qualities, had to emerge.

Chapter 27

The Workings of the Mind

THERE IS NO GREATER mystery in the known universe, except the universe itself, than the human mind. Born from the brain, critically dependent on the brain at every instant, crippled together with the brain in all sorts of weird fashions directly related to whatever brain area is maimed, the mind is without any doubt a product of polyneuronal functioning.

At the same time, the human mind is, collectively, the creator of the whole of technology, science, art, literature, philosophy, religion, and myth. The mind generates our thoughts, reasonings, intuitions, ponderings, inventions, designs, beliefs, doubts, imaginings, fantasies, desires, intentions, yearnings, frustrations, dreams, and nightmares. It brings up evocations of our past and it shapes plans for our future; it weighs, decides, and commands. It is the seat of consciousness, self-awareness, and personhood, the holder of freedom and moral responsibility, the judge of good and bad, the inventor and agent of virtue and sin. It is the focus of all our feelings, emotions, and sensations, of pleasure and pain, love and hate, rapture and despair. The mind is the interface between what we are wont to call the world of matter and the world of spirit. The mind is our window to truth, beauty, charity, and love, to existential mystery, the awareness of death, the poignancy of the human condition.

BRAIN AND MIND

There is no mind without brain, but much of the brain functions without mind. Consciousness is the tip of an iceberg. It emerges through the cortical layers of the brain above a vast and intricate network of highly active but unconscious centers and connections. Our nervous system operates to a large extent without our being aware of what is going on or our being able in any way to alter the chain of events. We are not conscious of, and have no control over, the myriad impulses that continually regulate our beating heart, the diameter of our blood vessels, the peristaltic movements of our intestine, the secretion of our glands, and our many other body functions. Even our conscious movements depend on largely unconscious nervous operations. When we grasp an object or take a brisk walk, we are totally unaware of the complex instructions that are being dispatched continually to many of our muscles to either contract or relax, of the equally complex signals whereby our eyes, muscles, and other parts of our body send information to our brain, and of the intricate interplay between sensory and motor impulses whereby coordination is achieved. Furthermore, many activities that require conscious attention when we learn them, like riding a bicycle or driving a car, become increasingly automatic and unconscious as we become more skilled.

This latter fact recalls another key feature, already mentioned in the preceding chapter: The brain wires a good part of itself in the course of learning. Interestingly, the conscious precedes the unconscious in this process. When I first learned to play the piano, I had to make excruciating efforts to get my fingers to do even the simplest of movements. Later, I could run scales or play small pieces without giving the matter any conscious attention, simply letting the program of coordinated impulses stored in my brain unfold. When I labored through new music, these automated schemata combined with conscious perceptions to guide my fingers through the unfamiliar score, in an interplay of stored and incoming informations. In skilled musicians, this interplay reaches an extraordinary degree of rapidity and efficacy. Such consciousness-directed structuring of networks that eventually become unconscious when "hard-wired" is typical of any sort of learning.

It is in this particular area of neuronal integration that modern brain research has made some of its most incisive advances. The visual system has been studied most extensively.[1] When we look at an object, we simultaneously perceive a number of elements of form, color, and motion that are somehow integrated into a unified, coherent image. It has been found that each element is perceived by a special subset of retinal cells and that each such subset projects its information on a separate group of neurons, or map, in the visual cortex. As many as thirty-two distinct areas have been identified in the visual cortex of the macaque monkey. A crucial problem, known as the "binding" problem, is how these multiple representations are correlated and integrated into a single image. The answer lies in what Edelman[2] has called reentry, a crisscross of impulses exchanged "horizontally" among the different areas within the visual cortex. What we "see" emerges from these interactions by a process akin to resonance—the physical matching of vibration frequencies that can cause a crystal glass to shatter upon a single musical note, or a bridge to collapse under a troop marching in step—that brings the activities in all the areas in phase. The product of this process is itself connected to an equally complex network of self-integrating motor impulses that direct our eye muscles when we focus on a given part of the image or track a moving object.

EMERGENCE OF THE MIND

The human mind is a product of evolution. This evolutionary process did not start with the ape-to-human transition; it was only completed in the course of this transition. This obvious truth long remained stifled in the stranglehold of the American behaviorist school of John Watson and B. F. Skinner,[3] who adamantly excluded any introspective or analogic approach from psychological investigation. Even the human mind, let alone any analogous attribute of animals, had to be viewed as a black box, with only entries and exits accessible to the observer. This view has come under increasing challenge from modern ethologists. Prominent among them is the American Donald Griffin,[4] formerly of the Rockefeller University in New York, now an associate of the Museum of Comparative Zoology at Harvard, renowned for his discovery that bats direct themselves by echolocation. In later years, Griffin has become the champion of animal "awareness," "thinking," and "minds," key words in the titles of the three books he has published on the subject in 1976, 1984, and 1992. In the face of often strong opposition on the part of mainstream animal ethologists, he has claimed consciousness and intentionality not only for the higher mammals but also for birds, fish, and even ants and honeybees. Not all researchers are ready to follow Griffin that far, but most now accept—what you no doubt knew all along—that your dog has feelings when he looks at you with "sorrowful" eyes, or that the chimpanzee, when it picks a small branch, strips it of twigs and leaves, pokes it into a termite burrow, pulls it out again, and licks clinging insects off it with evident relish, "knows" what it is doing and why, and actually has planned the whole thing with the purpose of regaling itself with a delicacy.[5]

Admittedly, such beliefs do not rest on scientifically established facts. Strictly speaking, we know only of our own consciousness and surmise it in others by analogy. Conversing with you, I may reasonably become convinced that something analogous to my consciousness exists in your brain, although your experiences are inaccessible to me. It is not too difficult to extend this kind of supposition to a dog or a chimpanzee, which are sufficiently close to us. Reading animal minds becomes increasingly hazardous, however, the wider the gulf between the animals and us. Whatever passes for a mind in the brain of an octopus or a honeybee must be very rudimentary and far removed from our own inner experiences.

We cannot know the details, and perhaps never shall. But we may take it as likely that animal awareness developed in successive stages as the structure of the brain increased in complexity. The final stages, from ape to human, were accomplished at an astonishing pace, driven by the advantages of strengthening social bonds and improved communication, culminating in the acquisition of language. This was a very circuitous evolutionary process. There are no genes for language. What the genes do is to prescribe certain rules that guide and limit the self-wiring of the brain in the course of development. A favorable mutation is one that alters the rules in such a way that additional bits of information can be wired in. Greater survival capacity and reproductive success are benefits of this modification and contribute to the spreading of the modified gene.

The late German-born physicist Max Delbrück, who worked at one time with the great Danish physicist Niels Bohr and later emigrated to the United States to become the acknowledged father of modern molecular biology, has drawn fascinating conclusions from the fact that the human mind is a product of evolution.[6] The way we view the world, he has pointed out, including our awareness of space, time, and matter, has been shaped by strictly utilitarian factors affecting survival and proliferation; it may have little to do with reality. Common sense and intuition may be very misleading in this respect. Only the exploration of nature by science has begun to provide us with glimpses of true reality, with its strange and bewildering concepts of quantized energy, wave-particle dichotomy and complementarity, probabilistic mechanics, indeterminacy, relativity, perhaps even logical inconsistency. Consciousness may be the ultimate frontier in this exploration.

The mind is not only the product of an evolutionary history; it is also the product of an epigenetic history, and it goes on molding itself throughout an individual's lifetime. What was said above about the wiring of the brain is crucially relevant to this history. So is memory, this remarkable ability of our brain to store information in a manner that allows the mind to retrieve it more or less at will. Indeed, continuity in time is a key property of the brain operations that generate the mind; it is essential to our feeling of being the same person while time flows past us.[7]

We thus see the mind as a late-emerging quality correlated with cortical expansion, arising from a brain that has been molded genetically by evolution and epigenetically by development into increasingly complex anatomical structures performing increasingly complex operations. But what *is* the mind? What is the nature of consciousness, of the inner self?

CONSCIOUSNESS EXPLAINED?

In the heading to this section, only the question mark is mine. The heading itself, in its confidently assertive form, is the title of a book published in 1991 by the American philosopher Daniel Dennett.[8] Not all readers of the book are likely to share its author's assurance. Some may feel instead that the author does not really explain consciousness but, rather, explains it away—by pulling it to pieces to the point that nothing remains to be explained. Consciousness does not exist. Dennett does not say so explicitly, but that is what his conclusion amounts to.

Dennett is only one among many contemporary philosophers who raise doubts about the reality of our mental experiences. The philosopher Patricia Smith Churchland[9] challenges the validity of what she calls "folk psychology," defined as "common sense psychology—the psychological lore in virtue of which we explain behavior as the outcome of beliefs, desires, perceptions, expectations, goals, sensations, and so forth." After comparing folk psychology to "folk physics," or "intuitive physics," that is, physics before Galileo, Newton, and Einstein, she reaches the conclusion that "folk psychology may be . . . dead wrong."

I would hesitate, as a philosophical illiterate, to question what are obviously the fruits of erudite disquisition. But I am encouraged to do so by another American philosopher, John Searle, who asks, citing Dennett, Churchland, and a host of others: "How is it that so many philosophers and cognitive scientists can say so many things that, to me at least, seem obviously false?"[10]

Readers not familiar with mainstream philosophy will no doubt be surprised to learn from Searle that "in the philosophy of mind, obvious facts about the mental, such as that we all really do have subjective conscious mental states and that these are not eliminable in favor of anything else, are routinely denied by many, perhaps most, of the advanced thinkers in the subject."[11] Searle then sets out to describe, dissect, and ultimately dismantle a number of contemporary systems, such as eliminative materialism, functionalism, physicalism, cognitivism, epiphenomenalism, "and so, depressingly, on," justifying Edelman's pithy characterization of philosophy as "a graveyard of Isms,"[12] or Crick's comment that "philosophers have had such a poor record over the last two thousand years that they would do better to show a little modesty rather than the lofty superiority they usually display."[13]

These few quotations should make it clear that mind research is still in an embryonic stage. This is not for want of study. Dozens of books have appeared on the subject in recent years, written by neuroscientists, psychologists, ethologists, anthropologists, sociologists, cognitive scientists, linguists, computer specialists, and philosophers, not counting theologians. Unfortunately, almost every one defends a different thesis, in good part because ideology plays a more important role in human psychology than in other scientific domains. I can hardly do justice to this mass of information and disputation, much of which is highly technical, but I shall abstract a few main themes.

THE RISE AND FALL OF DUALISM

An elegant solution to the mind-brain problem was proffered in the first half of the seventeenth century by the French physicist, mathematician, and philosopher René Descartes.[14] After cautiously proposing as a purely hypothetical concept—he did not want to share the fate of Galileo—that the human body be viewed as a machine functioning entirely according to the laws of physics, Descartes resolved the mind-body problem by postulating an independent soul interacting with the body by way of the pineal gland (which is situated in the middle of the brain—to Descartes a most revealing anatomical feature). Physiology has long dethroned the pineal gland as the abode of the soul, but Cartesian dualism has survived longer as an intellectually and emotionally satisfying way of reconciling the spiritual and material aspects of human nature. However, the dominant view among scientists today is that dualism violates the laws of energy conservation by assuming that a nonmaterial object could influence the behavior of material systems. A prominent exception is the Australian neurobiologist and Nobel laureate John Eccles, who has consistently and forcefully defended a dualistic conception of the mind-body relationship, partly in collaboration with Karl Popper, perhaps the most influential philosopher of science of our times.[15]

The views of these two venerable nonagenarians are generally dismissed with expressions of amused tolerance, if not derision, by the younger generation of neurobiologists and philosophers. Most workers advocate the "monist," or "materialist," view of the mind-brain interaction first proposed one century after Descartes by another French philosopher, Julien Offroy de La Mettrie (1709–1751), in a book uncompromisingly titled *L'Homme-Machine.* Another figure of the Enlightenment, the physician Pierre Jean Georges Cabanis[16] (1757–1808), summed up this view in a famous aphorism: "Le cerveau sécrète la pensée comme le foie sécrète la bile"—"The brain secretes thought the way the liver secretes bile"—a saying sometimes attributed to the Dutch physiologist Jakob Moleschott[17] (1822–1893) as: "The brain secretes thought as the kidney secretes urine."

Modern monism comes in many different forms. I have already referred to the most radical form, which goes so far as to deny the reality of subjective experiences. A milder form of monism accepts consciousness but views it as an irrelevant epiphenomenon, something that emerges from neuronal activity in the brain cortex but has no control over this activity. According to this view, our neurons do the whole job, including those parts, such as choosing, willing, and deciding, that we instinctively attribute to something we call our inner self and endow with the qualities of free will and responsibility. Our feeling of being in command is an illusion. We are mere spectators, provided with a tiny window through which we watch a minute fraction of what our brain is doing for us. Our watching makes no difference to the final outcome. We no more decide to run when setting a foot forward than we decide to accelerate the beating of our heart when running. The difference is simply that we are given a glimpse into the neuronal interactions that initiate running for us, whereas we are denied a similar glimpse into the mechanisms that cause our heart to beat faster.

The phrase "the difference is simply" hardly makes the difference simple. What is it that distinguishes a conscious from an unconscious polyneural event? We are offered no answer to that question, except the purely phenomenological statement, which is no answer at all, that neuronal events become conscious when they exceed a certain threshold of complexity or are organized according to wiring patterns peculiar to the brain cortex. In addition, we are still left with the problem of identifying the watcher, the so-called homunculus, the little man inside who looks at the screen in the "Cartesian theater," as Dennett[18] puts it, the "ghost in the machine," an expression derisively coined in 1949 by the materialist philosopher Gilbert

Ryle[19] and defiantly adopted as the title of a 1967 book by the controversial writer Arthur Koestler.[20] It is to avoid this difficulty that hard-core monists prefer to do without consciousness altogether. Dennett, for example, compares subjective experiences to a Russian-doll system of watchers-within-watchers that finally reduces to no watchers at all. The homunculus is made to vanish by the magic of infinite regress.

A theory that takes subjectivity one step beyond epiphenomenalism, while remaining true to orthodox, monist materialism, is the "central state" or "identity" theory, a favorite of many experts today. According to this theory, neuronal events and mental events are two aspects of one and the same thing. The self is not a watcher, it is a participant. Thoughts and sensations are undissociable facets of certain polyneuronal events taking place in the brain cortex. If, for example, I "debate" in my mind whether to shake hands with my enemy or to let him have it on the chin, different neuronal circuits in my brain play out different scenarios, each accompanied by conscious representations of their rationally deducible consequences and emotional concomitants. If I eventually settle on the pacific gesture, it is because the handshaking scenario has emerged as the most compelling inducer of motor behavior, linked with pleasurable feelings of relief, warmth, magnanimity, perhaps smugness and "holier than thou" complacency.

To explain this process, Darwin is once again invoked, though in a guise he might not have recognized. The different networks are seen as competing with each other, and the winning combination as selected according to its adaptation to the local (neuronal-mental) environment. The previous history of the brain plays a key role in this selection process. A Christian upbringing, by reinforcing networks conducive to "turning the other cheek," may confer a selective edge on forgiveness. Early exposure to street gangs, on the other hand, with its deeply etched patterns of violence, self-defense, and revenge, could give preeminence to the retaliation scenario.

An analogous Darwinian explanation is said to account for problem solving. Different neuronal networks display different conceptual scenarios that compete with each other within the ecological context of wired-in networks of acquired data, thought patterns, inclinations, prejudices, and other products of previous experiences. The winner in this competition, if there is a winner, is the adopted solution, its selection arising from the blending of many circuits by a sort of unifying resonance process similar to the unconscious mechanism whereby a coherent image arises from the integration of many distinct visual inputs. An intense feeling of pleasure—the flash of illumination, the thrill of discovery—appears as the subjective concomitant of this falling-in-phase of many circuits—something "clicking" in one's brain—whereas the preceding competitive stage, the often long-drawn succession of mismatchings that corresponds to "racking one's brain," is accompanied by agonizing torment. Hence, the delight often goes together with a sense of release. However, satisfaction is no guarantee that the solution is correct. Countless errors gestated in laborious suffering have sprung up in a gush of cathartic delight.

This reasoning can be extended to other enjoyable experiences, linking pleasure in each case to resonating neuronal networks and equating the intensity of exhilaration with the fullness and richness of the resonance phenomenon. Not only intellectual achievement, but also artistic emotion, musical delectation, religious fervor, and mystic ecstasy, could similarly arise from resonating neuronal circuits. So too could the strong feelings induced in us by passionate love, sexual climax, or hallucinogenic drugs. "Vibes" could be more than a colorful metaphor.

A distinctive feature of identity theory is that it implicates subjective experiences directly in the original wiring of the brain. When sensory inputs, including heard or written words, enter our brain for the first time, they do so laden with their emotional and conceptual significance and they are stored as such in our memory, so that summoning of the corresponding neuronal patterns automatically reawakens the sensations and concepts. The self is thus progressively molded in its indissociable physical and mental aspects by the barrage of inputs to which the brain is subjected from the moment of birth onward, or perhaps even earlier. From instruction to brainwashing, all forms of education, training, conditioning, programming, or imprinting depend on such simultaneous processing and recording of physical inputs and their mental correlates. Brain and mind are indissolubly wired together.

A major proponent of monism is Edelman, in the words of Dennett,[21] "one theorist who has tried to put it all together, from the details of neuroanatomy to cognitive psychology to computational models to the most abstruse philosophical controversies." I do not join Dennett in his conclusion: "The result is an instructive failure," but I must confess to my inability to follow Edelman all the way through, in spite of long, friendly discussions.

Edelman's central argument is that any acceptable theory of the mind must be compatible with and, preferably, derived from what is known of the brain processes from which consciousness arises.[22] Proceeding from this indisputable premise, which is ignored by many philosophers and psychologists who rely exclusively on introspection, Edelman takes as the starting point for his reflections his insights into the neuronal mechanisms underlying such key brain functions as perception, memory, and speech. In several books, he has explained in detail, but with few concessions to the nonspecialist, how he sees consciousness emerging from those functions, which are themselves the expression of a morphological organization historically produced by Darwinian evolution and epigenetic wiring. It is an impressive construction, which, however, addresses mainly the neuronal mechanisms underlying consciousness, not the nature of consciousness itself. Edelman does not discuss—deliberately, as he states in a note—"the classical disputes and categorizations related to the mind-brain problem and its surrounding philosophical hypotheses,"[23] except for expressing his unambiguous rejection of Cartesian dualism and his adhesion to monism-materialism. He tells us that "thinking about brain function in selectional terms relieves us of the horror of the homunculus"[24] and, like Dennett, disposes of this horror by endless regression. Also revealing is the following statement: "Mental properties cannot vary in the absence of variance in brain states, and thus are supervenient properties."[25] This sounds like epiphenomenalism, although it could be interpreted as compatible also with identity theory.

Crick is another molecular biologist who has directed his later interests to neurobiology. He has summed up his views in *The Astonishing Hypothesis,* a book largely devoted to contemporary research on vision. What Crick calls the "astonishing hypothesis" is an uncompromising adherence to monism: Neurons do it all, or, as Crick puts it in the mouth of Lewis Carroll's Alice, "You are nothing but a pack of neurons."[26] Consciousness is nothing but a correlate of certain polyneuronal activities.

Like Edelman, however, Crick takes a strictly empirical approach to the problem and refuses to speculate beyond the limits of scientific evidence. He acknowledges in the end that his book "has very little to do with the human soul,"[27] a somewhat unexpected conclusion to a book subtitled "the scientific search for the soul."

MIND POWER

All forms of monism have in common that they see consciousness only as a property arising from, or associated with, neuronal events, but do not grant consciousness the power to influence such events in turn. Even Searle, whom we have encountered as a major critic of modern materialism, shares this opinion. He also warns us against the "fallacy of the homunculus"[28] and defines consciousness as "an emergent feature of certain systems of neurons" that cannot itself influence the behavior of the neurons. As he writes: "The naive idea here is that consciousness gets squirted out by the behavior of the neurons in the brain, but once it has been squirted out, it then has a life of its own."[29]

Standing out by subscribing to this naive idea, while claiming to be a monist, is Roger Sperry, the late American neurobiologist who gained a Nobel Prize for his work on the functional specializations of the left and right cerebral hemispheres. Sperry has explained how his "long-trusted materialist logic was first shaken" in 1964, when he reached the conclusion "that emergent mental powers must logically exert downward causal control over electrophysiological events in brain activity."[30] In developing this notion, Sperry strongly insists on its monistic character. "I define this position," he writes, "and the mind-brain theory on which it is based as monistic and see it as a major deterrent to dualism."[31] However, this statement is preceded by the declaration: "In calling myself a mentalist, I hold subjective mental phenomena to be primary, causally potent realities as they are experienced subjectively, different from, more than, and not reducible to their physicochemical elements." This seems to me perilously close to dualism, which is what Eccles is said to have told Sperry.

The most determined attack against epiphenomenalism has been made by the

Oxford mathematician Roger Penrose, who, at the end of a wide survey of contemporary mathematics and theoretical physics, including algorithmic computation, the nature of mathematical truth, quantum mechanics, and relativity theory, concludes with an analysis of the mind-body problem. He distinguishes between the passive aspect of this problem—"How is it that a material object (a brain) can *evoke* a consciousness?"[32]—and its active aspect—"How is it that a consciousness, by the action of its will, actually *influence* the (apparently physically determined) motion of material objects?"[33]

To Penrose, there can be no doubt that consciousness does something. If it didn't, it had no selective value. Why, then, "did Nature go to the trouble to evolve *conscious* brains when non-sentient 'automaton' brains like cerebella would seem to have done just as well?"[34] Penrose finds in quantum mechanics a possible solution to the problem, in that "different alternatives at the quantum level are allowed to coexist in linear superposition."[35] He speculates that "the action of conscious thinking is very much tied up with the resolving out of alternatives that were previously in linear superposition."[36]

THE SELF AND FREE WILL

The question of active consciousness is intimately linked to the problem of free will. If, as claimed by all monist theories, neuronal events in the brain determine behavior, irrespective of whether they are conscious or unconscious, it is hard to find room for free will. But if free will does not exist, there can be no responsibility, and the structure of human societies must be revised. Very few among even the most uncompromising materialists are willing to drive this argument to its logical conclusions. They display remarkable inventiveness in trying to escape the snare.

Some take refuge in any loophole offered by quantum mechanics, uncertainty rules, chaos theory, probabilistic fluctuations, microscopic heterogeneity, or other forms of physical indeterminacy to loosen the shackles of neuronal determinism. Others invoke unpredictability. However, neither indeterminacy nor unpredictability accounts for the exercise of choice, design, or volition.

Some thinkers assert the lack of free will but recommend that we ignore this disturbing fact. Consider the following declaration by the German philosopher Bernhard Rensch,[37] a defender of the identity theory of brain function. In *Biophilosophy,* Rensch states that "the weight of evidence is far greater *against* freedom of the will than *for* it" (the italics are his).[38] He acknowledges that "to deny that the will is free involves grave consequences for our ethical, religious, and juristic ideas,"[39] and ends by stating that the important point is not that we should be free, but that we should behave as though we were: "The concept of freedom is itself an important determinant in our thought."[40] To me, this sounds much like saying that truth is important as long as we don't know or believe it.

Some, finally, face this truth unflinchingly and then, by intellectual legerdemain, argue that knowing we are not free actually increases our freedom—a curious application of the biblical promise "And you shall know the truth, and the truth shall make you free."[41]

We are told that our perception of personal freedom, like our concepts of space, time, and matter, is ingrained in our nature as a result of its evolutionary, adaptive value and could be equally deceptive. In order to function, human society must set rules obeyed by its members. Hence we, the members of such a society, emerged with the feeling of responsibility and freedom needed for society to function. We believe we are free for the same reason we believe matter to be solid and subject to the rules of causality, simply because such beliefs have helped our evolutionary success. They need in no way be true.

Those of us who have difficulty seeing ourselves as automatons, tricked by evolution into illusions of free will, are given stern warnings. We are reminded that a theory is not wrong simply because we don't understand it. Most of us don't understand the theory of relativity. This does not make the theory wrong. Many of us may similarly be baffled by the reasonings whereby some experts reconcile their objective reconstructions of the human mind with their subjective experiences. This may be our problem, not theirs. The second admonition, obviously true, is that a theory is not wrong simply because it conflicts with our inner convictions. Human history is replete with such misconceptions. We must learn to distrust those false certainties that our mind weaves for us when confronted with the inexplicable. As Patricia Churchland has pointed out, "folk psychology" could well go the way of "folk physics."[42]

Such arguments are difficult to refute. But they still leave unanswered the problem of *what* or *who* experiences the illusory feeling of freedom. Those who deny consciousness, or dismiss it as an irrelevant epiphenomenon or impotent participant, are enmeshed in contradictions when they speak and behave as though they are free to make decisions. They might reply that they use short cuts in order to avoid such tiresome periphrases as: "My brain neurons, molded by a combination of complex evolutionary and developmental factors, respond to the sum of external and internal impulses to which they are exposed by guiding my hand to write," in place of "I believe." They must find it hard to sustain this attitude in everyday life.

Does this mean that we must go back to Cartesian dualism? This is Eccles's opinion, and he offers an explanation, based on the uncertainty principle, of how the immaterial mind could, by affecting a probability field, modify without the expenditure of energy the likelihood of occurrence of a "microsite" synaptic event sufficient to influence neuronal firing.[43] He identifies this event as the discharge of a single neurotransmitter-containing synaptic vesicle. To most cell biologists, however, this explanation must appear totally unrealistic in terms of what is known of the mechanism of vesicular discharge. As a last stand, the microsite hypothesis can only herald the defeat of Cartesian dualism.

The philosopher Popper, another dualist, or "interactionist," does not bother about precise mechanisms and is content with pointing out that "the brain is an open system of open systems."[44] He and Eccles, in fact, join a number of monists in calling on some form of physical indeterminacy to allow consciousness to slip in without expenditure of energy; they differ from monists in that they see consciousness as a separate entity.

A problem common to dualism and to most forms of monism is that they rest on a preconceived definition of matter. As pointed out by Searle, those who refuse to face the facts of consciousness on the grounds of monism-materialism are led astray by an unacknowledged, residual dualism that opposes mental to material. They are caught in the snares of their own a priori definition of matter as something that excludes subjective mental experiences. "Materialism," he writes, "is thus in a sense the finest flower of dualism."[45]

On the subject of free will, however, Searle is evasive. He mentions free will only in passing, with the parenthetic comment "if there is such a thing,"[46] and argues that intentionality must be interpreted in the same way as we now interpret teleology in biology. It was common in the old days to see the heart as made for pumping blood through our arteries. Now we say that blood circulates through our arteries because the heart is constructed as it is, and we explain the apparently goal-directed structure of the heart as the product of natural selection. Those who happened to have a poorly functioning heart were eliminated. In the same vein, our visual neurons do not behave in a certain way in order to allow us to see things clearly. We see things clearly because our neurons were selected to behave the way they do. This "inversion of explanation"[47] conforms to perfect Darwinian orthodoxy and is unexceptionable. But I don't see in it an explanation of how I turn my eyes in one direction or in another as a result of what I experience as a free decision on my part.

Before concluding this cursory survey of what is said to go on in our brains when we feel or think, we must take a look at what this activity has produced. If we wish to understand the mind as a machine, we must consider also what the machine has wrought, that is, human culture.

Chapter 28

The Works of the Mind

HUMAN CULTURE is a product, collectively and cooperatively, of human minds. Rudiments of a culture exist in certain animal behaviors that are transmitted to a greater or lesser extent by imitative learning. Anathema to the behaviorist school, such a possibility is now accepted by many modern ethologists. Bird song, as studied, for example, by the American Peter Marler[1] from the Rockefeller University in New York, is only partly instinctive and depends, in part, on imitation. Birds are also involved in an amusing case recorded in England[2] in the 1930s, when some tits discovered the trick of pulling the tin-foil caps off milk bottles (traditionally left on doorsteps) and drinking the cream. This habit spread through a large part of the country, most likely by way of imitative learning. Among mammals, most prominently the primates, teaching plays an important role in the transmission, from generation to generation, of such behavioral activities as hunting, shelter building, using objects as tools, socializing, and communication, adding up to a sort of species "lore" that complements straightforward genetic transmission. With hominid evolution, this kind of cultural heredity became progressively more important as the means of communication improved, reaching a dramatic acceleration and expansion with the emergence of language and, especially, writing. Today, any cultural acquisition can immediately be stored for possible worldwide dissemination and retrieval.

CULTURAL EVOLUTION

Cultural evolution is very different from Darwinian evolution. It is much faster and resembles more the kind of evolution postulated at the beginning of the last century by the French naturalist Jean-Baptiste de Monet, Chevalier de Lamarck (1744–1829). One of the first evolutionists, Lamarck is known for his advocacy of the theory of the heredity of acquired characters. We no longer believe, with Lamarck, that giraffes acquired a long neck as a result of generations of giraffes stretching to reach higher leaves on a tree. Instead, we accept, with Darwin, that giraffes genetically endowed with a longer neck were able to survive better and produce more progeny because they had the advantage of reaching higher leaves. On the other hand, the ability to light fires, for example, which represented a major selective asset for humanoids who possessed it, was not transmitted by genes but by communication.

Although very different, cultural and biological evolution are not unrelated. Inventions were made by individuals who may have owed their talent to a particular combination of genes. On the other hand, once an invention was made—lighting a fire, for example, or sharpening a stone—it gave a selective advantage to those better endowed genetically with the appropriate skills. This subtle reciprocal interplay between the two kinds of evolution has been emphasized by many anthropologists and evolutionists. Even abstract concepts are subject to a Darwinian form of evolution. Beliefs, for example, compete on the strength of their ability to give those who hold them a reproductive advantage.

In the recent history of humankind, the Lamarckian process of cultural evolution has been overwhelmingly more important than the Darwinian process of biological evolution, largely because cultural acquisitions spread much faster than genetic modifications. Our genes and, therefore, our innate potentialities are hardly different from those of a Cro-Magnon human living 15,000 years ago. With luck—a winning number in the sexual lottery and a favorable family and social environment—a Cro-Magnon newborn transplanted into our century could give rise to an Einstein or a Picasso. In his day, instead, that individual would, perhaps, have become the inventor of a new tool or an avant-garde cave painter. The difference created by the 15,000-year interval is almost exclusively a matter of cultural heritage; it lies in the accumulated knowledge, technology, art, beliefs, mores, and traditions acquired and transmitted by the six-hundred-odd generations that have succeeded each other in the interval.

CULTURE AND MIND

Culture is produced and assimilated by human minds. It is transmitted between minds, either by direct communication or by books, art, and other media. Thus, culture has much to tell us about the human mind. In an illuminating approach to this question, Popper[3] divides reality into three parts, or "Worlds." World 1 is the "universe of physical entities." It comprises the material universe. Living beings, the human brain included, belong to World 1. Popper's World 3 is the world of culture. It encompasses all abstract notions and concepts, scientific theories, technological principles, aesthetic canons, ethical values, religious myths, and other creations of human minds. World 3 joins with World 1 in most humanmade objects, such as tools, machines, houses, clothes, books, disks, tapes, paintings, statues, and other artifacts, which are made of matter but bear a World 3 imprint. World 2 is the interface between Worlds 1 and 3. It is the realm of mind, serving as a two-way transducer between neuronal activities, which belong to World 1, and World 3 entities.

In this context, some contemporary views on the human mind seem ludicrous. The theory of relativity, the *Origin of Species,* the Sistine Chapel ceiling, the *Well-Tempered Clavier,* the *Discourse on Method,* the *Divine Comedy,* all products of irrelevant epiphenomena or of blindly interacting neuronal circuits in the brains of Einstein, Darwin, Michelangelo, Bach, Descartes, and Dante? It sounds unbelievable. Yet the facts are undeniable. Active neuronal circuits in the brain cortex are both necessary and sufficient for cultural creation.

The key to the mystery is found in Popper's World 2, though not, in my opinion, in his and Eccles's dualistic conception of this World as a matter-spirit interface. I would rather heed Searle's admonition that we must purge our definition of matter from its dualistic overtones.[4] If we want to escape from the dualism-monism quandary, we must enlarge our concept of matter to include the human mind with its full potential and access to World 3, instead of trying to shrink our view of the mind so as to make it fit inside our preconceived, "materialistic" definition of matter. We must view the mind as a special manifestation of matter.

If this means invoking a property not yet included in the description of physicists, this would not be the first time. In its history, physics has repeatedly been forced, sometimes reluctantly, to broaden its definition of matter. Gravitation, electromagnetism, relativity, quanta, the strong and weak intra-atomic forces all have had to be added to the Aristotelian concept of substance. One of the latest manifestations to be included in the definition of matter was life. Vitalism—the matter-life dichotomy once embraced as readily by physicists as by biologists—is another form of dualism that had to yield to modern insights. Living organisms are no longer viewed as made of matter "animated" by a (nonmaterial) vital spirit. Yet nobody disputes the fact that life exists as a special manifestation of matter organized in a special fashion. Why could we not think of mind in the same way?

Edelman is not ready to do so. He asks the question, "With such strangeness, why not get a little stranger and propose that additional, as-yet-undiscovered physical fields or dimensions might reveal the true nature of consciousness?" and emphatically rejects this use of physics as "surrogate spook."[5] Crick is less categorical and allows that "radically new concepts may indeed be needed—recall the modifications forced on us by quantum mechanics."[6] Penrose, while calling on quantum mechanics, adopts a resolutely Platonic view—"a point of view not to everybody's taste"[7] is Crick's terse comment—and sees consciousness as establishing direct and instantaneous "contact with Plato's world" of concepts and ideas.[8] This, to Penrose, is the essence of mathematical discovery; it is no more than "seeing" a truth that was always there. Whether mathematical truths are discoveries or inventions of the human mind is a much debated question. It is the object of a fascinating dialogue between two French scientists, the neurobiologist Jean-Pierre Changeux, a hardheaded adherent of the materialistic school, and the mathematician Alain Connes.[9] Changeux defends a relativistic, contingent view of mathematical truth, whereas Connes, like Penrose, believes in a universal mathematics existing independently of mathematicians.

ARTIFICIAL INTELLIGENCE

Before we end this schematic survey of the enormously complex field of brain science, one more question must be asked. Does the human brain operate like a computer? This question arises from the widely publicized achievements of what is known as AI (artificial intelligence), that is, the feats accomplished by increasingly sophisticated computers that not only perform calculations entirely beyond human capacity but are even beginning to beat their makers in other "mental" activities such as learning, translating, adapting to changing circumstances, solving problems, and playing chess. These achievements are impressive. Nevertheless, the answer to the question seems to be no. The human brain does not function like a computer, although computer simulations can be valuable in studies of the human brain.

This is the conclusion arrived at by the Harvard psychologist Howard Gardner in a 1985 book that provides an excellent historical account, sympathetic and critical at the same time, of the new field.[10] Edelman reaches the same verdict, pointing out that the brain's morphological and functional blueprint is entirely different from, and irreducible to, that of a computer. Brains and computers are also incongruent from the philosophical point of view, according to Searle, and from a mathematical point of view, according to Penrose, whose main thesis is that the brain, at least the conscious brain, operates by a nonalgorithmic process. On all accounts, the human brain is much more flexible and open-ended than any existing or imaginable computer, and there is nothing in what we know of the brain's construction that allows the suggestion that it includes a built-in programmer. Even the staunchest dualist would refuse to accept such a suggestion. The "cognitive revolution" has generated many important advances in information theory, symbolic logic, linguistics, and related fields, but it does not warrant some of the hopes conjured by the term "artificial intelligence."

Does this mean that machines operating like parts of the brain could not be constructed? Not necessarily, according to Edelman,[11] who has achieved considerable success with his family of Darwin robots, which are constructed so as to wire themselves by a selective mechanism. Crick also feels that "if we could build machines that had these [the brain's] astonishing characteristics . . . the mysterious aspects of consciousness might disappear."[12] Would these machines appear to possess consciousness? Crick believes that "in the long run this may be possible."[13] So do Edelman and several other proponents of identity theory. If we ever get that far, the problem will be how to find out whether the machine does or does not experience consciousness.

Chapter 29

Values

HUMANS, LIKE MANY of their forebears, are social animals. Their societies are ruled by laws based on a mixture of traditional customs, pragmatic measures, and shared values. How much of this organization is the product of biological evolution, how much of cultural evolution? Science has no definitive answer to this question, but it can contribute certain facts that anyone attempting an answer must take into account.

THE SHAPING OF HUMAN SOCIETIES

In 1975, the Harvard zoologist Edward O. Wilson, an expert in social insects, made headlines with a massive, lavishly illustrated opus challengingly titled *Sociobiology: The New Synthesis*.[1] In this ambitious, scholarly work, Wilson reviewed all forms of animal associations, from the lowest colonial invertebrates, such as corals and polyps, to the most advanced primates, paying special attention in each case to the evolutionary factors that might have favored the selection of certain genetically determined social behaviors.

In his approach, Wilson followed, albeit in more detailed and comprehensive fashion, a general tendency of modern ethology, which interprets animal sociality within the framework of Darwinian theory, guided by the principle that evolution selects any genetic trait that leads to a greater frequency of the relevant gene or genes in the gene pool of the species concerned. A lucid account of this approach is given in *The Selfish Gene,* by Britain's Richard Dawkins.[2] This book, published shortly after Wilson's *Sociobiology,* covers much the same ground, but more briefly and in a language entertainingly accessible to the general public. The key point is that animal societies usually consist of related individuals having many genes in common. Members of such societies can propagate their genes in two ways. They can do so by producing offspring, which faculty is said to account for the selection of such "selfish" traits as aggression, territoriality, male dominance, polygyny, female choosiness, parental care, and other behaviors that enhance an individual's reproductive success. But members of the societies can also help in propagating their genes by behaviors in which their own reproductive success is imperiled, or even forfeited, provided the payoff in terms of benefits for others is sufficient. Several ingenious models explaining the selection of such self-sacrificing behaviors have been proposed. A simple one is kin selection. Take siblings, for example, which share half their genes. If the death of one sibling saves more than two others that would otherwise have been unable to breed, the genetic balance sheet is positive and natural selection should favor the "altruistic" gene. Populations possessing this gene will outbreed those that lack the gene.

Considered within this framework, Wilson's book deserved to be acclaimed as an important contribution to an expanding field, with the usual spectrum of critical appraisals common for works in which theoretical explanations of observed facts play a major role. This might have been so,, were it not for the first and the last of the twenty-seven chapters, one titled "The Morality of the Gene" and the other "Man: From Sociobiology to Sociology," in which Wilson crossed the sacred border between the natural sciences and the humanities. This invasion of a traditionally "soft" science, sociology, by a "hard" science, biology, provoked an outcry, the major bone of contention being Wilson's assertion that much of human behavior is genetically determined. The smoldering nature/nurture controversy was rekindled, with predictable cleavage along ideological lines. The flare-up turned into a major conflagration through the involvement of a group of particularly vocal, openly Marxist political activists, several of whom, ironically, were colleagues of Wilson at Harvard. Fed by a barrage of books and articles pro and con, the conflict raged for several years.[3] Now that the dust has settled somewhat, what are we to make of the epic battle?

In its radical form, the conflict opposed two absolutes: genetic determinism—behavior is entirely inborn—and environmental determinism—the milieu has unlimited power in shaping behavior. The former corresponds to the doctrine known as social Darwinism, most vigorously defended by the nineteenth-century British philosopher Herbert Spencer,[4] in which social inequalities are the product of natural selection and, therefore, to be accepted as ordered by our nature and inevitable. The latter supports the Marxist view that human behavior is almost infinitely malleable and needs only the appropriate political, social, educational, and economic measures for a just, egalitarian society to be established. That the truth lies somewhere between these two extremes was admitted by even the most radical among the opponents. The question is one of degree. How much nature? How much nurture?

The answer lies in the relative impacts of biological evolution and cultural evolution on contemporary human behavior. Genetically, our species is the product of several billion years of evolution, culminating in the last crucial six million that witnessed the conversion of a chimplike ancestor into a modern human. What makes us specifically human, within the wider groups of primates, mammals, animals, and so on, is written into our genes. In all likelihood, this legacy includes certain behavioral patterns.[5] Wilson can hardly be wrong when he states that we still carry a number of the heritable traits that were useful to our hunter-gatherer ancestors and their hominoid forebears. He is right in insisting that we ignore the inborn nature of some of our social instincts only at our peril. What those instincts are, however, is a question that cannot be answered by simplistic reconstructions and naive extrapolations. The caveman image is a facile, inadequate justification of the macho bully.

We must take into account the overriding importance of cultural evolution and its ability to alter the course of biological evolution. The past history and present diversity of social structures enforce the conclusion that human genes prescribe few rules of social behavior. The most specifically human of our genes opened the mind to innovation, communication, intentionality, and choice, thereby helping to liberate human populations from the social straightjacket imposed by natural selection. Whether this liberation will be wisely exploited remains to be seen. It is our privilege and our burden that the way we use our evolutionarily gained freedom has become crucial for the future of our species and of much of the rest of the living world.

ALL CREATED EQUAL?

What has most inflamed the adversaries of sociobiology is its implication that there exist innate individual differences in intelligence, talent, and other psychological abilities that need to be taken into account for optimal education, professional orientation, and social integration. To admit the possibility of such differences and to attempt to measure them is viewed as opening the door to fascism, racism, sexism, capitalistic exploitation of the masses, and other forms of social discrimination. What does science have to say?

The answer is illustrated by the following anecdote. The great American dancer Isadora Duncan allegedly once suggested to George Bernard Shaw that they have a child together. "Just imagine," she said, "a child with my beauty and your intelligence." To which the celebrated Irish playwright, notorious for his ugliness and his wit, is said to have replied, politely declining, "What if it should have my beauty and your intelligence?"

Every child receives half of its genes from its mother and the other half from its father, but what these genes are is unpredictable. Because of the intricate, random chromosomal rearrangements that take place at meiosis, when diploid precursor cells are converted into haploid germ cells, each of the six-hundred-odd egg cells a woman produces in her lifetime, and even each of the billions of sperm cells a man may generate, is virtually guaranteed to contain a unique assortment of the progenitor's genes. Every child a couple may conceive—apart from identical twins, which arise from a single fertilized egg—is therefore guaranteed to have a unique combination of its parents' genes. Such is the basis of human individuality.

Only in simple cases, mostly involving "housekeeping" genes, can inheritance be predicted—and only statistically, at that—on the strength of Mendelian rules. A number of congenital diseases due to a single enzyme or other protein deficiency fall into this class and are now susceptible to prenatal testing and appropriate counseling. Examples are muscular dystrophy, sickle cell anemia, Tay-Sachs' disease, Gaucher's disease, among many others. Body build, facial features, and many other physical traits, as well as psychological traits such as intelligence, mathematical skill, musical ability, artistic talent, and linguistic aptitude, are not reducible to single genes, even though they may be genetically determined. They depend on a mosaic of genes and are influenced by the vagaries of meiosis and fertilization. Had another of Leopold Mozart's sperm cells won the race to the egg cell produced by Anna Maria Pertl sometime around May 1, 1755, some of the most beautiful music in the world would never have been written. The silly project, concocted some years ago, of propagating "Nobel ability" with the help of stored laureate sperm cells was based not only on a faulty sense of the factors that send a person to Stockholm, but also on ignorance of the laws of heredity. The only scientifically valid way of duplicating an individual is by cloning. In this procedure, first carried out on amphibians by the British biologist John Gurdon,[6] the nucleus of an unfertilized egg cell is replaced by that of a somatic (body) diploid cell. The "grafted" egg sometimes develops normally, giving rise to an individual genetically identical (apart from mitochondrial genes) to the donor of the nucleus. The similar cloning of mammals has not yet been successfully accomplished. Except for an imaginative writer's unauthenticated claim,[7] no millionaire has yet been perpetuated in this way. Such an operation may well become feasible in the future, however, which would prompt interesting ethical considerations. The debate has already been launched, with the recent announcement that cells derived from an in-vitro fertilized human egg cell had been "cloned" in a test tube, albeit to a very limited stage of development.[8]

The facts are clear. We are all born different and these differences include inequalities in innate abilities. In a rationally organized society, one would wish each individual to be given the opportunity for full realization of his or her genetic potential. How this goal is to be achieved is a question only society can answer. Granting that differences exist, the key issue, object of much debate, is whether they are best ignored or assessed and taken into account. Even scientists are divided on this question, despite their dedication to the truth, because they view differently the feasibility of a fair and accurate testing of the traits concerned.

THE BIOLOGY OF ETHICAL VALUES

"Scientists and humanists should consider together the possibility that the time has come for ethics to be removed temporarily from the hands of the philosophers and biologized."[9] This sentence, excerpted from Wilson's *Sociobiology,* has caused more turmoil among scientists and humanists than any other part of the book. The advocation is not new. It comes to us from Aristotle, by way of Jean-Jacques Rousseau—"man is good; society makes him bad"—and a host of other philosophers of the "naturalistic" school. The notion of a naturalistic ethics gained scientific support from evolutionary theory, especially through the efforts of the philosopher Herbert Spencer,[10] the father of social Darwinism, and the biologist Thomas H. Huxley,[11] popularly remembered for his defense of Darwin against Samuel Wilberforce, bishop of Oxford—"I would rather have a miserable ape for a grandfather than a man highly endowed by nature and possessed of great means and influence, and yet who employs these faculties and influence for the mere purpose of introducing ridicule into a grave scientific discussion." These two eminent Victorians did much to propagate the view that the sources of morality, and also its tenets and limitations, are to be sought in biology. Spencer forestalled Wilson by almost a century when, in 1892, he urged "the necessity of preluding the study of moral science by the study of biological science."[12]

In its strongest form, naturalistic ethics demands that the moral code be not only compatible with but also derived from biological and evolutionary knowledge. What comes naturally is good. Thus the inequalities of Victorian society were presented as products of natural selection and, therefore, morally just. Not only that, but to the extent current moral rules appear to conflict with human nature, they must be brought into line. As stated by Wilson and Charles J. Lumsden, the Harvard physicist enlisted by Wilson, "a sufficient knowledge of the genes and mental development can lead to the development of a form of social engineering that changes not only the likelihood of the outcome but the deepest feelings about right and wrong, in other words the ethical precepts themselves."[13] This philosophy, with its implicit menace of brain control, has been branded by much of the sociological establishment. It has also come under the attack of the philosophical establishment for its conceptual foundations.

According to many ethical theorists, sociobiologists are guilty of the "naturalistic fallacy"[14] denounced by the Scottish philosopher David Hume, a contemporary of Rousseau and his friend until they fell out. For Hume, the error consists in deriving prescriptions from descriptions, "ought's" from "is's." Ethics, the critics point out, deals with rules of conduct that are themselves based on general principles. Ethics should not be confused with the empirical observation of how people behave or are inclined to behave. Cheating by a majority does not make cheating good and commendable.

There are different forms of idealistic ethics, depending on the principles on which they are based. The strongest form is rooted in the philosophy of Plato. In modern times, the most illustrious and persuasive advocate of a—though definitely non-Platonistic—brand of idealistic values was the eighteenth-century German philosopher Immanuel Kant, who founded ethics on the transcendental idea of freedom and wanted ethical behavior to rest, ultimately, on one single "categorical imperative." A weaker form of idealistic ethics is relativistic and sees ethical principles as subject to choice and change. For example, an ethical code may be libertarian, utilitarian, or egalitarian, depending on whether individual freedom, the happiness of most, or social equality is considered the most desirable goal to be achieved in human intercourse.

At first sight, comparative anthropology and history offer strong support for a relativistic view of ethics. Different cultures obey different codes. What is seen as a basic courtesy in Eskimo society could get a person beheaded in Saudi Arabia. Even within a single culture, moral codes evolve, sometimes quite rapidly. In my lifetime, I have seen a major shift in the attitude toward contraception, abortion, and, most recently, euthanasia. Such changes concern not so much the ethical principles themselves as the manner in which these principles are interpreted and translated into precepts. What I have witnessed in my lifetime is not a modification of society's views on the "sanctity of human life," but rather a change in the definition of human life, its beginning, and its end. This change is influenced by the progress of scientific knowledge, imposed by the advances of medicine, and still subject to intense discussion.

A relativistic and evolutionary conception of ethics does not necessarily preclude the Kantian absolute. Morality developed progressively, together with the human mind. It may even have roots in animal behavior, as sociobiologists maintain. It is possible that ethical rules were fashioned and screened in the course of biological and, especially, cultural evolution, by a trial-and-error process in which their effects on individual fitness and social cohesion acted as selective factors. But this does not exclude the possibility that this evolution also reflects a progressive appreciation of absolute values, from dimly perceived and inadequately applied notions to more clearly apprehended and rationally argued imperatives. The two developments are not incompatible.

Regardless of the content of moral codes, the origin of our moral sense needs to be addressed. We may differ on what we call good and bad, but we all agree that there is a distinction between things good—and therefore to be commended—and things bad—and therefore to be avoided. A key question: Did we *invent* or did we *discover* the distinction between good and bad? The same discussion goes on concerning other abstract notions, for example, mathematical truth, rules of logic, and the concept of beauty. A Platonic view of mathematical truth, we have seen, is defended by the mathematicians Roger Penrose[15] and Alain Connes,[16] but rejected by the neurobiologist Jean-Pierre Changeux.[17] Beauty, no doubt, is in the eye of the beholder. But what about the abstract notion of beauty and our yearning for it?

In many societies, ethical precepts are formulated and imposed by organized religious bodies, often within the conceptual framework of a system of beliefs. Because of their partly irrational character, religions are targets of attacks by sociobiologists and materialistic neurobiologists. Yet religions survive and prosper in our increasingly materialistic and hedonistic world. As Wilson writes, "the enduring paradox of religion is that so much of its substance is demonstrably false, yet it remains a driving force in all societies. Men would rather believe than know."[18] Changeux echoes this disillusioned lament: "In spite of the unverifiable character of their content, of their physical as well as historical implausibility, creeds persist, and even spread."[19] Confronting the paradox, these biologists and many other scientists rationalize the "religious phenomenon" and the many mythologies it has generated as evolutionary concomitants of social development that happened to enhance the joint survival potential of the believers, as opposed to those societies whose members were not united by a belief or were united by a weaker belief. The winning ideology need not be true, it need only be stronger than the others. The growing appeal of fundamentalisms is a case in point.

We must ask, however, whether there is not in our nature, next to our intellectual hunger for rational explanations, a deep emotional need for religious belief. Even scientists, who are highly motivated to explain, cannot operate without a system of values. Most subscribe to what the French biologist Jacques Monod called the "ethics of knowledge," based on the "postulate of objectivity," a code of conduct that imposes intellectual integrity and respect for facts. The hue and cry raised by scientific fraud, as compared to the more lenient acceptance of fiscal cheating or fraudulent behavior in the business world, shows that society has particularly high expectations of its scientists. Yet the scientists themselves would be hard put to justify their self-imposed code rationally. As clearly stated by Monod, "the positing of the principle of objectivity as the condition of true knowledge *constitutes an ethical choice and not a judgment arrived at from knowledge*" (the italics are Monod's).[20] Is the choice gratuitous or a reflection of a Kantian precept of respect for truth? The question is at least debatable.

Ethics poses an even more fundamental problem for scientists. By definition, an ethical code implies the possibility of choice among conducts of different moral value. This condition is hardly compatible with strictly monist-materialistic theories of the brain, which imply a lack of free will and, therefore, of moral responsibility. I have mentioned in chapter 27 the unease of some neuroscientists and philosophers in the face of this problem. Sociobiologists are in the same quandary. In *Promethean Fire,* Lumsden and Wilson state their belief that "mental events are identical with physiological events in the brain"[21] and that "all of our behavior is indeed predestined."[22] This does not prevent them from speaking of free will as an essential quality of the human mind. Nowhere have I found an explanation of how these two kinds of statements are reconciled. Fortunately, the need to do so is not yet compelling. We still know too little about the human mind to affirm categorically that it is a mere emanation of neuronal activity lacking the power to affect this activity.

PART VII

THE AGE OF THE UNKNOWN

Chapter 30

The Future of Life

WE HAVE REACHED a crucial stage in the history of life. The face of the Earth has changed dramatically in the last few thousand years, a mere instant in evolutionary time, and it is changing ever faster. What would have taken one thousand generations in the past may now happen in a single generation. Biological evolution is on a runaway course toward severe instability.

In a way, our time recalls one of those major breaks in evolution signaled by massive extinctions. But there is a difference. The cause of instability is not the impact of a large asteroid or some other uncontrollable event. The perturbation is from life itself acting through a species of its own creation, an immensely successful species filling every corner of the planet with continually growing throngs, increasingly subjugating and exploiting the world. For the first time, also, in the history of life, natural selection has been replaced, be it only partly, by willful intervention on the part of a member of the biospheric community. The facts are before us, clear and unmistakable. Everybody can read the message and draw the obvious conclusions.

NATURAL SELECTION DERAILED

The human species is a product of natural selection, which put a premium on mutations that freed the hands and molded them for holding, signaling, weapon wielding, tool making, and other manual activities and on mutations that wired the brain into networks capable of directing the hands, planning the future, and communicating with other members of the species. The outcome is a being uniquely able to alter the course of the natural processes to which it owes its birth. Evidence of this power is all around us.

It all started some 40,000 years ago. Until then, early humans had coexisted fairly harmoniously with the rest of the biosphere. Organized in small, roving bands, they subsisted on fruits, berries, and other plant products and on the animals they were able to catch—mostly rodents, frogs, lizards, snails, insect larvae, and the defenseless young of birds and larger mammals. Theirs was a day-to-day, hand-to-mouth kind of living, totally dependent on what nature had to offer. But they learned to exploit every part of their biotope, moving from place to place in accord with seasonal and climatic changes, rarely disturbing natural balances to the point of endangering their food supply. Variously described as brutish or Arcadian, depending on personal viewpoint, the hunter-gatherer lifestyle may have been precarious but it was environmentally friendly.

Things changed when hunting became more successful, part of the general blossoming of human technology and social organization that started about 40,000 years ago. Included in this cultural explosion were the manufacture of better weapons and the development of efficient, cooperative hunting techniques that allowed successful slaying of even the largest mammals. Big game hunting progressively replaced the laborious catch of small animals, giving access to new, particularly rich sources of food. Plentifulness favored proliferation and also, perhaps, profligacy. Only choice morsels were selected for human consumption, the rest left to scavengers. Perhaps hunting even became a sport, practiced for the fun of it. Whatever happened sufficed to upset the cyclic balance of predator and prey. The first human assault on the biosphere was launched. In a matter of a few thousand years, the hunters wiped out aurochs, mammoths, giant sloths, and other prehistoric mammals in both Europe and the Americas. That hunting, rather than climatic or other environmental disturbances, was responsible for these mass extinctions is indicated by the fact that the extinctions took place at different times in different parts of the world, coinciding in each case with human invasion.[1]

The next assault started more insidiously, almost innocuously. Somewhere in the Middle East, around 10,000 years ago, a group of humans found ways of husbanding nutritious plants and animals, thereby discovering the benefits of staying in one place and letting the food come to them instead of going after it. These settlers built permanent dwellings that better protected them and their offspring against harsh climatic conditions, dangerous predators, and unfriendly neighbors. They forged stronger social bonds and knit closer networks of communal organization, labor division, and cooperative action. Thanks to these advantages, they bred and reared more young, expanding faster than the wandering hunter-gatherers. Whenever the number of settlers exceeded the carrying capacity of the surrounding land, a group split off, as from a beehive, to found a new colony some distance away, displacing or exterminating any roaming tribe that used the same land. According to the latest genetic and paleontological evidence, this is how the agrarian mode of life spread in those days, not by amicable melding and technology transfer.[2]

The spread moved at what looks like a snail's pace, about one kilometer (0.6 mile) per year. But it moved relentlessly. Acre by acre, the primeval forests gave way to pastures and arable land, of which parts, in turn, gave way to arid deserts, from overgrazing, biotope degradation, soil erosion, and other side effects of cultivation.[3] Acre by acre, the rich collectivities of interconnected microbes, plants, fungi, and animals that had been kept together under the life-bringing canopies of the great forests yielded everywhere to the same uniform fields of wheat and barley, the same pastures, the same herds of cattle, goats, sheep, and pigs, the same flocks of grounded hen and geese, the same cats and dogs, the same rodents, the same flies and cockroaches, the same parasites, symbionts, scavengers, decomposers, and other profiteers of the new, artificially created biotopes. Human ingenuity molded the adopted plants and animals into strange varieties that would never have been preserved by natural selection and that progressively replaced and supplanted their wild ancestors.

In the space of a few millennia, virtually the whole of humankind had adopted the agrarian mode of life and vast areas of forest had given way to farmland. At the same time, the growth of technology, the birth of modern science, the burgeoning of industry made further attacks on the biosphere through irrigation, fertilization, and other measures designed to increase agricultural yields and improve human living conditions. But hardly anyone worried. Nature seemed inexhaustible, which it almost was, in a way, until well into the nineteenth century.

The final blow was inflicted by the advances of medicine and sanitation, which increasingly saved human lives that would have been eliminated in the past by poor hygiene, malnutrition, and disease. Thanks to these advances, the curtailment of human population by natural selection was thwarted, resulting in the demographic explosion of the last 150 years. More mouths to feed has meant more food to produce, more acres of forest to destroy, more biotopes to disrupt, more impoverishment of the living world. Natural selection has been derailed, but not arrested. It continues to act within our artificially set boundaries, forced by these boundaries into directions that may, in the long run, prove more harmful to the biosphere and humankind than any direction it would have taken without interference. Unless we act soon to correct our course, natural selection could surge back with a vengeance.

SEVEN HEADS, ONE BODY

One of the twelve labors the legendary Hercules was challenged to perform was the slaying of the Hydra of Lerne, a monstrous sea snake said to inhabit the depths of the marshy lake of Lerne in southern Greece. The Hydra had seven heads—some say nine, fifty, or even one hundred. The heads grew back immediately when severed, a property, incidentally, that is shared by the monster's namesake, a tiny polyp celebrated for its regeneration ability. After vainly trying to cut off one head after the other, Hercules finally killed the beast with a mighty blow that shattered all the heads at the same time. Brute force won. But we may wonder why Hercules did not instead use his own head and attack the monster's body.

Humankind now confronts a multiheaded monster of its own devising—deforestation, loss of biodiversity, exhaustion of natural resources, overconsumption of energy, environmental pollution, and humankind's own deterioration.[4] Fighting each head separately is ineffective. Fighting them all together may prove too Herculean a task. The only workable solution: go for the body—alter the behavior of the human species itself.

A major threat to our future is deforestation. More than 50 percent of the forests of the world have disappeared (more than half this amount in the last century) and they continue to shrink at an alarming rate as a result of land development, farming, lumbering, burning, and damage from drought, acid rain, blights, and other ills. Large areas, shorn of protecting trees, fail to resist the combined onslaughts of the elements and of overexploitation by human and animal populations. Everywhere, the desert moves on to reclaim the devitalized land. The Sahara, which once was green, is gaining millions of acres every year. In the Sahel region, the desert advances at the rate of more than three miles a year. The tropical rain forests are faring worse, loosing some 40 million acres (about the size of the state of Florida), or almost 2 percent of their present surface area, every year. At this rate, no tropical rain forests will be left in fifty years' time.

Television has brought home to us, to the point of trivialization, the tragic consequences of deforestation. The emaciated bodies of infants clinging to flattened breasts, the look of despair in the eyes of the mothers, the gaunt frames of old men and women awaiting death have become so familiar that we have almost ceased to be moved by these images. But famines are local dramas. Most of the world is well fed. Surplus food is produced in many areas, silos and warehouses overflow, farmers are paid to leave land fallow. Hunger is a political and economic problem, as well as a biological one. But the disappearance of forests has other consequences, less visible but more global and irreparable.

One is the loss of biodiversity. Nobody knows how many species exist on Earth. In a recent plea to save endangered species, Edward O. Wilson lists a total of 1,402,900 catalogued species.[5] More than half of these are insects (751,000), to which must be added an additional 123,400 noninsect arthropods and 106,300 other invertebrate animals. In contrast, there are only 42,300 known vertebrate species, of which less than 10 percent are mammalian. Plants are represented by 248,400 species, fungi by 69,000, protists by 57,700 (among them 26,900 phototrophic algae), and bacteria by a mere 4,800. Everybody agrees that some of these figures are grossly underestimated. The bacterial world is almost entirely unexplored. Millions of insect species may await characterization. Wilson believes that "the total number [of species] alive on earth is somewhere between 10 and 100 million."[6] The Oxford specialist Robert M. May offers a more modest but still considerable guess: around five to eight million.[7]

More than half the known species inhabit tropical rain forests, also the main repositories of the millions of uncatalogued species. The toll paid by these depositories to deforestation is staggering. According to Wilson's most optimistic estimate, at the present rate of eradication of the rain forests, "the number of species doomed each year is 27,000. Each day it is 74, and each hour 3."[8] This figure—

between one thousand and ten thousand times the estimated prehistoric extinction rate—dwarfs, without reducing it to insignificance, the number of endangered species in temperate parts of the world, such as the giant panda, the spotted owl, or the snail darter, on which most efforts by conservationists are spent.

The disappearance of living species is not just a blow to orchid growers, butterfly collectors, and beetle buffs. It is an irremediable loss of precious information, the biological equivalent of the burning of the library of Alexandria in 641. It is the destruction of a large part of the book of life before it can be read, the irreplaceable loss of vital clues to biological evolution and our own history. Resources of potentially great practical benefit may also be lost. With each daily shrinking of the biosphere, a valuable source of food or a molecule that could have cured malaria, AIDS, or some other scourge may be vanishing forever.

Another threatening effect of deforestation is to weaken the world's defense against excessive carbon dioxide production. The forests are the lungs of the world, or, rather, its "reverse lungs." Trees (and other plants), when illuminated, remove carbon dioxide and produce oxygen. They thereby compensate the use of oxygen and production of carbon dioxide by much of the rest of the living world and, increasingly, by the burning of fossil fuel and, also, of living trees, as has been practiced on a large scale in the Amazon region. For the last 150 years and in a steadily increasing fashion, more oxygen has been used than produced, and more carbon dioxide produced than used. This imbalance hardly affects the oxygen content of the atmosphere, but appreciably influences the level of atmospheric carbon dioxide, which is three orders of magnitude lower than that of oxygen. The carbon dioxide content of the atmosphere has risen from 0.0315 to 0.0360 percent between 1958 and 1993 and it is rising ever faster; it is expected to reach 0.060 percent between the years 2050 and 2100.[9] Many believe that this change may cause a significant warming of the Earth's temperature through what is known as the greenhouse effect.

This phenomenon is due to the presence of a one-way filter—the glass of a greenhouse or the carbon dioxide in the atmosphere—that lets in visible light from the sun but partly blocks exit of the higher-wavelength infrared radiation emitted back. Thus, sunlight energy is retained and causes the temperature to rise inside a greenhouse or on the surface of this planet. It is generally agreed that the increase of atmospheric carbon dioxide—and of methane, a gaseous side product of cattle breeding formed by anaerobic microbes in the stomachs of ruminants—will produce global warming. The extent of the warming and its significance in a world inevitably exposed to alternative periods of warmth and glaciation are vigorously debated. Some scientists predict disastrous climatic change, major alteration in vegetation and biotope composition, melting of polar ice, catastrophic coastal flooding, and other planetary upheavals. Others minimize the dangers and trust Gaia to respond. Even Gaia's father, however, England's James Lovelock, admits that her response, albeit adequate from her long-range vantage point, may be too late or too little to avert major disasters for our species.[10]

Forests play an important role in the cycling of water; they store huge quantities of it, regulate its loss from the soil, and create atmospheric instabilities that draw rain-bringing clouds to the lands. The droughts that have followed deforestation in many parts of the world dramatically illustrate the water-saving role of the Earth's "green cathedrals." Artificial irrigation may bring back fertility to the desert, but at an enormous cost of energy, which brings us to another of the Hydra's heads: energy overconsumption.

Humankind is the only living species to spend more energy than it requires for sustenance and reproduction. This extra energy was long derived exclusively from the living world, mostly from wood to provide heat and from animal (or human) muscles to supply work. Eventually, sails were spread to make use of the winds, and the power of falling water was harnessed to drive mills. However, the major change came when ways were found to convert heat into work—the invention of the steam engine. Almost symbolically, the first such engine was used to pump water up from coal mines.

Intimately linked from the start, steam and coal established a partnership that spawned the industrial age, ploughed the seas with steamships, and crisscrossed continents with railways, spurring geologists to find more coal, miners to dig deeper for power-giving seams. Skies darkened, rains soured, façades blackened and crumbled, underground floods and explosions killed thousands; silicosis and tuberculosis killed many more. Few cared. Humankind had entered the age of energy debauchery.

Then came the second wave, more sweeping than the first. The internal combustion engine was invented; oil fever flared, sending new armies drilling for the gusher that would make their fortune. Thus was inaugurated the Ford-Rockefeller era that now has hundreds of millions of gas guzzlers scurrying like ants all over the planet.

Because immense stores of fossilized biomass accumulated when plant photosynthesis exceeded heterotrophic breakdown, the new machines could be developed far beyond the capacity of the biosphere to support them. However hungry the mechanical slaves grew to be, there was always enough coal and oil to feed them. A further giant step was taken when devices were invented for the conversion of mechanical energy into electric energy, the transport of electric current, and the conversion of electric energy to mechanical energy, heat, and light. These advances made accessible to every home and factory the power produced by combustion of fossil fuels in huge, centralized plants.

Today in the United States, every man, woman, and child spends for transport and comfort one hundred times the energy required to support their metabolism.[11] Many other countries in the industrialized world approach this consumption. The developing world lags far behind but is striving to improve its standard of living at the cost of greater energy expenditure. Throughout the world, most of the energy consumed by humankind comes from the combustion of fossil fuels: coal, petroleum, and natural gas. But it does not come free.

Overconsumption of fossil fuels is largely responsible for the overproduction of carbon dioxide and, thus, for the greenhouse threat. In addition, fuel combustion, especially of coal, releases into the atmosphere toxic oxides of sulfur and nitrogen, main causes of the acid rain that is killing forests and lakes, damaging human health, and upsetting ecological balances in many parts of the world. Furthermore, fossil fuels, especially petroleum, provide the main materials used by petrochemical industries for the manufacture of plastics and chemicals that, in turn, add to pollution of the planet. The extraction and transport of products needed to meet growing demands are themselves not without hazards—landscapes disfigured by strip mining, the heavy toll exacted on human lives and health by coal mining and oil drilling, the damage inflicted on the environment by burning oil wells, and, especially, the ecological disasters that have flowed from gutted tankers. Most serious of all, fossil fuels are a nonrenewable resource that will one day cease to be available. It is estimated that production of petroleum and natural gas will reach its peak in a few years' time and will then decline to complete depletion one century from now. Coal is more plentiful but also limited. Its exploitation should peak in the middle of the twenty-first century and fall to exhaustion sometime between 2400 and 2600.[12]

These facts are well known and have been brought to the attention of the world's leaders by the energy crisis of the 1970s. Already in 1976, the environmental and political activist Barry Commoner, from Washington University in St. Louis, published a penetrating analysis of energy production and consumption in the United States.[13] One need not accept his condemnation of the economic structure of the country to appreciate the validity of his objections to present energy consumption, which, in the light of thermodynamic principles, appears grossly wasteful.

The mistake, according to Commoner, is to calculate efficiency only in terms of the fraction of consumed energy (fuel burned) converted into useful work (First Law efficiency, that is, derived from the first law of thermodynamics). This is not unsound in itself and has, in fact, inspired the design of fuel-economizing motorcars, better insulated houses, and other energy-saving devices. But we do not pay enough attention to the Second Law efficiency (derived from the second law of thermodynamics) of our energy converters, that is, to the energy they consume relative to the minimum amount needed to perform the desired work under the best conditions. According to the second criterion, mass transit is superior to private transportation, whatever the First Law efficiency of engines. Also, it is nonsensical to use electric power for heating, which involves the wasteful conversion of heat into electricity, only to convert the electricity back into heat. More surprising, it would be more energy-sparing to heat our houses with heat pumps than with burners. Heat pumps are machines such as refrigerators that use energy to pump out heat from a colder environment (the inside of the refrigerator or the outside of the house) into a warmer one (the inside of the house in both cases).

Much research has gone into harnessing alternative, inexhaustible, and nonpolluting sources of energy, such as the winds, natural or artificial waterfalls, tidal movements, waves, oceanic currents, temperature differences between surface and deeper water levels, hot springs, subterranean heat, and, of course, the sun. Efforts are also being made to develop a safe, transportable fuel that could someday replace fossil fuels, the ideal one being hydrogen, which is nonpolluting and burns to water. There is progress on all fronts but no major breakthrough. It is generally believed that solar power would be the ideal source of energy for our planet, but the large-scale exploitation of this source is hindered by many practical obstacles. One of the numerous problems to be solved arises from the diffuse and irregular manner in which sunlight falls on the Earth's surface. Immense areas need to be devoted to light collection (0.4 percent of the surface of the country to satisfy the total needs of the United States[14]) and huge storage facilities must be provided to ensure a supply of energy while the sun is hidden. Perhaps one day vast deserts will come back to life with thousands of square miles of solar screens feeding energy into giant hydrogen-generating plants that supply fuel for the whole world. At present, this is a dream, but it is worth pursuing.

In the meantime, only nuclear power can provide energy at an affordable price and in essentially nonpolluting form (except for the heating of local waters). Several countries, prominently France and Belgium, now cover an important fraction of their needs with this form of energy. Many others, including Austria, Sweden, and the United States, have been forced under pressure of public opinion to scrap or curb their development of nuclear power.

Nuclear power has a bad name, not entirely undeserved. Hideously manifested in the infernos of Hiroshima and Nagasaki, it remains linked to apocalyptic visions of war and massive death. It is also intimately associated in the public mind with radioactivity, an invisible and frightening force known to produce cancer, leukemia, and genetic defects. Furthermore, the generation of nuclear power is not as safe as some scientists and engineers have maintained, a perception now symbolized by the names of Windscale, Three Mile Island, and Chernobyl. Objective observers point out that nothing real can be totally risk-free and that the toll of peaceful nuclear power since it was inaugurated is remarkably small—only a fraction of the yearly fatalities due to motorcar accidents or cigarette smoking in the United States alone. But nuclear fear will not be cured with cold figures. It will take demonstrated engineering competence of a compelling sort.

Viewed rationally, nuclear technology has some serious drawbacks besides the relatively small risk of major accidents. It produces highly radioactive waste products that need to be stored safely for millennia, a problem not yet solved. In addition, the world's supply of uranium, which provides the main fuel of nuclear reactors, is limited. Breeder reactors, which generate more fuel than they consume, theoretically offer a lasting solution to this problem, but they produce plutonium, the ideal material for do-it-yourself bombs, which evoke frightening vistas of nuclear piracy and terrorism.

All nuclear reactors presently in use rely on atomic fission, the phenomenon exploited in the first atom bombs. Hydrogen fusion, which powers the sun and hydrogen bombs, would be much safer and almost nonpolluting, but requires very high temperatures within adequately contained regions of space, conditions increasingly approached in experimental devices but not yet realized despite enormous research effort. It will take many years, if not decades, before this form of energy is mastered, if it ever is.

Pollution is perhaps the most vicious and unvanquishable of the Hydra's heads. There is almost no industry that does not generate byproducts that pollute the air, streams, oceans, and the soil and its precious water. Lakes made lifeless by heavy metals, rivers suddenly covered with dead fish, pipes pouring foul fluids, skies laden with fumes and smog, woods reduced to silence except for the croaking of crows, fields rendered barren but for a few hardy weeds, all used to be common sites around factories and sometimes reappear, by accident or neglect.

Many of the products of industry also do harm, unintended or accepted as the price of derived advantages. Pesticides were spread around the world to benefit farmers until Rachel Carson sounded the alarm. Chlorofluorocarbons (CFCs) were hailed as ideal refrigerator fluids and spray propellants but now are branded as greenhouse gases and destroyers of the ozone layer that protects the planet against excessive ultraviolet radiation from the sun. Antibiotics, rightly acclaimed as miracle drugs when they were first produced, turn out to be powerful selectors of dangerous pathogens. The list seems endless.

Even harmless materials, by their sheer bulk, create environmental problems. The plastic containers and wrappings that envelop almost every item in a supermarket; the "disposable" equipment, from laboratory ware to lighters, cameras, dishes, and cutlery; the artificial materials that now replace wood or metals in countless applications; the synthetic fabrics that substitute for wool, cotton, or fur, all add to the "plastosphere." Resistant to biodegradation, almost impossible to burn cleanly, this stuff piles up, transforming lands and seas into garbage dumps. Garbage is a major problem in itself, completely overwhelming communities and Nature herself.

Shrinking forests, expanding deserts, vanishing species, dwindling resources, rising greenhouse gases, a punctured ozone shield, polluted skies, poisoned waters, stinking landfills, radiating wastes, mountains of garbage. The Hydra has too many heads.

A few months before his death, Lord (Solly) Zuckerman, one of Britain's most influential scientists, who attended both the 1972 Stockholm and 1992 Rio conferences on the global environment, summed up his impressions. "Like Stockholm," he wrote, "the significant lessons that Rio has left behind are that national interests differ; that there is a difference between national and global environmental problems; that long-term and short-term social and environmental issues do not belong in the same category; and that development in the poorer countries almost inevitably brings in its train not only financial and managerial, but also new environmental problems. Another major lesson is that, in practice, the solutions to most environmental problems lie almost entirely in the political and economic domains."[15] The tone is more resigned than sanguine.

In 1992, the officers of the United States National Academy of Sciences and of the Royal Society of London, the two most august scientific bodies in the Anglo-Saxon world, issued a joint, solemn appeal to the world's conscience.[16] The statement opens with an ominous warning: "If current predictions of population growth prove accurate and patterns of human activity on the planet remain unchanged, science and technology may not be able to prevent either irreversible degradation of the environment or continued poverty for much of the world." Unfortunately, the document's conclusions sound more like wishes than forceful recommendations: "Global policies are urgently needed to promote more rapid economic development throughout the world, more environmentally benign patterns of human activity, and a more rapid stabilization of world population." And the authors add: "Sustainable development can be achieved, but only if irreversible degradation of the environment can be halted in time."[17]

This mood of guarded optimism is shared by a number of experts, especially economists, sociologists, and political scientists. They sound the alert but trust humankind to muddle through, as it has so often done. They recall Thomas Malthus, who, two centuries ago, in his celebrated *Essay on the Principle of Population as It Affects the Future Improvement of Society,*[18] claimed to prove that arithmetically increasing resources could not possibly keep pace with the needs of a geometrically increasing population. But Malthus did not foresee the Industrial Revolution, which developed resources far beyond the needs of the growing population of England and considerably augmented its standard of living. Technology, it is maintained, has not said its last word and may once again save humankind from the Malthusian predicament.

The situation now is not the same as it was in the time of Malthus. In spite of all efforts, the world has worsened in the last twenty years, and the future looks somber. The reason is simple. The Hydra's heads draw strength from a body that grows relentlessly. There are too many of us on Earth; too many are added every year. If expansion of the human population continues, our species will be trapped by its own success.

THE HEART OF THE PROBLEM

Every second, three human beings die and twice this number are born. One century ago, infant mortality would have evened the score. Thanks to modern medicine and sanitation, this no longer happens. As a result, 100,000,000 humans are added to the planet's population every year.[19]

The globe held one billion people in 1825. This figure doubled in one century, doubled again in the next half-century, and now stands at 5.6 billion in 1994; it is expected to exceed 10 billion by 2050 if it continues apace. Whether our fragile world can stand this hypertrophy of a single species is by no means certain. The question is not whether we can avoid further degradation of the world, which a resolute believer in the power of science and technology might consider an achievable goal. The question is whether so many humans can live together in harmony.

We are not a pacific lot. Through conflicts, invasions, conquests, crusades, holocausts, and wars, history reveals an uninterrupted succession of battles among human groups. The rest of the animal world holds no equivalent of such intraspecies fighting. There is nothing ritualistic about our contests. We fight for real. Mahatma Gandhi and Mother Teresa are less representative of our species than Alexander the Great, Genghis Khan, Napoleon, Hitler, or the Godfather. How much of this aggressivity is genetic, how much acquired, remains unclear. Whatever the answer to this question, the fact stands out. We are aggressive animals. Will we submit passively to the pressures of our growing numbers? Not likely.

One type of conflict already under way will pit the poor against the affluent, especially in large cities. By the year 2000 the world will hold twenty-three cities of more than 10 million inhabitants each, seventeen of them in developing countries.[20] Anybody who has visited São Paulo, or even New York, is aware of the explosive mixture of lawlessness and violence that poverty and despair can breed. Another form of impending conflict is between the young and the old,[21] perhaps one day between the healthy and the sick, especially in the medically advanced countries with stabilized or falling demographies. Most disquieting of all are the widening gulf between North and South and the resulting rising tensions between the two.[22] As a biologist, I observe two facts. With few exceptions, the populations of developing countries increase faster than their ability to satisfy the elementary needs of their people. Famine, epidemics, abject poverty, social unrest, political instability, and other ills, all are on the uprise, largely because of the growing imbalance between numbers and means. By contrast, in the developed countries, relatively sparse, aging populations enjoy a standard of living undreamt of by the poorer populations of the south. The consequence of the imbalance is predictable and clearly evident in many developed countries.

In the short view that most politicians must adopt if they want to stay in power, this trend may not appear too worrisome. In the long view, informed by contemplating the history of life and humankind on Earth, as we have done in this book, the prospect is more than worrisome; it is terrifying. With 100 million people added to the world population every year, mostly in the south, pressure is bound to build and the counterreaction cannot lag far behind. Major conflicts cannot be excluded. Remember the past. We have not changed. Not yet.

The conclusion is inescapable. If we do not, in the near future, succeed in curtailing our population growth with rational care, *natural selection will do it for us the hard way,* at the cost of unprecedented hardship to human populations and irreparable harm to the environment. Such is the lesson of four billion years of the history of life on Earth.

THE ROLE OF SCIENCE

Perhaps the most problematic factor affecting our future is the role to be played by science, accused of being both the best and the worst in the world, depending on how it is used. The benefits of science are all around us. Health, comfort, security, freedom from drudgery, worldwide communication, unlimited information storage, and the immense development of culture these advances have permitted are all products of science and technology that were not available to our hunter-gatherer ancestors or even, to a large extent, to our forebears of a few centuries ago. But so are all the ills that threaten the future of humankind and the planet.

Science good or bad? The question is increasingly raised. After the triumphant scientism of the nineteenth century, with its confident belief that science can explain, and technology solve, everything, the twentieth century has witnessed a rising tide of antiscientism.[23] There is no room in this book to discuss the many social, political, economic, and ideological issues raised by these questions, but a few points deserve attention here.

All except a radical fringe would agree that the benefits of science and technology far outweigh their drawbacks. Who would want to go back to the "good old days," when half the children never reached the age of two; when one-third of women died in childbirth; when smallpox, typhus, cholera, and plague decimated populations; when tuberculosis took its toll; when pneumonia, diphtheria, meningitis, and polio killed or crippled millions; when nutritional deficiencies stunted growth; when epileptic seizures were viewed with terror and the victims of ergot poisoning were burnt at the stake as witches? Who would willingly return to the times when humans had to toil day and night simply to keep alive? Surely not the millions who still hover near enough to this precarious way of life to know by experience what it means.

Even if such a return to earlier times were desirable, it is not feasible. There is no way to erase the acquisitions of science and technology. They are here to stay. It is not just a question of burning a library. Knowledge has become planetary, stored in millions of "memories," virtually indestructible. Even thwarting human inquisitiveness, although possibly enforceable within certain confines of space and time, could not be imposed globally or persistently without a radical change in human nature. Science is a product of intelligence, itself a product of higher brain function, itself a product of human brain development, itself a product of a genetically determined program, itself a product of evolution and natural selection: a direct outcome of qualities, including the urge to understand, that have shaped our evolutionary success. We cannot possibly, even if we wanted to, suddenly renounce our heritage and cease looking at the world in wonder, asking questions, and using our ingenuity to solve them. Only the end of the human species can accomplish this. If science is to be our downfall, then it will be because we as a species are flawed, the victims of a lethal mutation.

Such an outcome is not to be lightly discounted. Objective analysis of our predicament does, indeed, reveal a flaw that could be fatal. We owe our success to our intelligence. The lack of corresponding wisdom could spell our ultimate ruin. We are capable of acquiring knowledge but not of using it wisely. This statement hardly needs documentation. Just turn on your television set and watch the news.

The way to survival is not less science, but more wisdom. In this respect, scientists are not the only guides, perhaps not even the best ones. Wisdom is not a necessary correlate of knowledge or understanding, or even of intelligence. Neither, however, is wisdom to be sought in ignorance, stupidity, prejudice, or superstition. There is a tendency in some circles to refuse to seek the truth so that it may comfortably be ignored. This attitude has gone so far as to cause certain lines of research to be banned because the results may conflict with some preconceived opinion or ideology. Such moves are understandable. But they are doubly perverse. They are insulting, by treating humans as immature children who have to be protected from the truth, and they are futile. Truth, whatever we may do to deny or ignore it, is bound to catch up with us.

The great advances of the last decades in understanding the nature and evolution of life, including our own species, deserve to be heeded. There is urgent necessity that our leaders be better acquainted with the "facts of life," in the real sense of that expression. We are part of the biosphere and have become its custodians. Whatever other preoccupations we may have, we cannot afford to ignore our own nature. We must learn to think biologically.

Science is also our best chance to solve our present and future problems. To be true, the choice of means poses great difficulties. Even those who agree that anything that can be known should be known are not ready to claim that anything that can be done should be done. Here is where wisdom will be required most urgently. Present trends allow a glimmer of hope.

The last twenty years have witnessed a remarkable rise in global responsibility. The ecological movement, in spite of excesses, deserves to be praised. Also notable is the development of bioethics. Hospitals and research centers all over the world are now controlled by multidisciplinary committees that are consulted whenever a potentially hazardous research or treatment is contemplated. The existence of world organizations concerned with problems of health, environment, economic development, and population growth on a planetary scale is also encouraging, even though their functioning often remains cumbersome and ineffectual. In less than a single human lifetime, the biosphere has acquired a conscience, a necessary condition for the exercise of collective and informed wisdom.

THE NEXT FIVE BILLION YEARS

In 1953, Sir Charles Galton Darwin, grandson of the great Charles Darwin, published a book called *The Next Million Years*.[24] It so happens that in the same year

Watson and Crick announced the double helix, while the endocrinologist Gregory Pincus at the Worcester Foundation near Boston was busy testing the biological effects of a substance newly synthesized in the laboratories of the Syntex Company in Mexico City by the Austrian-born American chemist Carl Djerassi—a substance that was to become "The Pill."[25] Thus should we be warned against making predictions that may be completely upset by future discoveries which, by their nature, cannot be included in our evaluations.

Sir Charles's view of the next million years is surprisingly tame, little different from an extrapolated version of the present but for a few corrections. He mentions the population problem, but without panic and in deploring but resigned admission that population and food supply will continue to balance each other at the expense of an inevitable "starving margin." He even bemoans the low birth rate of developed countries as undermining the ramparts of civilization. He acknowledges the energy problem but offers few solutions. He shows little confidence in nuclear power, thinks solar energy will be difficult to harness, and tends to favor hydraulic power. This leads to the quaint notion of mountaineers possessing the source of power, pitted in some sort of bargaining situation against the food-producing dwellers on the plain. There is no mention of pollution, overcrowding, or loss of biodiversity, no hint of possible catastrophes except those caused by climatic or cosmic factors beyond human control.

Can we do better with predictions today? The Princeton astrophysicist Richard Gott thinks so. In a recent article,[26] he has used what he calls the Copernican principle to predict our future prospects. By Copernican principle, Gott means the assumption that there is nothing special about us, whether in time or space. That is, we are just random representatives of our kind. We may not be, of course, but the extent to which we actually differ from this idealized mean obeys a statistical function that can be taken into account. On the basis of this assumption, Gott calculates our probable future as a simple function of our past. According to his formula, if we have been around for a time t, then there is a fifty-fifty chance that we shall still be here in $t/3$, and not in $3t$. At the 95 percent confidence level, the limits become $t/39$ and $39t$. For the human species, assuming it arose 200,000 years ago, this calculation gives limits of 67,000 and 600,000 years at the 50 percent confidence level and 5,100 and 7.8 million years at the 95 percent level. Put simply, there is a 95 percent chance that the human species will survive more than 5,100 years and will be extinct 7.8 million years from now. The chances are already one in two that our species will have disappeared 600,000 years from now. Note that the 95 percent lower limit falls to a value of 12 years if the rise in population number is taken into account. Doomsday may indeed be around the corner, as some street preachers warn us!

Life itself fares better in Gott's arithmetic. Having been present for about 3.8 billion years, life is 95 percent assured still to exist some 100 million years from now, and it has a 50 percent chance to last more than 11 billion years, that is, longer than the Earth will be able to sustain life. According to the latest cosmological estimates, the Earth should become engulfed in the explosion of the sun in about six billion years and is expected to become uninhabitable long before.[27]

Life will continue as long as there is a niche on Earth capable of supporting it. We don't need Gott's calculations to make this prediction confidently. Its 3.8-billion-year history tells us that life will not merely hold on but will flourish and evolve toward greater variety and complexity, despite major geographic and climatic changes, even planetary cataclysms. It actually seems to thrive on catastrophes. Every time a mass extinction occurred in the history of life, a riotous upsurge of new living forms followed.

Life will continue. But with us or without us? As Gott points out, the average longevity for most species is between one million and eleven million years, and for mammals it is two million years. His prediction for the life span of our species thus falls within the right ballpark. One may, however, wonder whether Gott's methodology—which has not gone unchallenged[28]—can be applied to the unique case of a species that covers the whole surface of the planet and has amassed a vast, powerful, and commonly shared cultural heritage stored in virtually imperishable form. However, there is no foreseeing what can happen over very large lengths of time—about 40,000 generations per million years—once a degenerative process, perhaps triggered by a major holocaust, has set in. Entire populations have been wiped out in our times. Why not the whole world population sometime in the future?

If *Homo sapiens* becomes extinct, what will replace it? Perhaps nothing comparable, in which case human intelligence may be just a flicker in the darkness, never to be reignited again, "only an afterthought," as Stephen Jay Gould puts it, "a kind of cosmic accident, just one bauble on the Christmas tree of evolution."[29] As I shall explain in the next chapter, my reading of the book of life is different and sees at work throughout animal evolution a strong selective pressure favoring the creation of neuronal networks of increasing complexity. If our species disappears, I am inclined to predict its replacement by another intelligent species with perhaps greater powers than we have, notably more wisdom. This species could be a direct *Homo* offshoot, or it could arise through a separate pathway from some other animal species. Life on Earth has the time to recapitulate one thousand times the emergence of the human species from its last common ancestor with chimpanzees, and some twenty times the whole history of mammals. Many wonderful things can still happen in the next five billion years, and no doubt will.

Another possibility: Our species may evolve into some kind of planetary superorganism, a society in which individuals would abandon some of their freedom for the benefit of the whole society. Something of the kind happened to the protists that joined to form the first multicellular organisms, also, to some extent, to the social insects, although I would expect the "humanhive" to be much less stereotyped and more sophisticated than a beehive or an ant nest. Such a transformation could be already under way.[30]

Chapter 31

The Meaning of Life

WE HAVE COME to the end of our voyage. We have retraced step by step, to the extent present knowledge allows, the history of life on Earth, from the first biomolecules to the vast panoply of existing microbes, fungi, plants, and animals, including our own species. We have used our understanding of the forces and constraints that molded the course of this adventure to hazard guesses concerning its future. In this last chapter, I contemplate the biospheric tissue of our planet within a cosmic context, to see whether it has anything to tell us regarding a question all of us have asked: What does it all mean? I start by presenting two contrasting viewpoints, both of which happen to come to us from France.

A TALE OF TWO FRENCHMEN

Our first protagonist is Pierre Teilhard de Chardin,[1] born in 1881 in the mountainous province of Auvergne, not far from the source of the Dordogne River on whose banks Cro-Magnon remains were discovered thirteen years before Teilhard's birth. He grew up in one of those small chateaux that guard almost every French village and make up, together with the church steeple, the twin pillars of traditional virtues. Thus are explained the two imperatives of Teilhard's life. From the surrounding ranges, he gained an early passion for geology and paleontology, which led him to a scientific career digging for human vestiges in Asia and Africa. From his sheltered, deeply religious family environment, he drew an enduring faith that propelled him into the priesthood under the austere discipline of the Jesuit order.

Throughout his life, Teilhard strove to reconcile the conflicting demands of science and faith. He elaborated a kind of naturalistic theology that evoked the distrust of his superiors, who forbade him to publish his ideas. Only after his death in semi-exile in New York in 1955 were his works published. His major opus, *Le Phénomène*

Humain, was written in the years 1938–40, further reworked in 1947 and 1948, and finally published in 1955. Its English translation appeared in 1959.[2]

Teilhard's philosophy is influenced by Henri Bergson (1859–1941), a confirmed evolutionist, winner of the 1927 Nobel Prize in literature. In *L'Evolution Créatrice,* published in 1907, Bergson defended a vitalistic, spiritualistic view of biological evolution, which he depicted as obligatorily driven toward increasing complexity by a creative force, the *élan vital.* Teilhard adopted this vision, substituting for *élan vital* the seemingly more orthodox—though equally nebulous—"radial energy" (the conventional energy of physics being designated "tangential").

According to Teilhard's vision, mind and matter coexist in elementary form from the beginning of the universe and are jointly driven toward increasing complexity by the combined and complementary powers of these two forms of energy. Life emerged naturally from this "complexification" and went on to weave an increasingly elaborate fabric of living organisms around our planet. The evolution of this "biosphere" culminated in the appearance of humankind and consciousness. The next step, which he considered to be going on at present, is the creation of a "noosphere" (from the Greek word for spirit or soul), a planetary spiritual entity destined finally to converge, perhaps in association with other noospheres produced elsewhere in space, into point Omega, which is Teilhard's "scientific" name for the God of his religion.

Our second Frenchman is Jacques Monod (1910–1976).[3] A descendant of a prominent Huguenot family that produced several influential philosophers and churchmen, Monod did not grow up in the traditional Calvinistic environment of his forebears. He spent his youth in the sunny Mediterranean city of Cannes, where his father, a well-known painter, had settled with his American wife. He was thus exposed early in his life to both Latin and Anglo-Saxon influences. His wife, whom he married just before the war, was Jewish, the granddaughter of a rabbi who had been the "Grand Rabbin de France," the head of the French-Jewish community, at the time of the infamous Dreyfus trial.

Monod's early interests were divided between biology and music. He was an excellent cellist, directed a choir, and might have embarked on a musical career had he been sure of making it to the top. His elder brother, upon learning that Jacques was very good but perhaps not a prodigy, commented: "So he won't be a Bach; he will be a Pasteur instead"—a status Jacques Monod came close enough to achieving. He shared the 1965 Nobel Prize in medicine with his mentor, André Lwoff, and his younger coworker François Jacob, for work they performed at the Pasteur Institute in Paris, of which he was the director at the time of his death.

Like Teilhard, Monod sought to put biology within a wider philosophical context, but his ideological framework was totally different. He unequivocally rejected his religious heritage and later also renounced Marxism just as vehemently, after a brief flirtation with the Communist party, in whose ranks he fought in the Resistance during the Second World War. Like many of his generation in France, he eventually found a congenial intellectual climate in the existentialist philosophy of Jean-Paul Sartre and, especially, of Albert Camus, for whom he had a veritable veneration. Monod's scientific background was also very different from that of Teilhard. Trained in the rigorous discipline of biochemistry, he had been led by his studies of microbial adaptation to become one of the founders of modern molecular biology, the science that gave concrete substance to the abstract concepts of genetics and evolutionary biology.

Monod's "essay on the natural philosophy of modern biology," *Le Hasard et la Nécessité,* appeared in 1970. Its English translation, *Chance and Necessity,* was published in 1971.[4] The main philosophical message of this work is that biological evolution, far from being in any way directed by some sort of *élan vital,* radial energy, or other mystical force, depends entirely on random mutations (chance) screened by natural selection (necessity). There is no meaning, purpose, or design to be read in the appearance and evolution of life, even intelligent life. "The universe," Monod writes, "was not pregnant with life, nor the biosphere with man."[5] And he concludes with a mixture of austere grandeur and stoic romanticism: "The ancient covenant is in pieces: man knows at last that he is alone in the universe's unfeeling immensity, out of which he emerged only by chance. His destiny is nowhere spelled out; nor is his duty. The kingdom above or the darkness below: it is for him to choose."[6] Monod's kingdom has nothing to do with heaven, of course; it is what he calls the "ethic of knowledge," the freely chosen, self-imposed rule of the scientist, based on the "postulate of objectivity." His darkness is any form of "animism," an all-encompassing term that includes myths, superstitions, religious creeds, vitalistic and teleological explanations of life, and Marxist ideologies. The "ancient covenant" is the age-old alliance between man and nature under the aegis of one sort of animism or another.

No two books on the same general subject could be more dissimilar in form and content than *The Phenomenon of Man* and *Chance and Necessity,* but they have one thing in common. They evoked an inordinate number of passionate reactions, some favorable, many more adverse. Teilhard, in particular, while being acclaimed by a small, progressive fraction of the Catholic laity, suffered the sharp rebuke of scientists. Not without good reason. I remember reading his book when it came out and feeling intensely irritated by its bombastic style and scientific woolliness. Monod dismissed him in a few cutting words. Teilhard's philosophy, he wrote, "would not merit attention but for the startling success it has encountered even in scientific circles."[7] He further professed himself "struck by the intellectual spinelessness of this philosophy."[8] Peter Medawar, one of Britain's most distinguished and philosophically literate scientists, blasted Teilhard's book in terms that were even more contemptuous. "The greater part of it," he charged, "is nonsense, tricked out with a variety of metaphysical conceits . . . [it] cannot be read without a feeling of suffocation, a gasping and flailing around for sense."[9] Stephen Jay Gould, America's popular biology writer, went so far as to accuse Teilhard of fraud and complicity in the Piltdown forgery, the planting of a chemically treated primate jaw purported to have belonged to some prehuman "missing link."[10]

This is uncommonly harsh punishment for a man who was highly regarded in the paleontological circles of his day and who, on all accounts, was a mild-mannered, likable, and unassuming person. Teilhard is hardly read, let alone heeded, by biologists today. An exception is the Yale biochemist Harold J. Morowitz, who has adopted Teilhard's vision and incorporated it within a mystical, pantheistic picture of the universe.[11]

Surprisingly, the philosophy of Teilhard has found a sympathetic ear among a number of physicists and cosmologists preoccupied with the eternal conundrum: Why is the universe built the way it is? Their answer: So that it should be known. In 1974, the American physicist Brandon Carter identified this argument as the "anthropic principle" (from the Greek word for man). The "weak" form of this principle allows other universes, but then they would not be known: Only a universe built like ours can produce the intelligent life needed to know it. The "strong" wording of the anthropic principle states that the universe must be such as to produce intelligent life. The "participatory" version of this principle, put forward by the American physicist John A. Wheeler, asserts that observers are actually necessary to bring the universe into being, a definition that seems to call for some sort of retroactive creation by the human mind after it emerged.

Such speculations have become a hot topic of contemporary discussion, as witnessed by the massive, vastly documented opus—700 pages, 600 mathematical equations, 1,500 notes and references—published in 1986 by the British astronomer John D. Barrow and the American physicist and mathematician Frank J. Tipler, under the title *The Anthropic Cosmological Principle.*[12] In this book, the authors marshal evidence from history, philosophy, religion, biology, physics, astrophysics, cosmology, quantum mechanics, and biochemistry in support of their general viewpoint. Of Teilhard, they do not hesitate to state: "The basic framework of his theory is really the only framework wherein the evolving cosmos of modern science can be combined with an ultimate meaningfulness to reality."[13] In considering the future of the universe, they see life and mind invading the whole cosmos and converging into Omega Point. They conclude (the emphases are theirs): "At the instant the Omega Point is reached, life will have gained control of *all* matter and forces not only in a single universe, but in all universes whose existence is logically possible; life will have spread into *all* spatial regions in all universes which could logically exist, and will have stored an infinite amount of information, including *all* bits of knowledge which it is logically possible to know. And this is the end."[14]

Heady stuff. One wonders what words Medawar, who unfortunately died in 1987, would have conjured from his rich vocabulary to qualify this sort of prophetic announcement in a scientific context. Physicists, however, have been driven into such weird territories by their explorations that they are now far ahead of the most imaginative science fiction writers in the kind of cosmological scenarios they can invent. After Einstein's relativity, Planck's quanta, and Heisenberg's uncertainties, after black holes, cosmic strings, antimatter, and the dizzying dance of quarks and gluons, they know better than anyone that the real world is stranger than the representations conceived by brains adapted to human constraints of time and space.

Another distinguished physicist attracted by the anthropic principle, although he does not mention the term, is Freeman Dyson, an English-born American member of the Institute for Advanced Study in Princeton, New Jersey (where Einstein spent the last part of his life). In *Disturbing the Universe,* Dyson wrote: "The more I examine the universe and study the details of its architecture, the more evidence I find that the universe in some sense must have known that we were coming. There are some striking examples in the laws of nuclear physics of numerical accidents that seem to conspire to make the universe habitable."[15] By "numerical accidents," Dyson refers, as do the defenders of the anthropic principle, to the values of a number of physical constants that, had they been different, would have made matter unsuitable for the development of life. Although more cautious than his anthropic-principle colleagues, Dyson nevertheless allows his imagination to soar. He dreams of a future when humankind will invade and colonize space with the help of "green technology" and appropriate evolutionary adaptations, and he hints at the possible existence of "a universal mind or world soul which underlies the manifestations of mind that we observe."[16]

Not all physicists and cosmologists agree with such an idealistic, "anthropic" view. Witness the following quotations out of *The First Three Minutes,* the 1977 "modern view of the origin of the universe" by the American theoretical physicist and Nobel laureate Steven Weinberg: "It is almost irresistible for humans to believe that we have some special relation to the universe, that human life is not just a more-or-less farcical outcome of a chain of accidents reaching back to the first three minutes, that we were somehow built in from the beginning. . . . It is very hard to believe that all this [the Earth, as seen from aboard an aircraft] is just a tiny part of an overwhelmingly hostile universe. . . . The more the universe seems comprehensible, the more it also seems pointless."[17] Weinberg may not have read Monod when he wrote this (he does not cite Monod). Yet, except for substituting "farcical" for "chance," and "hostile" and "pointless" for "unfeeling"—or rather "indifferent," the original word used in the French text—his conclusion is surprisingly close to that of Monod—including the projection of human anxiety into a mindless cosmos.

Monod's book did not draw the same kind of fire as did Teilhard's. Much of it is straightforward science, made accessible to a wider public in clear, unexceptionable terms. His excursion outside the realm of strict science did, however, render Monod open to a broadside of attacks, even on the part of some scientists who denied the validity of his claim that modern biology enforces his conclusions concerning, for example, the low probability of the emergence of life or consciousness. Such was my case, but the lengthy critique I wrote at the time (in French) remained buried in an obscure journal[18] and perhaps was not even read by Monod, even though I sent him a copy. Another critic was Gunther Stent, a German-American molecular neurobiologist with an interest in the history of science and philosophy, who has the distinction of having worked with both Max Delbrück, the father of modern molecular biology, and Jacques Monod. In *Paradoxes of Progress,*[19] a strange work prophesying "the end of the arts and science," Stent points out that Monod's "ethic of knowledge" rests on a Kantian a priori just as do "animist" ethical systems—Monod did not dispute this—the difference being that the ethic of knowledge is self-imposed, whereas the others are derived from some religious or ideological system of beliefs.

The philosophers reacted to Monod's intrusion into their domain with greater vigor and almost unanimous opposition, although for a variety of reasons. Their scorn, incidentally, is not restricted to Monod. It extends to most other scientists who venture to cross boundaries and trespass upon philosophy's pastures. In a 1992 book, for example, the British philosopher Mary Midgley evenhandedly executes Monod, Dyson, and the defenders of the anthropic principle.[20] The lesson is clear: Scientists should stick to science and leave philosophy to the professional philosophers.

A NECESSARY DIALOGUE

Science and philosophy are each so demanding that it is virtually impossible for a person to be proficient in both, let alone in one, except in a bookish, observer capacity that serves as a poor substitute for an insider's intimate familiarity. Yet scientists and philosophers must talk to each other. Unless philosophy is merely to be an exercise in pure thought, it must take account of science. And scientists ask metaphysical questions, like everyone else, but they seek answers from the vantage point of their expertise.

Traditionally, the dialogue with philosophers has been held mainly by theoretical physicists and mathematicians, probably because of a common meeting ground in abstraction. The resulting cosmological picture comprised all facets of the physical world, from elementary particles to galaxies, but either ignored life or had life and mind tagged on to the picture as separate entities by some implicit, sometimes explicit, recourse to vitalism and dualism. This is wrong. Life is an integral part of the universe; it is even the most complex and significant part of the known universe. The manifestations of life should dominate our world picture, not be excluded from it. This has become particularly mandatory in view of the revolutionary advances in our understanding of life's fundamental processes.

Throughout this book, I have recounted the history of life on Earth in a manner designed to reveal underlying causalities and driving forces. A pattern emerges, dominated at the start by deterministic factors, increasingly affected by contingency as evolution progresses, though within constraints more stringent than is often assumed. Looking at the broad picture, we may now ask the question: Who, of Teilhard or Monod, was closer to the truth?

THE LIVING COSMOS

The universe is a hotbed of life. When I was a student, organic chemistry was viewed as esoteric, a chemistry practiced only by living organisms, including the organic chemists themselves. We did not believe in vitalism, of course, but some residue of the vitalist mystique colored our thoughts. Space chemistry has shattered this fantasy. Organic chemistry is just carbon chemistry, no more mysterious than any other chemistry, just immensely richer because of the unique associative properties of the carbon atom. Organic carbon compounds are everywhere. They make up 20 percent of interstellar dust, and interstellar dust makes up 0.1 percent of galactic matter.

In this organic cloud, which pervades the universe, life is almost bound to arise, in a molecular form not very different from its form on Earth, wherever physical conditions are similar to those that prevailed on our planet some four billion years ago. This conclusion seems to me inescapable. Those who claim that life is a highly improbable event, possibly unique, have not looked closely enough at the chemical realities underlying the origin of life. Life is either a reproducible, almost commonplace manifestation of matter, given certain conditions, or a miracle. Too many steps are involved to allow for something in between.

If I am right, there are about as many living planets in the universe as there are planets capable of generating and sustaining life. As mentioned in chapter 13, even a conservative estimate puts this number in the trillions. Unless this estimate is totally off the mark, we may take it that trillions of foci of past, present, or future life exist. Trillions of biospheres coast through space on trillions of planets, channeling matter and energy into the creative fluxes of evolution. In whatever direction we turn our eyes when we look into the sky, there is life out there, somewhere. This fact completely alters the cosmological picture. The Earth is not a freak speck around a freak star in a freak galaxy, lost in an immense "unfeeling" whirlpool of stars and galaxies hurtling in time and space ever since the Big Bang. The Earth is part, together with trillions of other Earth-like bodies, of a cosmic cloud of "vital dust" that exists because the universe is what it is. Avoiding any mention of design, we may, in a purely factual sense, state that the universe is constructed in such a way that this multitude of life-bearing planets was bound to arise. Among the billions of stars that make up each galaxy, many are bound to be circled by planets, a few of which, at least, are bound to be of the right size and in the right spatial orientation with respect to their sun (perhaps a large moon may be needed as well, to cause tides) to offer a cradle for life. The universe is not the inert cosmos of the physicists, with a little life added for good measure. The universe *is* life, with the necessary infrastructure around; it consists foremost of trillions of biospheres generated and sustained by the rest of the universe.

THE THINKING COSMOS

Evolution is at work in each biosphere, following the same universal principles. Because of the incessant interplay of chance mutations and environmental circumstances in determining the course of natural selection, no two biospheres could have the same history. The entire cloud of vital dust forms a huge cosmic laboratory in which life has been experimenting for billions of years. What it may have produced challenges imagination. The Earth's biodiversity, staggering as it is, may represent only a small sample of the diversity of life throughout the cosmos. What, in this overall pattern, is the likelihood of other biospheres besides our own giving rise to conscious, thinking organisms?

According to many evolutionists, this likelihood is very small, perhaps so small that the event happened only once in the whole universe and then only by an extraordinary stroke of luck. The German-born American biologist Ernst Mayr, from Harvard University, does not hesitate to write, in a work summarizing the fruits of a lifelong scrutiny of biological evolution: "An evolutionist is impressed by the incredible improbability of intelligent life ever to have evolved."[21]

Stephen Jay Gould echoes this view in a book devoted to the fossils found in the Burgess Shale, a geological site in the Canadian Rockies about 530 million years old and rich in fossils of bizarre animals. "Wind back the tape of life to the early days of the Burgess Shale," he writes, "let it play again from an identical starting point, and the chance becomes vanishingly small that anything like human intelligence would grace the replay."[22] He returns to this theme at the end of the book, exclaiming: "Biology's most profound insight into human nature, status, and potential lies in the simple phrase, the embodiment of contingency."[23]

Jared Diamond also believes humans to be unique. His reason: There are no woodpeckers in Australia, New Guinea, New Zealand, or Madagascar. "If woodpeckers had not evolved that one time in the Americas or Old World, a terrific niche would be flagrantly vacant over the whole Earth."[24] Diamond's idea is that one cannot trust convergent evolution. Woodpeckers, although ideally adapted, have emerged only once. So must it also be with humans. Hence, "for practical purposes, we are unique and alone in a crowded universe."[25] To which the chronicler of "the third chimpanzee" adds a disenchanted "Thank God."

The sociobiology advocate Edward O. Wilson likewise singles out woodpeckers, but for the opposite reason. "Woodpeckers and woodpecker-like forms," Wilson writes in *The Diversity of Life,* "illustrate the dual patterns of adaptive radiation and evolutionary convergence. During radiations of birds in different parts of the world, separate lines evolved to fill the woodpecker niche."[26] Wilson does not, however, extrapolate from woodpeckers to humans, although he does, in another book written with Charles J. Lumsden, express the view that "there is life around some of the stars" and that "there are probably also advanced civilizations."[27]

The reasons why many evolutionists believe in the uniqueness of intelligent life are easy to understand. Humankind is the outcome of a succession of many key events in which chance played an important role. To mention only a few of those events, some prokaryote first had to evolve into a primitive eukaryotic cell, a transformation that took more than one billion years, required a large number of genetic innovations as well as a special environmental setting, and, for all we know, may well have happened only once. At least, no trace of more than one eukaryotic line has so far been found among present-day organisms. While this prokaryote-eukaryote transition went on, other major events had to take place in the bacterial world, in particular the evolution of oxygen-generating phototrophy and oxygen-utilizing aerobic life, followed later by the endosymbiotic adoption of the ancestors of mitochondria and chloroplasts. Then, starting from some eukaryotic protists, animals of increasing complexity had to arise, preceded by the plants on which they were to feed. Also, sexual reproduction had to be inaugurated. Several body plans of increasing sophistication were to emerge in succession: two-layered with radial symmetry, three-layered with bilateral symmetry, inversion of mouth-to-anus polarity, creation of a coelom, formation of a notochord and a neural tube, adaptation to terrestrial life, with the sequential appearance of amphibians, reptiles, and mammals, development of primates, and, finally, the fateful step from ape to man.

Each of these decisive steps, as well as the many others that link them, depended crucially on the right genetic changes occurring at the right time under the right conditions. Let a single small event be different and the whole history of life might have been different. It is a typical example of historical contingency, as pictured by the familiar "if" stories: If Cleopatra's nose had been shorter; if a grain of sand had not obstructed Cromwell's urethra; if . . . the fate of the world might have been different. This reasoning cannot be faulted without dismantling the solid edifice of modern evolutionary theory. Its conclusion is incontrovertible. Should things start all over again, here or elsewhere, the final outcome would not be the same. But how different would it be?

To answer this question, we must consider what I have called the constraints of chance. Evolution does not operate in a world of infinite possibilities in which only a throw of dice decides which possibility will become reality. Let me list some of the limitations.

1. Mutations are not truly random events, in the sense of being ruled entirely by chance. Some areas in the genome are more sensitive to mutagenic influences than others, and this sensitivity itself varies according to genetic and environmental influences. Gene mutability has been woven by natural selection within a network of responses that all, one way or the other, facilitate or hamper the mutability of

certain genes in a manner favorable to the organism. This complex control has even led some researchers to speculate about the possibility of "selective" or "adaptive" mutations.[28]

2. Not all genetic changes are equally significant. It is now generally agreed that simple point mutations, those that result merely in the replacement of one nucleotide by another in a nucleic acid or of one amino acid by another in a protein, rarely play a role in what is known as macroevolution, the kind of evolution we are talking about. Most often, the truly creative mutations involve whole chunks of DNA that are duplicated, inverted, transposed, or otherwise reshuffled.

3. Not all genes are equally significant targets of mutation. The genes concerned in nontrivial evolutionary steps most often belong to the small class of regulatory genes, such as homeotic genes. Run-of-the-mill "housekeeping" genes are rarely involved. It is striking that loss of enzymes, rather than gain of enzymes, has often accompanied evolutionary "progress." We humans are particularly indigent in this respect, which is why we must find in our food so many vitamins and essential nutrients manufactured by so-called "lower" forms of life, which are, in fact, biochemically richer than we are.

4. Then, there is the organism in which the mutation occurs. Only in a given organismic context can a given genetic change be evolutionarily influential. Preexisting body plans limit the possibilities of viable change. Once a direction has been taken, the possible scope of future changes narrows, and it narrows even further with each subsequent evolutionary step. This explains the occurrence of evolutionary lines, as well as the increasing pace at which evolution often proceeds along such lines. The evolution of the horse is a textbook example. The ape-to-human transformation is a particularly striking one.

5. Related to the preceding condition, there is a historical, hierarchical factor. At each level of complexity, a different kind of genetic change becomes relevant. To take just a few examples, very early in the development of life, the most decisive changes were those that increased the accuracy of RNA replication. In the course of the prokaryote-eukaryote transition, the emergence of structural proteins, such as actin and tubulin, was an essential condition of cellular enlargement. In the formation of multicellular animals, cell adhesion molecules (CAMs) and substrate adhesion molecules (SAMs) became pace-setting innovations. And so on. To each evolutionary phase its own type of mutations.

6. To these numerous internal conditions must be added the crucial importance of the environment. Certain genetic changes can be beneficial and, therefore, be retained by natural selection only under certain conditions. Quite often, it is an environmental change that makes the difference. When we surveyed the evolution of plants and animals, we encountered many instances of major changes triggered by climatic or geological upheavals. Again, a relevant example in our own history is the transition from forest to savannah, which may have played a key role in hominization.

7. Finally, not every genetic change retained by natural selection is equally deci-

sive. Most changes, in fact, have but a marginal impact on the unfolding of evolution, contributing only to secondary biodiversity. Most outer branches and twigs on the tree of life are no more than that, variations on the same basic theme, the sources of taxonomic delights or nightmares, depending on your temperament, but of no significance for the course of evolution. Darwin's famous finches, which evolved differently on each island of the Galapagos archipelago, are a case in point. Key mutations are those that determine a major fork in the tree of life, for example, those that led to segmentation in the body plan of a worm, to polarity inversion in some annelid, or to neoteny in a hominoid ape.

These facts are illustrated by the uneven course of evolution, highlighted under the name of punctuated equilibrium by Stephen Jay Gould and his colleague Niles Eldredge, from the American Museum of Natural History in New York.[29] Evolution proceeds by short phases of rapid change separated by long periods of stasis or slow drift. The rapid change takes place when all the conditions listed above are met simultaneously. Chance still plays a role, but within a set of constraints so stringently limiting the range of possibilities open to its exploration that evolving systems sometimes have to wait many millions of years for the dice to fall inside the allowed boundaries.

Viewed within this context, the history of life on Earth allows less leeway to contingency and unpredictability than current fashion claims. Just as the proverbial forest is hidden by its trees, the tree of life itself is hidden by its luxurious canopy (see figure 31.1). The realm of contingency lies mostly in the millions of outer twigs and small branches, where chance was given full scope to play innumerable variations with whatever blueprint was provided by the parent limb. Trim the tree of this outer diversity and you are left with a stark trunk delineated by a relatively small number of major forks, each of which introduced a significantly altered blueprint.

Here is where contingency was channeled most stringently by inner and outer constraints, sometimes causing evolution almost to grind to a standstill until the right kind of mutation occurred. I am comforted in this view by the fact that the right kind of mutation—or, at least, *a* right kind of mutation—did indeed occur when circumstances called for it. To recall only one example, when the need arose for a protein molecule capable of building strong intracellular props by self-assembly, appropriate mutations happened not once but twice, to produce actin and tubulin. Please, do not misunderstand me. I am not invoking the intervention of an agency that produced the right kind of mutation *because* it was needed. What I am saying is that the kind of mutations that produced actin and tubulin presumably were fairly banal events, which happened to hit the jackpot when the time was ripe. The same may have been true of other key evolutionary events, though by no means all. Other mutations, if they had happened first, could have initiated other forks, branching out in other directions. There is plenty of room in my reconstruction for the development of differently shaped evolutionary trees on the other planets where life has taken hold. But certain directions may carry such decisive selective advantages as to have a high probability of occurring elsewhere as well.

The direction leading to polyneuronal circuit formation is likely to be specially privileged in this respect, so great are the advantages linked with it. Let something like a neuron once emerge, and neuronal networks of increasing complexity are almost bound to arise. The drive toward larger brains and, therefore, toward more consciousness, intelligence, and communication ability dominates the animal limb of the tree of life on Earth, and could well do so on many other life-bearing planets. On the other hand, the bodies serving the brains and controlled by them need not be similar to human bodies, although they would likely possess appropriate means for sensing, acting, and communicating.

My conclusion: We are not alone. Perhaps not every biosphere in the universe has evolved or will evolve thinking brains. But a significant subset of existing biospheres have achieved intelligence, or are on the way to it, some, perhaps, in a form more advanced than our own. In the 1960s, a number of scientists seriously entertained this possibility. They even succeeded in convincing NASA to invest considerable sums in the search for extraterrestrial intelligence (SETI).[30] This highly iffy project now enjoys less favor, to some extent because of growing doubts about the existence of extraterrestrial intelligence and, especially, because it is felt that priority should be given to urgent "Earth" problems such as environmental pollution, AIDS, and other social ills. I agree with the second reason but not with the first one.

According to the view I defend, it is in the nature of life to beget intelligence wherever and whenever conditions allow. All around us, in distant space, little islands exist where thinking beings use their minds individually and cooperatively, as we do, to create cultures. Conscious thought belongs to the cosmological picture, not as some freak epiphenomenon peculiar to our own biosphere, but as a fundamental manifestation of matter. Thought is generated and supported by life, which is itself generated and supported by the rest of the cosmos.

Is there any chance of those little islands of thought ever communicating with each other and sharing cultures? According to the laws of physics, only a tiny part of cosmic thought could do so in a reasonable amount of time. Using the fastest available means of communication, that is, some kind of radio signals moving at the speed of light, we could not in our lifetime communicate—that is, send a message and receive an answer—with civilizations more than some thirty light-years away. We could, by enlisting future generations, extend this range to a few hundred light-years, but surely not much farther. Compare this distance with the diameter of our galaxy (about 100,000 light-years) and you can see that our access is perforce limited to only a small number of star systems in our own galaxy, a minuscule fraction of the universe, which contains billions of other galaxies between millions and billions of light-years away from us. With enormous luck, we might establish contact with one or two outside civilizations—which, in itself, would be an achievement of extraordinary significance; hence the SETI project—but most of the think-

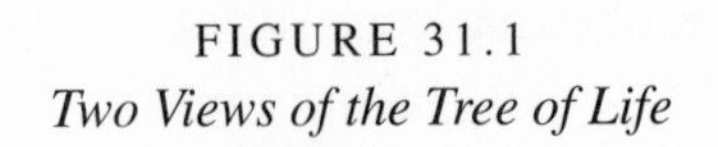

FIGURE 31.1
Two Views of the Tree of Life

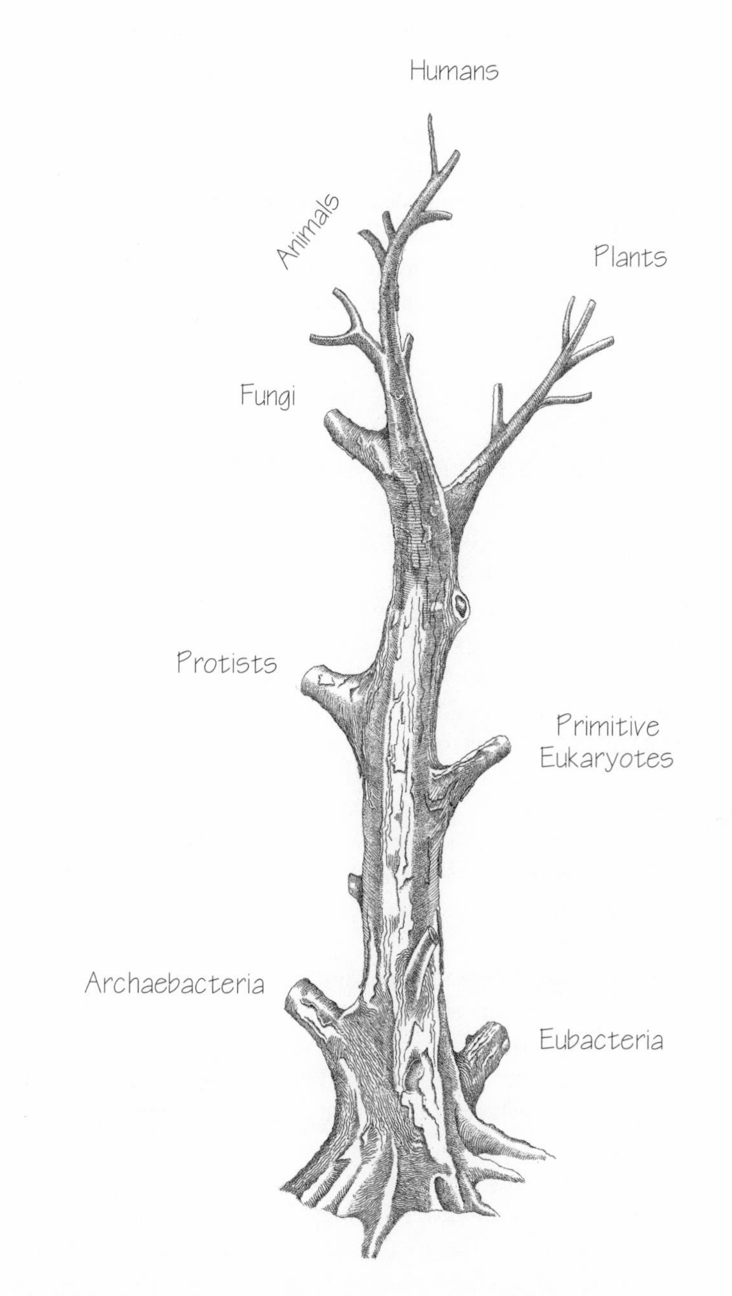

On the left is shown the tree of life, as it appears to us, with its rich canopy of greatly diversified ramifications hiding the trunk. Above is seen the same tree shorn of its canopy. The structure of the trunk, with its progressive rise toward greater complexity, is clearly evident. (Drawings by Ippy Patterson.)

ing cosmos, even within our own galaxy, would remain forever beyond our reach. For widespread cosmic communication we need an exception to the basic postulate of Einstein's relativity theory—that nothing can move faster than the speed of light.

Physicists, ever imaginative, have played with possible ways of getting around Einstein's rule, talking of things such as tachyons, hyperspace, and space warp, but these fantasies still belong to the realm of science fiction. Platonists like Penrose hint at instantaneous contact between minds and "Plato's world,"[31] but do not mention the possibility of two minds communicating in this way. A merger of minds into Teilhard's "noosphere" or Dyson's "world soul" remains no more than poetic images at the present time. But so would the notion of satellite television have appeared to Lucy if she had been at all capable of conceiving this possibility. Who can tell what the future has in store?

A MEANINGFUL UNIVERSE

It may appear that I have opted for Teilhard against Monod, but this is not so; I feel scientifically much closer to Monod than to Teilhard. I have, however, opted in favor of a meaningful universe against a meaningless one. Not because I want it to be so, but because that is how I read the available scientific evidence, which includes much that was unknown to Monod, who himself knew much more than did Teilhard. In addition, I have tried to encompass the whole history of life, not just the part best known to me, a dangerous but necessary annexation of unfamiliar territories.

My reasons for seeing the universe as meaningful lie in what I perceive as its built-in necessities. Monod stressed the improbability of life and mind and the preponderant role of chance in their emergence, hence the lack of design in the universe, hence its absurdity and pointlessness. My reading of the same facts is different. It gives chance the same role, but acting within such a stringent set of constraints as to produce life and mind obligatorily, not once but many times. To Monod's famous sentence "The universe was not pregnant with life, nor the biosphere with man,"[32] I reply: "You are wrong. They were."

Even if life and mind were rare, they would still be awe-inspiring manifestations of matter. Their products—the whole of biodiversity, the whole of human culture—would still be such as to induce feelings of reverence and wonder.

The notion of cosmic absurdity may appeal to scientists because of a "once bitten, twice shy" syndrome. For centuries, we gloried in the conviction of being the masters of a world that existed only for us. Copernicus shook us from our pedestal; every subsequent advance of science reduced our importance further. Our Earth was displaced from its central cosmological position to that of a mere planet circling the sun. The sun itself was shown to be but one of hundreds of billions of stars in one of hundreds of billions of galaxies, a speck lost in the vastness of the universe. The most stunning blow came with Darwin. Humankind was displaced from its self-appointed peak position to the tip of a twig among millions of other twigs sprouting from the tree of life. It has been a lesson in humility, engendering distrust of external authority and of internal certainty. We have become skeptics.

The lesson can be overstressed. It should not be converted into "human-bashing," calling on science to downgrade the human species and to vindicate the view that we humans have no place in the scheme of things. That there is, in fact, no scheme of things for us to fit in. The philosopher William Barrett, America's major expositor of European existentialism, has denounced what he views as "one of the supreme ironies of modern history: the structure that most emphatically exhibits the power of mind nevertheless leads to the denigration of the human mind."[33] Before him, the Hungarian-born British scientist and philosopher Michael Polanyi had written: "It is the height of intellectual perversion to renounce, in the name of scientific objectivity, our position as the highest form of life on earth and our own advent by a process of evolution as the most important problem of evolution."[34]

If the universe is not meaningless, what is its meaning? For me, this meaning is to be found in the structure of the universe, which happens to be such as to produce thought by way of life and mind. Thought, in turn, is a faculty whereby the universe can reflect upon itself, discover its own structure, and apprehend such immanent entities as truth, beauty, goodness, and love. Such is the meaning of the universe, as I see it.

What is important in this view is not absolute truth, probably inaccessible at our level of development, but the search for truth. In the same way, there is no absolute beauty, but a shared yearning for beauty; no absolute good, but a shared striving after goodness. Just see what different peoples have held or hold for beautiful or for good in different parts of the world and at different times. To me, the main message should be one of tolerance for others, and of humility for oneself. I have tremendous faith in modern science and have devoted my life to it. But I feel that science should not be arrogant. The human mind may be only a link—perhaps even a side branch—in an evolutionary saga that is far from completed and may well some day produce minds much more powerful than ours. According to the predicted lifetime of the sun, on our planet alone the thinking biosphere has another five billion years to go, one thousand times the duration of the step from ape to man. We must bow to mystery.

EPILOGUE

I have opposed two paradigmatic personalities, Monod and Teilhard; two philosophies, one featuring absurdity and the other meaningfulness. It is up to each of us to choose. Shall we proclaim with Macbeth that life is "a tale told by an idiot, full of sound and fury, signifying nothing"? Or shall we heed Hamlet: "There are more things in heaven and earth, Horatio, than are dreamt of in your philosophy"?

I have stated my choice and my reasons for it. Although based on science, these reasons are not unbiased. Teilhard, the devout Jesuit, wanted with all his strength to discover in the living world objective evidence in support of his faith. Monod, the proudly despairing existentialist, wanted just as passionately for the living world to bolster his feeling of isolation and absurdity. Neither of them, surely, saw himself as other than totally honest and intellectually rigorous. It would be foolish on my part to claim immunity from bias when such great minds failed to overcome their own prejudices.

I shall leave the last words to two other Frenchmen. Here, out of Pascal's *Pensées,* is God's message to the researcher: "Be consoled; you would not look for me if you had not already found me."[35] Not all may find solace, or even meaning, in the message. Those who prefer to steer clear of metaphysics may, with Medawar,[36] be content to follow the concluding advice Voltaire put in the mouth of his Candide: "We must cultivate our garden."[37]

Notes

Reference numbers in the notes refer to works in the additional reading list.

PREFACE

The epigraph is from a text by Einstein originally published in the October 1930 issue of *The Forum* and reprinted in M. Hill, ed., *Wise Men Worship* (New York: E. P. Dutton, 1931), p. i.

INTRODUCTION

1. Data on the comparative sequencing of cytochrome *c* are from R. E. Dickerson and I. Geis, *Structure and Action of Proteins* (Menlo Park, Calif.: Benjamin/Cummings, 1969), pp. 64–65.
2. In his landmark work, *On the Origin of Species by Natural Selection* (London: Murray, 1859), Charles Darwin defines natural selection as the "preservation of favorable variations and the rejection of injurious variations" (p. 81), that is, redefined in modern terms, the process whereby, in a genetically diverse population of organisms competing for the same limited resources, organisms endowed with the hereditary ability to produce the largest number of similarly endowed offspring—whatever the reason—progressively outnumber the others. This process is considered the driving force of biological evolution. For additional information, see M. Ruse, *Darwinism Defended* (Reading, Mass.: Addison-Wesley, 1982).
3. Popular accounts of the evidence on the antiquity of life provided by microfossils and stromatolites are to be found in reference 7 and in L. Margulis and L. Olendzenski, eds., *Environmental Evolution* (Cambridge, Mass.: MIT Press, 1992). For more technical, detailed reviews, see reference 1.
4. Isotopes are atoms of the same element with different atomic masses. They have the same number of protons and electrons, but differ by the number of neutrons in their nucleus. With six peripheral electrons, ^{12}C has six protons and six neutrons

in its nucleus, whereas ^{13}C has six protons and seven neutrons. Carbon assimilation by phototrophic organisms discriminates in favor of the lighter isotope. Measurements of the relative abundance of the two isotopes in old carbon deposits can thus provide information on the occurrence of biological activity at the time the deposits were made. Details on this technique and on the results it has yielded can be found in reference 1.

5. Estimates of the conditions during the first 800 million years of the Earth's existence depend very much on the model adopted for the Earth's formation. See G. Arrhenius in *Earth, Moon and Planets* 37 (1987): 187–99.
6. In reference 5, R. Shapiro gives an informative, critical, and entertaining overview of the main theories of the origin of life.
7. F. Hoyle and N. C. Wickramasinghe have described their theory in *Lifecloud* (New York: Harper & Row, 1978).
8. The hypothesis of directed panspermia is proposed by F. H. Crick in *Life Itself* (New York: Simon & Schuster, 1981).
9. Alan Truscott, the bridge columnist of the *New York Times,* has reported two accounts of complete suits being dealt, but he tells me that the possibility that the hands were prearranged cannot be excluded.
10. Reference 64, p. 145.
11. Hoyle's Boeing 747 analogy is cited in reference 5, p. 127.

CHAPTER 1

1. The formation of planet Earth is discussed in reference 2. Conditions on the primitive Earth are examined in references 1 and 4. The subject is also dealt with by a number of contributions in reference 8. See also the introduction, note 5.
2. An alternative possibility is that phosphate was present in more soluble forms on the prebiotic Earth. This is the opinion defended by G. Arrhenius, B. Gedulin, and S. Mojzsis, who believe that apatite formation depends on the presence of living organisms, in C. Ponnamperuma and J. Chela-Flores, eds., *Proceedings: Conference on Chemical Evolution and the Origin of Life* (Hampton, Va.: Deepak Publishing, 1993).
3. The thesis that life must have originated at low temperature has been repeatedly defended by the abiotic chemistry pioneer S. L. Miller. See his contribution in reference 7, pp. 1–28.
4. Information on deep-sea hydrothermal vents and the various forms of life they harbor is collected in N. G. Holm, ed., *Marine Hydrothermal Systems and the Origin of Life* (Norwell, Mass.: Kluwer, 1992). See also the review of Holm's book by A. Lazcano in *Science* 260 (1993): 1154–55, and the intriguing suggestion by T. Gold that a vast biosphere may be hidden in the hot depths of the Earth (*Proc. Natl. Acad. Sci. USA* 89 [1992]: 6045–49). A number of new organisms belonging to this special, hot biosphere have been isolated by the German microbiologist Karl O. Stetter. A summary of his early findings are published in reference 8, pp. 195–219.
5. This quotation is excerpted from a letter written by Charles Darwin to his botanist

friend Joseph Dalton Hooker and reproduced in F. Darwin, ed., *The Life and Letters of Charles Darwin,* vol. 2 (New York: D. Appleton, 1887).

6. See A. I. Oparin, *The Origin of Life on the Earth,* 3d ed. (New York: Academic Press, 1957).
7. This historic paper appeared in *Nature* 171 (1953): 737–38. Contrasting personal accounts of this discovery are to be found in J. D. Watson, *The Double Helix* (New York: Atheneum, 1968; see also the re-edition, together with reviews, edited by G. S. Stent [New York: Norton, 1980]), and in F. Crick, *What Mad Pursuit* (New York: Basic Books, 1988).
8. Stanley Miller's classic paper was published in *Science* 117 (1953): 528–29.
9. See H. Urey, *The Planets: Their Origin and Development* (New Haven: Yale University Press, 1952).
10. An excellent review of abiotic chemistry is given by J. Oró, S. L. Miller, and A. Lazcano in *Annu. Rev. Earth Planet. Sci.* 18 (1990): 317–56. For a summary survey of the field, see the article by S. L. Miller in reference 7, pp. 1–28.
11. Extraterrestrial chemistry is reviewed by S. Green in *Annu. Rev. Phys. Chem.* 32 (1981): 103–38. Several contributions in reference 8 deal with space chemistry.
12. A. H. Delsemme has summarized his views in *Orig. Life Evol. Biosp.* 21 (1992): 279–98.
13. Born in 1892, J. B. S. Haldane is remembered as one of England's most brilliant and versatile scientists and thinkers. He made important contributions to population genetics and evolution theory and wrote a large number of incisive essays on a great variety of subjects. He was also a militant member of the Communist party and became a great admirer of India, where he spent the last years of his life. He wrote only one paper, but a very influential one, on the origin of life. It was published in 1929 in *The Rationalist Annual* (see J. B. S. Haldane, *On Being the Right Size and Other Essays,* ed. J. M. Smith [Oxford and New York: Oxford University Press, 1985], pp. 101–12).
14. In an early paper on protein synthesis in *Symp. Soc. Exp. Biol.* 12 (1958): 138–63, Crick defines the Central Dogma as follows: "This states that once 'information' has passed into protein, it cannot get out again. In more detail the transfer of information from nucleic acid to nucleic acid, or from nucleic acid to protein may be possible, but transfer from protein to protein, or from protein to nucleic acid is impossible."
15. For information on ribozymes, see T. R. Cech in *Sci. Am.* 255, no. 5 (1986): 64–75.
16. W. Gilbert in *Nature* 319 (1986): 618.
17. W. Gilbert in *Cold Spring Harbor Symp. Quant. Biol.* 52 (1987): 903.
18. On RNA-like molecules, see L. E. Orgel in *Nature* 358 (1992): 203–9.
19. PNA is described in papers by D. Y. Cherny et al. in *Proc. Natl. Acad. Sci. USA* 90 (1993): 1667–70; and by P. Wittung et al. in *Nature* 368 (1994): 561–63.
20. Several quotations from experts referring to the difficulty of a prebiotic synthesis of RNA are reproduced in reference 6, p. 129. See also G. F. Joyce in *New Biologist* 3, no. 4 (1991): 399–407; and, for a technically more demanding collection of papers, R. F. Gesteland, ed., *The RNA World* (Cold Spring Harbor, N.Y.: Cold Spring Harbor Laboratory Press, 1993).

21. The argument in support of congruence is explained in greater detail in my paper "The RNA World: Before and After?" in *Gene* 135 (1993): 29–31.

CHAPTER 2

1. J. Oró in *Biochem. Biophys. Res. Commun.* 2 (1960): 407–12.
2. R. Shapiro in *Orig. Life Evol. Biosp.* 18 (1988): 71–85.
3. J. D. Bernal, an eminent British physical chemist and crystallographer, is the author of *The Origin of Life* (London: Weidenfeld and Nicholson, 1968), one of the first books on the topic.
4. J. P. Ferris and G. Ertem in *Science* 257 (1992): 1387–89.
5. For reference, see the introduction, note 5.
6. G. Wächtershäuser has described his model in *Microbiol. Rev.* 52 (1988): 452–84, and in *Proc. Natl. Acad. Sci. USA* 87 (1990): 200–4. A critique of the model by C. de Duve and S. L. Miller has appeared in *Proc. Natl. Acad. Sci. USA* 88 (1991): 10014–17. Wächtershäuser has written a rebuttal in *Proc. Natl. Acad. Sci. USA* 91 (1994): 4283–87.
7. R. E. Eakin in *Proc. Natl. Acad. Sci. USA* 49 (1963): 360–66.
8. A possible participation of protein catalysts in protometabolism is evoked by L. Dillon in *The Genetic Mechanism and the Origin of Life* (New York: Plenum Press, 1978); F. Dyson in *Origins of Life* (Cambridge: Cambridge University Press, 1985); and R. Shapiro in reference 5.
9. The formation of "proteinoids" was first reported by S. W. Fox and K. Harada in *Science* 128 (1958): 1214. A recent account of Fox's work is given in his book *The Emergence of Life* (New York: Basic Books, 1988).
10. T. Wieland has reviewed his work in H. Kleinkauf, H. von Döhren, and L. Jaenicke, eds., *The Roots of Modern Biochemistry* (Berlin: Walter de Gruyter, 1988), pp. 213–21.
11. The discovery of acetyl-coenzyme A was announced by F. Lynen and E. Reichert in *Angew. Chem.* 63 (1951): 47–48.
12. An overview of the role of group transfer in biosynthesis is given in chapter 8 of reference 12.
13. T. Wieland and W. Schäfer in *Angew. Chem.* 63 (1951): 146–47, and in *Liebigs Annal. Chem.* 576 (1952): 104–9.
14. The synthesis of bacterial peptides from thioesters is reviewed by H. Kleinkauf and H. von Döhren in *Annu. Rev. Microbiol.* 41 (1987): 259–89. F. Lipmann has reflected on the evolutionary significance of this mechanism in *Science* 173 (1971): 875–84.
15. The synthesis of key thiols under plausible prebiotic conditions is described in two papers by S. L. Miller and G. Schlesinger in *J. Mol. Evol.* 36 (1993): 302–7, 308–14.
16. The possible involvement of catalytic multimers in protometabolism is discussed in reference 6.
17. Catalytic effects of mixtures of amino acids are reported by A. Bar-Nun, E. Kochavi, and S. Bar-Nun in *J. Mol. Evol.* 39 (1994): 116–22.

CHAPTER 3

1. F. M. Harold, *The Vital Force: A Study of Bioenergetics* (New York: W. H. Freeman, 1986), p. 168.
2. C. E. Folsome, *The Origin of Life: A Warm Little Pond* (San Francisco: W. H. Freeman, 1979). See also H. J. Morowitz, *Cosmic Joy and Local Pain: Musings of a Mystic Scientist* (New York: Scribner's, 1987), and his *Beginnings of Cellular Life* (New Haven: Yale University Press, 1992).
3. See P. S. Braterman, A. G. Cairns-Smith, and R. W. Sloper in *Nature* 303 (1983): 163–64, and Z. K. Borowska and D. C. Mauzerall in *Orig. Life* 17 (1987): 251–59.
4. Details on banded iron-formations are to be found in reference 1.
5. E. Drobner et al. in *Nature* 346 (1990): 742–44.
6. On iron-sulfur proteins, see the review by R. Cammack in *Chem. Scripta* 21 (1983): 87–95. Also to be noted is the paper by R. V. Eck and M. O. Dayhoff in *Science* 152 (1966): 363–66, which indicates that an ancient bacterial iron-sulfur protein may have originally descended from a tetrapeptide. This fact points to the possibility that very small peptides may have been involved in protometabolism. See also note 15 in this chapter.
7. The distinction between high-energy and low-energy bonds was made by F. Lipmann in a landmark paper published in *Adv. Enzymol.* 1 (1941): 99–162.
8. The role of inorganic pyrophosphate in extant organisms and its possible involvement in prebiotic energy transfer have been reviewed by H. G. Wood (pp. 581–602) and by H. Baltscheffsky, M. Baltscheffsky, and M. Lundin (pp. 917–22) in H. Kleinkauf, H. von Döhren, and L. Jaenicke, eds., *The Roots of Modern Biochemistry* (Berlin: Walter de Gruyter, 1988).
9. Concerning the volcanic production of polyphosphates, see Y. Yamagata et al. in *Nature* 352 (1991): 516–19. G. Arrhenius has recently described mineral-catalyzed mechanisms of polyphosphate synthesis that could have played a prebiotic role. For reference, see the introduction, note 5.
10. The prebiotic role of thioesters is discussed in detail in reference 6. See also my paper in reference 8, pp. 1–20.
11. A. Weber in *J. Mol. Evol.* 18 (1981): 24–29, and *BioSystems* 15 (1982): 183–89.
12. See chapter 1, note 5.
13. A good survey of hyperthermophilic bacteria and their phylogeny pointing to a very ancient origin is given by K. O. Stetter in his contribution in reference 8, pp. 195–219. See also chapter 1, note 4.
14. A. Weber in *Orig. Life* 15 (1984): 17–27.
15. L. Kerscher and D. Oesterhelt in *Trends Biochem. Sci.* 7 (1982): 371–74. See also note 6 in this chapter.
16. The association of iron with sulfur also characterizes Wächtershäuser's model (see chapter 2, note 6). Membranes made of iron sulfide have recently been suggested as possible prebiotic energy transducers by Michael J. Russell and coworkers from the University of Glasgow in *Terra Nova* 5 (1993): 343–47, and *J. Mol. Evol.* 39 (1994): 231–43.

CHAPTER 4

1. See chapter 2, note 1.
2. P. G. Stoks and A. W. Schwartz in *Geochim. Cosmochim. Acta* 45 (1981): 563–69.
3. A. Eschenmoser and E. Loewenthal in *Chem. Soc. Rev.* 21 (1992): 1–16.
4. The reaction whereby ATP could originally have arisen with the help of a thioester writes as follows:

$$\text{R-S-CO-R' + AMP + PP}_i \rightleftharpoons \text{R-SH + R'-COOH + ATP}$$

   From right to left, this reaction now serves universally in the activation of acids, with coenzyme A as R-SH.
5. See chapter 1, note 10.
6. See the chapter by H. B. White III in J. Everse, B. Anderson, and K.-S. You, eds., *The Pyridine Nucleotide Coenzymes* (New York: Academic Press, 1982), pp. 1–17.
7. S. L. Miller and L. E. Orgel, *The Origins of Life* (Englewood Cliffs, N.J.: Prentice-Hall, 1973), p. 185.
8. I have heard Miller use the term "robust" several times in lectures. I have not seen it in print.
9. In his chapter in reference 7 (pp. 1–28), S. L. Miller also criticizes the notion that the emergence of life must have taken a very long time. He mentions a figure of 10,000 years as plausible. But this is for the whole process. My estimate of a few years or less refers to the time needed to reach RNA.

CHAPTER 5

1. See chapter 1, note 14.
2. The key sentence, which went unheeded at the time, appears in a short paper by E. Chargaff in *Experientia* 6 (1950): 201–9: "It is, however, noteworthy—whether this is more than accidental, cannot yet be said—that in all deoxypentose nucleic acids examined thus far the molar ratios of total purines to total pyrimidines, and also of adenine to thymine and guanine to cytosine, were not far from one." In his autobiography, *Heraclitean Fire* (New York: Rockefeller University Press, 1978), Chargaff recounts how, being a "terrible complexifier," he "missed the opportunity of being enshrined in the various halls of fame of the science museums" (p. 98).
3. See chapter 1, note 7.
4. Reference 11, p. 180.
5. S. Spiegelman's pioneering work is reviewed in *Amer. Sci.* 55 (1967): 221–64.
6. L. E. Orgel in *Proc. R. Soc. London B* 205 (1979): 435–42, and *J. Theor. Biol.* 123 (1986): 127–49.
7. M. Eigen et al. in *Sci. Am.* 244, no. 4 (1981): 88–118.
8. Reference 22.

9. M. Eigen and R. Winkler-Oswatitsch in *Naturwissenschaften* 68 (1981): 282–92.
10. A. M. Weiner and N. Maizels in *Proc. Natl. Acad. Sci. USA* 84 (1987): 7383–87.
11. Most striking is the interaction between RNA and arginine, described by M. Yarus in *Science* 240 (1988): 1751–58.
12. Most likely, D- and L-amino acids coexisted in prebiotic days. In particular, the catalytic multimers of my model (see chapter 2) are assumed to have included both kinds of amino acids (as do the bacterial peptides, such as gramicidin S, that are made from thioesters in present-day organisms).
13. In the words of F. Crick (*J. Mol. Biol.* 38 [1968]: 367–79), it was "not impossible to imagine that the primitive machinery had no protein at all and consisted entirely of RNA" (p. 50).
14. H. F. Noller, V. Hoffarth, and L. Zimniak in *Science* 256 (1992): 1416–19.

CHAPTER 6

1. Eigen has made a theoretical study of such feedback loops, which he calls "hypercycles" (see chapter 5, note 7). In reference 62, S. A. Kauffman has extended these studies to highly complex systems.
2. F. Lipmann in *Essays in Biochemistry* 4 (1968): 1–24.
3. For reviews, see D. D. Buechter and P. Schimmel in *Crit. Rev. Biochem.* 28 (1993): 309–22, and M. E. Saks, J. R. Sampson, and J. N. Abelson in *Science* 263 (1994): 191–97.
4. I have described these primeval correspondences between amino acids and RNA molecules under the name "The Second Genetic Code" (*Nature* 333 [1988]: 117–18).
5. See chapter 5, note 9.

CHAPTER 7

1. See chapter 5, note 7.

CHAPTER 9

1. See chapter 1, note 6.
2. A. L. Herrera has summarized "forty-three years of experimental investigation" in a short note in *Science* 96 (1942): 14.
3. See chapter 2, note 9.
4. See chapter 3, notes 1 and 2.
5. The "myth of the primeval soup" has been denounced by C. R. Woese (*J. Mol. Evol.* 13 [1979]: 95–101); C. B. Thaxton, W. L. Bradley, and R. L. Olsen (*The Mystery of Life's Origin* [New York: Philosophical Library, 1984]); R. Shapiro (reference 5); and G. Wächtershäuser (*Microbiol. Rev.* 52 [1988]: 452–84), among others.

6. See chapter 5, note 7 (p. 101).
7. In opposition to van der Waals forces, which govern attractions between nonpolar molecules, electrostatic attraction and repulsion forces are often designated Coulomb forces, from the name of the French physicist who discovered them. Both forces play essential roles in the structures of macromolecules and in all sorts of intermolecular interactions. Hydrogen bonds, mentioned in chapter 5, depend on a special kind of electrostatic attraction.
8. G. Blobel in *Proc. Natl. Acad. Sci. USA* 77 (1980): 1496–1500.
9. T. Cavalier-Smith in *Ann. N.Y. Acad. Sci.* 503 (1987): 55–71.
10. The different kinds of membranes that are present in the trillions of cells that compose a complex organism, such as the human body, for example, all originate by accretion, broken by successive cell division, from the corresponding membranes present in the fertilized egg cell. There is thus a maternally transmitted membrane continuity from generation to generation. There are reasons for believing that a certain amount of topological information—membranes serving as templates for like membranes—is transmitted in this way.
11. Readers interested in the penicillin story may enjoy D. Wilson, *Penicillin in Perspective* (London: Faber & Faber, 1976), which tries to discriminate between fact and myth.

CHAPTER 10

1. In anaerobic fermentations—for example, alcoholic fermentation, the conversion of glucose to ethanol; or lactic fermentation, the conversion of glucose to lactic acid—glucose gives rise, by way of a level-A-to-level-B, thioester-dependent, ATP-generating electron-transfer process, to a substance—pyruvic acid—that serves as the electron acceptor at the B level, to produce either ethanol and carbon dioxide or lactic acid, which are the final products of fermentation. Thanks to this doubling-up of the metabolic sequence through the A and B levels, the fermentation process can proceed without an outside acceptor (oxygen) to pick up the released electrons.

CHAPTER 12

1. C. F. Amabile-Cuevas and M. E. Chicurel in *Amer. Scient.* 81 (1993): 332–41.
2. C. R. Woese in *Sci. Am.* 244, no. 6 (1981): 98–122, and *Microbiol. Rev.* 51 (1987): 221–71.
3. C. R. Woese, O. Kandler, and M. L. Wheelis in *Proc. Natl. Acad. Sci. USA* 87 (1990): 4576–79.
4. According to a recent review by O. Kandler (*Progr. Botan.* 54 [1993]: 1–24), *all* of the most ancient prokaryotes, among both archaebacteria and eubacteria, are thermophilic. This amounts to a very strong case in support of a thermophilic ancestor.
5. P. Forterre et al. in *BioSystems* 28 (1993): 15–32.
6. M. Sogin in *Curr. Opin. Genet. Dev.* 1 (1991): 457–63.

CHAPTER 13

1. For information on the possible existence of life elsewhere in the universe, see G. Feinberg and R. Shapiro, *Life Beyond Earth* (New York: William Morrow, 1980); R. T. Rood and J. S. Trefil, *Are We Alone?* (New York: Scribner's, 1981); and R. Breuer, *Contact with the Stars* (San Francisco: W. H. Freeman, 1982). The first book ends with the conclusion that "the universe is full of life." The other two are distinctly less sanguine. Frank Drake, a pioneer in the search for extraterrestrial intelligence (SETI), has recently given a partly autobiographical account of this search, written in collaboration with D. Sobel, in *Is Anyone Out There?* (New York: Delacorte Press, 1992). See also the review of this book by R. N. Bracewell in *Science* 258 (1992): 1012–14.
2. Readers interested in the new science of computerized modeling of living processes and in the mecca of such research, the Santa Fe Institute, have a choice between two recent books written in a chatty, journalistic style (references 60 and 61), to which may be added S. Levy, *Artificial Life: The Quest for a New Creation* (New York: Cape/Pantheon, 1992). For a meatier account, readers should turn to reference 62. If in a hurry, they may fall back on papers by S. A. Kauffman in *Sci. Am.* 265, no. 2 (1991): 78–84, or by R. Ruthen in *Sci. Am.* 268, no. 1 (1993): 130–40.
3. Reference 62.
4. Reference 62, p. 232. For a history of the "edge of chaos" concept, see chapter 3 in reference 61.

CHAPTER 14

1. The repressor hypothesis is presented in a landmark paper by F. Jacob and J. Monod in *J. Mol. Biol.* 3 (1961): 318–56. Jacob has given a personal account of this discovery in his autobiography, *La Statue Intérieure* (Paris: Editions Odile Jacob, 1987), translated into English by F. Philip under the title *The Statue Within* (New York: Basic Books, 1988).
2. F. M. Burnet, *Clonal Selection Theory of Acquired Immunity* (Cambridge: Cambridge University Press; and Nashville, Tenn.: Vanderbilt University Press, 1959).
3. For information on archaebacteria, see the review by W. J. Jones, D. P. Nagle, and W. B. Whitman in *Microbiol. Rev.* 51 (1987): 135–77; C. Edwards, ed., *Microbiology of Extreme Environments* (New York: McGraw-Hill, 1990); M. J. Danson, D. W. Hough, and G. G. Lunt, eds., *The Archaebacteria* (London: Portland Press, 1992); and the review by O. Kandler in *Progr. Botan.* 54 (1993): 1–24.
4. Ibid.
5. See the introduction, note 3.
6. J. W. Schopf's latest data are described in *Science* 260 (1993): 640–46.

CHAPTER 15

1. W. Zillig, P. Palm, and H.-P. Klenk in reference 8, pp. 181–93.
2. The hypothesis of a eukaryote ancestor with a DNA genome anteceding prokary-

ote has been proposed by P. Forterre et al. (chapter 12, note 5). The hypothesis of a similar ancestor with an RNA genome has been presented by H. Hartman in *Specul. Sci. Technol.* 7 (1985): 77–81, and, in a somewhat different form, by M. Sogin (chapter 12, note 6).

3. The endosymbiont theory goes back more than a century. It was revived by L. Margulis (then L. Sagan) in *J. Theor. Biol.* 14 (1967): 225–74. She has since expanded the theory in *Origin of Eukaryotic Cells* (New Haven: Yale University Press, 1970); reference 23; and, in collaboration with D. Sagan, *Micro-Cosmos* (New York: Summit Books, 1986). In collaboration with M. McMenamin, she has recently edited an English translation of *Concepts in Symbiogenesis* (New Haven: Yale University Press, 1992), a book published in the Soviet Union in 1979 in which L. N. Khakhina reviews the work of Russian biologists who pioneered the endosymbiosis concept in the early twentieth century. Many interesting contributions to the topic are to be found also in J. J. Lee and J. F. Fredrick, eds., *Endocytobiology III* (*Ann. N.Y. Acad. Sci.* 503 [1987]); P. Nardon et al., eds., *Endocytobiology IV* (Paris: Institut National de la Recherche Agronomique, 1990); and H. Hartman and K. Matsuno, eds., *The Origin and Evolution of the Cell* (Singapore: World Scientific, 1992).
4. A microfossil believed to be of a eukaryotic alga has been found by T.-M. Han and B. Runnegar (*Science* 257 [1992]: 232–35) in a 2.1-billion-year-old iron formation in Michigan. Should the eukaryotic nature of this organism be confirmed, the time of endosymbiont adoption would have to be put back at least 500 million years.
5. My description of *Giardia* is based largely on the work of K. S. Kabnick and D. A. Peattie, reviewed in their article in *Amer. Sci.* 79 (1991): 34–43. For a different viewpoint, see M. E. Siddall, H. Hong, and S. S. Desser in *J. Protozool.* 39 (1992): 361–67. The ancientness of *Giardia* has been reported by M. L. Sogin et al. in *Science* 243 (1989): 75–77.
6. Although designated by the same name, bacterial and eukaryotic flagella are totally different chemically, structurally, and functionally. To avoid confusion, L. Margulis (see reference 23) has long advocated that eukaryotic flagella and cilia be grouped under the name "undulipodia," a term borrowed from the old German literature. This proposal has not been widely adopted so far.
7. There is no description in the literature of *Giardia* actually taking up bacteria. However, there is clear evidence that the organism can engulf extracellular materials and digest them intracellularly. Furthermore, several of *Giardia*'s close relatives are known to be voracious eaters of bacteria. My description of *Giardia* on the hunting trail is within the limits of poetic license.
8. The expression "fierce predator" is from *Micro-Cosmos* (see note 3, above), p. 129.
9. M. McCarty, the last surviving member of the famous trio, has given a fascinating account of this epoch-making discovery in *The Transforming Principle* (New York: Norton, 1985).
10. It was stated by the diplomonad expert G. Brugerolle (*Protistologia* 11 [1975]: 111–18) that *Giardia* has no detectable Golgi, suggesting that this cell structure had not yet formed at the time *Giardia*'s ancestor branched from the eukaryotic line. However, a characteristic Golgi-associated secretion system becomes de-

tectable in *Giardia* when this organism prepares to encyst and builds an external carbohydrate-protein shell in the process. See F. D. Gillin, D. S. Reiner, and M. McCaffery in *Parasitology Today* 7 (1991): 113–16.

11. On the ancientness of microsporidia, see C. R. Vossbrinck et al. in *Nature* 326 (1987): 411–14.
12. On the evolution of the mitotic apparatus, see D. F. Kubai in *Int. Rev. Cytol.* 43 (1975): 167–227, and I. B. Heath in *Int. Rev. Cytol.* 64 (1980): 1–80.

CHAPTER 16

1. Growth associated with surface pleating can occur in walled cells. Giant bacteria encased in a cell wall have been described. They include a free-living sulfur bacterium isolated by E. Fauré-Frémiet and C. Rouiller (*Exp. Cell Res.* 14 [1958]: 29–46), and a fish intestinal parasite described by K. D. Clements and S. Bullivant (*J. Bacteriol.* 173 [1991]: 5359–62) and unambiguously identified as a prokaryote by E. R. Angert, K. D. Clements, and N. R. Pace (*Nature* 362 [1993]: 239–41). Both organisms have a highly pleated cell membrane.
2. The theory linking the development of a primitive eukaryote to the acquisition of phagocytosis was first proposed by C. de Duve and R. Wattiaux (*Annu. Rev. Physiol.* 28 [1966]: 435–92). It has been put forward independently by R. Y. Stanier (*Symp. Soc. Gen. Microbiol.* 20 [1970]: 1–38), and elaborated further by T. Cavalier-Smith (*Ann. N.Y. Acad. Sci.* 503 [1987]: 17–54). See also reference 6.
3. K. S. Kabnick and D. A. Peattie in *Amer. Sci.* 79 (1991): 34–43. For a contrasting opinion, see M. E. Siddall, H. Hong, and S. S. Desser in *J. Protozool.* 39 (1992): 361–67, who make no mention of haploidy and attribute the presence of two nuclei to delayed cell division.
4. The key experiment describing the first production of a monoclonal antibody was reported by G. Köhler and C. Milstein in *Nature* 231 (1975): 87–90. A general overview of the topic is given in J. W. Goding, *Monoclonal Antibodies: Principles and Practice,* 2d ed. (New York: Academic Press, 1987).

CHAPTER 17

1. See, however, chapter 15, note 4.
2. *Lorenzo's Oil* tells the story of the frantic search for a cure by the parents of a boy afflicted with adrenoleukodystrophy. The movie is brilliant, but its message may be dangerously misleading, coming at a time when medical research is particularly promising and, paradoxically, is subjected to many irrational attacks. See F. Rosen's review in *Nature* 361 (1993): 695.
3. See the review by M. Müller in *J. Gen. Microbiol.* 129 (1993): 2879–89. See also T. Fenchel and B. J. Finlay in *Amer. Sci.* 82 (1994): 22–29.
4. See chapter 15, note 3, and the chapter by L. Margulis in L. Margulis and L. Olendzenski, eds., *Environmental Evolution* (Cambridge, Mass.: MIT Press, 1992), pp. 173–99.
5. An endosymbiont origin of the eukaryotic nucleus is proposed by H. Hartman in

Specul. Sci. Technol. 7 (1985): 77–81, and, in a somewhat different form, by M. Sogin (see chapter 12, note 6).

6. F. Jacob in *Science* 196 (1977): 1161–66
7. Two distinct building blocks, α–tubulin and β–tubulin, actually combine to make microtubules. But they are closely similar and descendants of a common molecular ancestor.

CHAPTER 18

1. Some bacterial colonies behave like true organisms, according to J. A. Shapiro in *Sci. Am.* 258, no. 6 (1988): 82–89.
2. G. M. Edelman, *Topobiology* (New York: Basic Books, 1988).

CHAPTER 19

1. There is some evidence that bacteria may have taken hold on land as early as 1.2 billion years ago. See R. J. Horodyski and L. P. Knauth in *Science* 263 (1994): 494–98.
2. An excellent overview of plant and mold evolution is given in reference 16. For a simplified account, reference 24 is both informative and entertaining.
3. The great Permian crisis is described in reference 21. Additional data are given by I. H. Campbell et al. in *Science* 258 (1992): 1760–63.
4. Reference 16, pp. 556–57.
5. P. O. Wainright et al. in *Science* 260 (1993): 340–42.
6. See chapter 9, note 11.

CHAPTER 20

1. As a general reference book to animal evolution, I recommend reference 18. A simplified and entertaining account is given in reference 24.
2. In reference 17, S. J. Gould gives a comprehensive and critical analysis of the recapitulation law.
3. Hollow, spherical, multicellular arrangements, one cell thick, are formed by some bacteria (see chapter 18, note 1) and by some algae. *Volvox* is a textbook example of the latter.
4. The phylogenetic relationship between diploblasts and triploblasts is still uncertain. It is generally agreed that the two groups separated very early, but whether from a single protist ancestor or from distinct protist ancestors is not clear. Results reported by R. Christen et al. in *EMBO J.* 10 (1991): 499–503, "do not exclude the possibility that triploblasts and diploblasts arose independently from different protists." According to P. O. Wainright et al. in *Science* 260 (1993): 340–42, all animals are derived from a single choanoflagellate ancestor, which they share with fungi.
5. For an attractively presented summary of the work on the nematode *Caenorhabditis elegans,* see the article by M. Pines in reference 25, pp. 30–38.

6. For recent data on the Cambrian explosion, see A. H. Knoll in *Sci. Am.* 265, no. 4 (1991): 64–73; J. S. Levinton in *Sci. Am.* 267, no. 5 (1992): 84–91; A. H. Knoll and M. R. Walter in *Nature* 356 (1992): 673–78; A. H. Knoll in *Science* 256 (1992): 622–27; S. C. Morris in *Nature* 361 (1993): 219–25; and S. A. Bowring et al. in *Science* 261 (1993): 1293–98.
7. Knoll's theory linking the Cambrian explosion to a rise in atmospheric oxygen is outlined in articles cited above (see note 6).
8. See the introduction, note 3.
9. Claude Bernard's major work is his *Introduction à l'Etude de la Médecine Expérimentale,* published in Paris in 1865. Its English translation by H. C. Greene has appeared under the title *An Introduction to the Study of Experimental Medicine* (New York: Macmillan, 1927; New York: Dover, 1957).

CHAPTER 21

1. A simple introduction to the complexity of homeotic genes can be found in the article by P. Radetsky in reference 25, pp. 18–29.
2. See chapter 20, note 2.

CHAPTER 22

1. The saga of the discovery of the coelacanth is told in reference 24.
2. Zallinger's mural is reproduced, with a number of interesting comments, in V. Scully, R. F. Zallinger, and J. H. Ostrom, *The Age of Reptiles* (New York: Abrams, 1990).
3. Early descriptions of the iridium anomaly can be found in references 21 and 24. L. W. Alvarez, in his autobiography, *Adventures of a Physicist* (New York: Basic Books, 1987), and W. Alvarez and F. Asaro, in their chapter in J. Bourriau, ed., *Understanding Catastrophe* (Cambridge and New York: Cambridge University Press, 1992), pp. 28–56, give a personal account of their investigations of the anomaly. Recent evidence on Chicxulub as the site of the asteroid impact that presumably caused the extinction of the dinosaurs and many other species is given by V. L. Sharpton et al. in *Science* 261 (1993): 1564–67.
4. Pasteur's complete sentence is: "Dans les champs de l'observation, le hasard ne favorise que les esprits préparés."

CHAPTER 23

1. Alfred Lotka and Vito Volterra were two mathematical biologists who, in the early 1920s, made a theoretical study of the mutual influences of predator and prey populations. This study remains a classic. See reference 26.
2. J. F. Kasting in *Science* 259 (1993): 920–26.
3. J. Lovelock has developed his Gaia hypothesis in reference 26. In reference 61, the science writer R. Lewin recounts his visit and conversations with Lovelock.
4. For a portrait of L. Margulis and her endorsement of the Gaia hypothesis, see the vignette by C. Mann in *Science* 252 (1991): 378–81.

5. A scientist, physician, and administrator, Lewis Thomas, who passed away in December 1993, will be remembered mostly for his unique brand of scientifico-poetical essays collected in *The Lives of a Cell* (New York: Viking, 1974) and several other books. The quotation is from his foreword to reference 26, p. x.
6. F. Dyson, *From Eros to Gaia* (New York: Pantheon, 1992), p. 344.
7. Reference 26, p. 236.

CHAPTER 24

1. W. Gilbert in *Nature* 271 (1978): 501, and in *Science* 228 (1985): 823–24.
2. L. E. Orgel and F. H. C. Crick in *Nature* 284 (1980): 604–7.
3. Borrowed from the title of a book by R. Dawkins (reference 22), the concept of DNA selfishness was developed further by W. F. Doolittle and C. Sapienza in *Nature* 285 (1980): 601–3, and by Orgel and Crick (see note 2, above). See also chapter 9 of reference 19.
4. Reference 22, p. 47.
5. Evidence that genes contain intervening sequences that are removed from their RNA transcripts was reported simultaneously by S. M. Berget, C. Moore, and P. A. Sharp in *Proc. Natl. Acad. Sci. USA* 74 (1977): 3171–75, and by L. T. Chow et al. in *Cell* 12 (1977): 1–8. See also J. A. Witkowski in *Trends Biochem. Sci.* 13 (1988): 110–13.
6. See note 1, above.
7. There is increasing evidence that proteins are the products of modular construction. Some of the evidence is reviewed in E. M. Stone and R. J. Schwartz, eds., *Intervening Sequences in Evolution and Development* (New York: Oxford University Press, 1990). See also the article by M. Gō and M. Mizutani in H. Hartman and K. Matsuno, eds., *The Origin and Evolution of the Cell* (Singapore: World Scientific, 1992), and the review by R. F. Doolittle and P. Bork in *Sci. Am.* 269, no. 4 (1993): 50–56. Several such modules are shared between prokaryotes and eukaryotes and, therefore, may go back to the distant times when the genes were first assembled (see P. Green et al., in *Science* 259 [1993]: 1711–16). Interestingly, the modules, also called domains or motifs, often consist of a protein part involved in binding or recognition of some molecular entity, for example DNA or NAD, which explains their evolutionary conservation. The extent to which exons are module-coding sequences is, however, variously appreciated.
8. W. Gilbert, M. Marchionni, and G. McKnight in *Cell* 46 (1986): 151–54.
9. See the review by B. Dujon in *Gene* 82 (1989): 91–114.
10. B. McClintock has reviewed her work in her Nobel lecture. See *Les Prix Nobel 1983* (Stockholm: Almquist & Wiksell, 1984), pp. 174–93.
11. According to a recent paper by A. Stoltzfus et al. in *Science* 265 (1994): 202–7, "the exon theory of genes is untenable."
12. The estimate of about seven thousand for the number of exons is reported in a paper by R. L. Dorit, L. Schoenbach, and W. Gilbert in *Science* 250 (1990): 1377–82. For a critique of the method used, see R. F. Doolittle in *Science* 253 (1991): 677–79, and the authors' rebuttal (ibid., 679–80).

CHAPTER 25

1. Mary Leakey has told the story of the discovery of the Laetoli footprints in *Nation. Geogr.* 155 (1979): 446–57. In this article, she tells of only two individuals. She mentions a third in an account contributed to a Dutch collection, *De Evolutie van de Mens* (Maastricht: Natuur en Techniek, 1981).
2. The discovery of Lucy is recounted by D. Johanson in reference 32.
3. For eminently readable accounts of human origins, see references 31, 32, and 34. As an introduction to the topic, readers may turn to articles by B. Wood in *Nature* 355 (1992): 783–90; K. S. Thomson in *Amer. Sci.* 80 (1992): 519–22; and P. Andrews in *Nature* 360 (1992): 641–46. More details on the recent evolution of the human species are given in M. C. Corballis, *The Lopsided Ape* (New York: Oxford University Press, 1991); C. Willis, *The Runaway Brain* (New York: Basic Books, 1993); and reference 33. A pleasantly romanticized and nicely illustrated narrative of the origin and evolution of early humans and of their way of life is given in P. Angela and A. Angela, *The Extraordinary Story of Human Origins,* translated from the Italian by G. Tonne (Buffalo, N.Y.: Prometheus Books, 1993). Written largely in the style of a historical novel, this book is based on good scientific data and includes conversations with many leaders in the field.
4. The ups and downs of African Eve can be followed in articles by R. L. Cann, M. Stoneking, and A. C. Wilson in *Nature* 325 (1987): 31–36; C. B. Stringer in *Sci. Am.* 263, no. 6 (1990): 98–104; L. Vigilant et al. in *Science* 253 (1991): 1503–7; A. C. Wilson and R. L. Cann in *Sci. Am.* 266, no. 4 (1992): 68–73; A. G. Thorne and M. H. Wolpoff in *Sci. Am.* 266, no. 4 (1992): 76–83; A. Gibbons in *Science* 257 (1992): 873–75; and J. Klein, N. Takahata, and F. Ayala in *Sci. Am.* 269, no. 6 (1993): 78–83. See also reference 34 and *The Runaway Brain* by C. Wills (see note 3, above). Confirmation of a single African origin of modern humans has been given by D. M. Waddle in *Nature* 368 (1994): 452–54; A. M. Bowcock et al., ibid., 455–57; and Y. Coppens in *Sci. Am.* 270, no. 5 (1994): 88–95.
5. P. Lieberman, E. S. Crelin, and D. H. Klatt in *Amer. Anthropol.* 74 (1972): 287–307, and P. Lieberman, *On the Origin of Language* (New York: Macmillan, 1975).
6. The discovery of a fossil hyoid bone in Israel has been used as evidence against Lieberman's claim that the Neanderthals lacked the ability to speak. See B. Arensburg et al. in *Nature* 338 (1989): 758–60. In recent years a number of authors have rallied in defense of the Neanderthals, said to be wrongly accused of brutishness. See, for example, reference 34 and *The Runaway Brain* by C. Wills (see note 3, above).
7. Reference 33.

CHAPTER 26

1. An excellent introduction to the structure and evolutionary development of the human brain can be found in reference 42. For readers in a hurry, reference 46 gives a good overview of the topic.

2. On Paley and his theology, see reference 20.
3. Ibid.
4. No more than 1,829 successive steps, spread over some 400,000 generations, could have sufficed to transform a flat patch of light-sensitive cells sandwiched between a transparent protective layer and a layer of dark pigment into a respectable camera eye with refractive lens, according to a computer simulation study by D.-E. Nilsson and S. Peiger, published in *Proc. R. Soc. B* 256 (1994): 53–58. See also the commentary on this work by R. Dawkins in *Nature* 368 (1994): 691–92.
5. Darwin's religious preoccupations are discussed by J. H. Brooke in J. Durant, ed., *Darwinism and Divinity* (Oxford and New York: Basil Blackwell, 1985), pp. 40–75.
6. See chapter 20, note 5.
7. D. H. Hubel and T. N. Wiesel in *Sci. Am.* 241, no. 3 (1979): 150–62.
8. Reference 38.
9. G. M. Edelman has given a detailed account of his theory in *Neural Darwinism* (New York: Basic Books, 1987). In reference 47, he gives an outline of the theory directed to a wider readership.
10. F. Crick in *Trends Neurosci.* 12 (1989): 240–48.
11. Reference 38, p. 272.
12. H. Keller, *The World I Live In* (New York: Century, 1908).
13. Neoteny is discussed in detail in reference 17.
14. Reference 33, p. 48.

CHAPTER 27

1. See *The Mind's Eye,* a collection of articles originally published in *Scientific American,* with introductions by J. M. Wolfe (New York: W. H. Freeman, 1986), and reference 48.
2. In *The Remembered Present: A Biological Theory of Consciousness* (New York: Basic Books, 1989), G. M. Edelman explains in detail his theory of consciousness. This book follows his *Neural Darwinism* (see chapter 26, note 9). Edelman summarizes his main ideas for a wider readership in reference 47. See also the review of the last book (reference 47), really a thoughtful essay, by O. Sacks in the *New York Review of Books,* April 8, 1993, pp. 42–49. Readers may further enjoy Steven Levy's article "Dr. Edelman's Brain" in the *New Yorker,* May 2, 1994, pp. 62–73.
3. For a brief history of behaviorism and its replacement by a new cognitive science, see reference 39. Readers may also be interested in a recent biography, *B. F. Skinner: A Life,* by D. W. Bjork (New York: Basic Books, 1993).
4. D. Griffin has exposed his ideas in references 35, 37, and 44. See also M. S. Dawkins, *Through Our Eyes Only? The Search for Animal Consciousness* (New York: W. H. Freeman, 1993).
5. Jane Goodall tells this story in *The Chimpanzees of Gombe: Patterns of Behavior* (Cambridge, Mass.: Harvard University Press, 1986).
6. Reference 40.
7. In *A History of the Mind* (New York: Simon & Schuster, 1992), N. Humphrey

emphasizes the distinction between physical time, a mere succession of instants, and subjective time, a succession of overlapping segments of a certain duration.

8. *Consciousness Explained* (Boston: Little, Brown, 1991), by D. C. Dennett, dismantles popular concepts of consciousness to the point that little is left.
9. Patricia Smith Churchland has the rare merit of attempting to join neurobiology and philosophy, two disciplines that tend to keep very much to themselves. In *Neurophilosophy* (Cambridge, Mass.: MIT Press, 1986), she defends the doctrine of "eliminative materialism," also advocated by her husband, Paul M. Churchland (*Matter and Consciousness: A Contemporary Introduction to the Philosophy of Mind* [Cambridge, Mass.: MIT Press, 1984]). Allusions to "folk psychology" come from her book (pp. 299ff.).
10. Reference 45, p. 3.
11. Ibid.
12. Reference 47, p. 157.
13. Reference 48, p. 284.
14. Descartes delayed publication of his *Discours de la Méthode* from 1633 to 1637 after hearing of Galileo's condemnation by the Holy Office. To avoid the same fate, he took great precautions to explain that his theory was a purely "as if" account and in no sense a claim to absolute truth. A key tenet of this theory is that the bodies of both humans and animals can be viewed as entirely automatic machines, coordinated by the brain by means of "animal spirits" circulating through the nerves. Humans differ from animals by possessing an immortal soul, the site of intelligence and consciousness, which interacts with the animal spirits in the pineal gland.
15. Reference 36.
16. P. J. G. Cabanis has summed up his philosophical ideas in *Traité du Physique et du Moral de l'Homme* (1802).
17. Born in Holland, J. Moleschott taught mostly in Germany. His materialistic views are expounded in his book *Der Kreislauf des Lebens* (1852).
18. See note 8, above.
19. G. Ryle, *The Concept of Mind* (London: Hutchinson, 1949).
20. A. Koestler, *The Ghost in the Machine* (New York: Macmillan, 1967).
21. *Consciousness Explained* (see note 8, above), p. 268.
22. See note 2, above.
23. *The Remembered Present* (see note 2, above), p. 308.
24. Reference 47, pp. 79–80.
25. *The Remembered Present* (see note 2, above), p. 260.
26. Reference 48, p. 3.
27. Ibid., p. 285.
28. Reference 45, p. 212.
29. Ibid., p. 112.
30. R. Sperry, *Science and Moral Priority* (New York: Columbia University Press, 1983), p. 79.
31. Ibid.
32. Reference 43, p. 404.
33. Ibid., p. 408.
34. Ibid.

35. Ibid., p. 399.
36. Ibid., p. 438.
37. In *Biophilosophy,* trans. C. A. M. Sym (New York: Columbia University Press, 1971), B. Rensch develops a "panpsychistic, identistic, polynomistic" (p. 296) theory of the mind.
38. Ibid., p. 217.
39. Ibid., p. 233.
40. Ibid., p. 310.
41. John 8:32.
42. *Neurophilosophy* (see note 9, above), pp. 299ff.
43. Reference 42, pp. 187–92.
44. Reference 36, p. 565.
45. Reference 45, p. 26.
46. Ibid., p. 227.
47. Ibid., p. 228.

CHAPTER 28

1. P. Marler in *Amer. Sci.* 58 (1970): 669–73, and P. Marler and S. Peters in *Science* 146 (1981): 1483–86.
2. Recounted by D. R. Griffin, in reference 44, p. 41, the behavior of the cream-eating tits was originally reported by J. Fisher and R. A. Hinde in *Brit. Birds* 42 (1949): 347–57, and R. A. Hinde and J. Fisher in *Brit. Birds* 44 (1951): 393–96.
3. Reference 36, pp. 36–50.
4. Reference 45.
5. Reference 47, p. 216.
6. F. Crick and C. Koch in *Sci. Am.* 267, no. 3 (1992): 152–59.
7. Reference 48, p. 314.
8. Reference 43, pp. 158ff. and 426ff.
9. Reference 41.
10. Reference 39. For an updated history of artificial intelligence, see reference 63. For a good survey of neuronal network simulation and its relation to artificial intelligence, see J. D. Cowan and D. H. Sharp in *Daedalus* 117, no. 1 (1988): 85–121.
11. Edelman has described his Darwin robots in *The Remembered Present* (see chapter 27, note 2).
12. Reference 48, p. 282.
13. Ibid., p. 283.

CHAPTER 29

1. Reference 50.
2. Reference 22.
3. An excellent overview of the battlefield is offered in reference 51, a collection of essays pro and con the new doctrine. Major works on the sociobiology side are

references 52 and 53, as well as the more technical *Genes, Mind, and Culture,* by C. J. Lumsden and E. O. Wilson (Cambridge, Mass.: Harvard University Press, 1981); and on the opposition side, references 54 and 55.

4. Herbert Spencer's main work on ethics is *The Principles of Ethics* (New York: D. Appleton, 1892).
5. The relationship between genes and behavior in both animals and humans is discussed in detail in a special issue of *Science* 264 (1994): 1685–1739.
6. J. B. Gurdon in *Sci. Am.* 219, no. 6 (1968): 24–35.
7. D. Rorvik's *In His Image* (Philadelphia and New York: Lippincott, 1978) tells the story of a millionaire's allegedly successful attempt to get himself cloned. The author "protects the identities of those involved." On the other hand, he props his story with extensive scientific documentation, which makes the book interesting reading, whether true or not.
8. The successful cloning of human embryos—to a very limited stage of development—has been reported by J. L. Hall and coworkers from George Washington University School of Medicine in Washington, D.C. See R. Kolberg in *Science* 262 (1993): 652–53.
9. Reference 50, p. 562.
10. See note 4, above.
11. T. H. Huxley is the author of *Evolution and Ethics* (New York: D. Appleton, 1894). Huxley tells of his historic encounter with Bishop Wilberforce in a letter to a friend, cited by G. de Beer in *Charles Darwin* (Garden City, N.Y.: Doubleday, 1964), p. 167.
12. Reference 51, p. 23.
13. Reference 53, p. 179.
14. A. Flew, reference 51, pp. 142–62.
15. Reference 43.
16. Reference 41.
17. Ibid.
18. Reference 50, p. 561.
19. My translation from reference 41, p. 254.
20. Reference 64, p. 176.
21. Reference 53, p. 76.
22. Ibid., p. 174.

CHAPTER 30

1. The role of human hunting in the extinction of the large mammals is discussed in references 21, 29, and 33.
2. L. L. Cavalli-Sforza, P. Menozzi, and A. Piazza in *Science* 259 (1993): 639–46.
3. For a view of agriculture as a "two-edged sword," see reference 33.
4. References 57, 58, and 59.
5. Reference 29, p. 136.
6. Ibid., p. 346.
7. R. M. May in *Sci. Am.* 267, no. 4 (1992): 42–48.
8. Reference 29, p. 280.

9. The atmospheric content of carbon dioxide has been recorded since 1958 at the Mauna Loa Observatory in Hawaii. The graph showing the steady rise in this content is reproduced in reference 58, p. 5.
10. Reference 26, pp. 156–59.
11. Reference 57, p. 164.
12. Ibid., p. 176.
13. B. Commoner, *The Poverty of Power* (New York: Knopf, 1976).
14. Reference 57, p. 185.
15. Lord Zuckerman in *Nature* 358 (1992): p. 273.
16. The joint appeal by the Royal Society of London and the U.S. National Academy of Sciences was issued in 1992 under the title *Population Growth, Resource Consumption, and a Sustainable World.*
17. Fortunately, the appeal has been followed by more energetic action, though still in the form of recommendations. Representatives from fifty-eight academies of science worldwide, gathered in New Delhi in October 1993, issued a joint statement calling for government policies and initiatives that will help "achieve zero population growth within the lifetime of today's children." See *Population Summit of the World's Scientific Academies* (Washington, D.C.: National Academy Press, 1994).
18. Malthus's book, which had a major influence on Darwin, was first published in 1798.
19. The demographic problem is discussed in a number of the books already cited. It is the central topic of P. Ehrlich's 1968 book, *The Population Bomb* (reference 49), followed by *The Population Explosion* (reference 56), which opens with a chapter entitled "Why Isn't Everyone as Scared as We Are?"
20. Reference 57, p. 293.
21. S. J. Olshansky, B. A. Carnes, and C. K. Cassel in *Sci. Am.* 268, no. 4 (1993): 46–52.
22. T. F. Homer-Dixon, J. H. Boutwell, and G. W. Rathjens in *Sci. Am.* 268, no. 2 (1993): 38–45.
23. Starting with J. Rifkin's *Algeny* (New York: Viking, 1983), an increasing number of books have attacked science, especially biology, as both a perverse myth and a dangerous power. See, for example, M. Midgley, *Evolution as a Religion* (London and New York: Methuen, 1985); R. C. Lewontin, *Biology as Ideology* (New York: HarperCollins, 1991); B. Appleyard, *Understanding the Present* (London: Picador, 1992); R. Hubbard and E. Wald, *Exploding the Gene Myth* (Boston: Beacon Press, 1993); as well as references 54, 55, 66, and 67. For balancing viewpoints, see B. D. Davis, *Storm Over Biology* (Buffalo, N.Y.: Prometheus Books, 1986); M. F. Perutz, *Is Science Necessary?* (New York: Dutton, 1989); J. E. Bishop and M. Waldholz, *Genome* (New York: Simon & Schuster, 1990); B. D. Davis, ed., *The Genetic Revolution* (Baltimore: Johns Hopkins University Press, 1991); R. Shapiro, *The Human Blueprint* (New York: St. Martin's Press, 1991); G. Holton, *Science and Anti-Science* (Cambridge, Mass.: Harvard University Press, 1993); and P. R. Gross and N. Levitt, *Higher Superstition: The Academic Left and Its Quarrels with Science* (Baltimore: Johns Hopkins University Press, 1994). See also references 28 and 30.
24. C. G. Darwin, *The Next Million Years* (London: Rupert Hart-Davis, 1953).

25. C. Djerassi, *The Pill, Pigmy Chimps, and Degas' Horse* (New York: Basic Books, 1992).
26. J. R. Gott III in *Nature* 363 (1993): 315–19.
27. Reference 3.
28. In an article entitled "Horoscopes for Humanity?" published on the op-ed page of the July 14, 1993, issue of the *New York Times,* the physicist and author Eric J. Lerner, from Lawrenceville, New Jersey, has written: "Mr. Gott's forecast is, like astrological forecasts, pseudo-science, a mere manipulation of numbers to disguise an implausible argument." See also the exchange of correspondence in *Nature* 368 (1994): 106–8.
29. S. J. Gould, *Wonderful Life* (New York: Norton, 1989), p. 44.
30. This is the opinion defended by G. Stock in *Metaman: The Merging of Humans and Machines into a Global Superorganism* (New York: Simon & Schuster, 1993).

CHAPTER 31

1. Details on Teilhard's life and philosophy may be found in reference 65 and in H. J. Morowitz, *Cosmic Joy and Local Pain* (New York: Scribner's, 1987). For a biography of Teilhard, see C. Cuénot, *Teilhard de Chardin: A Biographical Study,* trans. V. Colimore (Baltimore: Helicon, 1965).
2. *Le Phénomène Humain,* by P. Teilhard de Chardin, was published in Paris in 1955 by Editions du Seuil, and in New York in 1959 as *The Phenomenon of Man* by Harper & Row.
3. Biographical details on Jacques Monod can be found in reference 11 and in *The Statue Within* (New York: Basic Books, 1988), the autobiography of Monod's coworker François Jacob.
4. Reference 64.
5. Ibid., pp. 145–46.
6. Ibid., p. 180.
7. Ibid., p. 31.
8. Ibid., p. 32.
9. Sir Peter Medawar (1915–1987), who gained the 1960 Nobel Prize in medicine for his immunological work, is the author of several books on science and philosophy. His critical review of Teilhard's book appeared in *Mind* 70 (1961): 99–106, and is reproduced in his *Pluto's Republic* (New York: Oxford University Press, 1982), pp. 242–51. The sentence quoted is from the first page of this review.
10. S. J. Gould in *Natural History* (August 1980): 8–28. According to the veteran South-African paleoanthropologist Phillip V. Tobias, the accusation is unfounded. The villains in the Piltdown conspiracy have now been identified with reasonable certainty as Charles Dawson, the local amateur archeologist who "discovered" the faked bones, and Sir Arthur Keith, a respected anatomist who provided the expertise for the faking. See *The Sciences* 34, no. 1 (1994): 38–42.
11. H. J. Morowitz, *Cosmic Joy and Local Pain* (see note 1, above).
12. Reference 65.
13. Ibid., p. 204.

14. Ibid., p. 677.
15. Reference 68, p. 250.
16. Ibid., p. 252.
17. S. Weinberg, *The First Three Minutes* (New York: Basic Books, 1977), p. 148.
18. My critique of Monod's book appeared as "Les Contraintes du Hasard" in the second 1972 issue of *Revue Générale* (a Belgian cultural magazine), pp. 15–42.
19. G. Stent, *Paradoxes of Progress* (San Francisco: W. H. Freeman, 1978).
20. Reference 67.
21. E. Mayr, *Toward a New Philosophy of Biology* (Cambridge, Mass.: Harvard University Press, 1988), p. 69.
22. S. J. Gould, *Wonderful Life* (see chapter 30, note 29), p. 14.
23. Ibid., p. 320.
24. Reference 33, p. 192.
25. Ibid., p. 195.
26. Reference 29, p. 99.
27. Reference 53, p. 53.
28. An enlightening overview of "genetic intelligence" is given by D. S. Thaler in *Science* 264 (1994): 224–25.
29. S. J. Gould and N. Eldredge have recently celebrated the twenty-first anniversary of their theory, which substitutes an uneven course of evolution for the gradual course allegedly postulated by Darwin (*Nature* 366 [1993]: 223–27). Offering a contrasting viewpoint, one of the chapters in reference 20 is entitled "Puncturing Punctuationism."
30. The SETI project has been mentioned in chapter 13.
31. See chapter 28, note 8.
32. Reference 64, pp. 145–46.
33. Reference 66, p. 75.
34. M. Polanyi, *The Tacit Dimension* (New York: Doubleday, 1966), p. 47.
35. "Console toi, tu ne me chercherais pas si tu ne m'avais trouvé" is from Blaise Pascal's *Pensées,* section VII, 553.
36. P. B. Medawar, *The Limits of Science* (New York: Harper & Row, 1984), p. 99.
37. "Il faut cultiver notre jardin" is the last sentence of Voltaire's *Candide.*

Glossary

Acid. A substance that releases hydrogen ions in aqueous solution. Organic acids are characterized by a carboxyl group. See also **Carboxyl group, Proton.**

Actin. A structural protein, found only in eukaryotes, that polymerizes into helical double-stranded fibers that are major components of the cytoskeleton and, in association with myosin, of motile systems. See also **Myosin.**

Active transport. A process, usually powered by the hydrolysis of ATP, sometimes by protonmotive force, whereby substances are transported across membranes against an opposing concentration gradient and/or membrane potential. See also **ATP, Membrane potential, Protonmotive force, Pump.**

Adenine. A purine base found in both DNA and RNA. See also **Base pairing, Deoxyribonucleic acid, Guanine, Purines, Ribonucleic acid.**

ADP. Adenosine diphosphate, a combination of AMP and phosphate linked by a pyrophosphate bond. See also **AMP, ATP, Pyrophosphate.**

Aerobiosis. A mode of life that requires oxygen. See also **Anaerobiosis.**

Agonist. A substance that exerts a biological effect. See also **Receptor.**

Alcohol. A substance bearing a hydroxyl group. See also **Hydroxyl group.**

Allantois. An embryonic membrane of amniotic eggs, involved in gas exchanges and excretion.

Allele. One of two or more variants of the same gene. See also **Gene.**

Alternation of generations. In plants, a mode of development in which haploid, gamete-producing organisms derived from spores alternate with diploid, spore-producing organisms derived from fertilized eggs. See also **Diploid, Egg cell, Fertilization, Gamete, Haploid, Spore.**

Amino acid. A substance possessing both an amino group and a carboxyl group, and serving as a building block for the formation of peptides and proteins. See also **Amino group, Carboxyl group, Peptide, Protein.**

Amino group. $-NH_2$, a characteristic group of amino acids.

Amnion. An embryonic membrane enclosing a fluid-filled sac within which the embryo develops in the eggs of reptiles, birds, and mammals.

AMP. Adenosine monophosphate, a ribonucleotide containing adenine as the base. See also **Adenine, ADP, ATP, Ribonucleotide.**

Amphipathic. Same as **Amphiphilic.**

Amphiphilic. The property of molecules possessing both a hydrophilic and a hydrophobic end. See also **Hydrophilic, Hydrophobic.**

Anaerobiosis. Life in the absence of oxygen. See also **Aerobiosis.**

Angiosperms. Flowering, fruit-producing plants. See also **Flower, Fruit, Gymnosperms.**

Antibody. A protein produced by the immune system in response to a substance, called antigen, and having the property of specifically combining with the antigen. See also **Antigen.**

Anticodon. The triplet of bases whereby, in protein synthesis, an amino acid–carrying transfer RNA, properly positioned on the ribosome surface, binds by specific base pairing to a codon specifying the amino acid in a messenger RNA. See also **Amino acid, Base pairing, Codon, Messenger RNA, Protein, Ribosome, Transfer RNA, Translation.**

Antigen. A substance, usually foreign to the organism, able to elicit an immune response. See also **Antibody.**

Archaea. In the new classification, proposed by the American microbiologist Carl Woese of the University of Illinois, the synonym of archaebacteria. See also **Archaebacteria, Bacteria.**

Archaebacteria. A major group of prokaryotes, recognized as a separate kingdom or domain. See also **Eubacteria.**

ATP. Adenosine triphosphate, a combination of ADP and phosphate linked by a pyrophosphate bond. The central conveyer of metabolic energy, thanks to its two terminal pyrophosphate bonds. See also **ADP, AMP, Phosphorylation, Pyrophosphate.**

Autotrophy. The ability to grow and develop from simple mineral nutrients such as carbon dioxide, molecular nitrogen or nitrate, sulfate, and so on. See also **Chemoautotrophy, Heterotrophy, Phototrophy.**

Axon. The filament whereby a neuron exerts an effect. See also **Dendrite, Neuron, Synapse.**

Bacteria. In common usage, adopted in this book, the synonym of prokaryotes. In Woese's new classification (see **Archaebacteria**), the synonym of eubacteria. See also **Eubacteria, Prokaryotes.**

Base pairing. The phenomenon, based on chemical complementarity, whereby adenine binds to uracil or thymine, and guanine to cytosine, in nucleic acids.

Bilayer. See **Lipid bilayer.**

Biosphere. The sum total of living organisms on the Earth.

Blastopore. The opening whereby the early embryonic stage called gastrula communicates with the outside. It becomes the mouth in protostomes and the anus in deuterostomes. See also **Blastula, Deuterostomes, Gastrula, Protostomes.**

Blastula. An early stage in embryonic development consisting of a spherical sac, which subsequently flattens and folds to form a double-walled pouch, the gastrula. See also **Gastrula.**

Blue-green algae. See **Cyanobacteria.**

Carbohydrates. A major group of natural substances including simple sugars and related molecules, as well as more complex substances made mostly of such molecules.

Carboxyl group. -COOH, characteristic of all organic acids.

Carrier. A metabolic cofactor serving in electron transfer or group transfer. See also **Coenzyme.**

Carrier-level phosphorylation. A coupled process whereby ATP is assembled from ADP and inorganic phosphate by way of protonmotive force generated by a downhill electron-transfer process involving membrane-embedded electron carriers. See also **ADP,**

ATP, Electron-transfer chain, Phosphorylation, Protonmotive force, Substrate-level phosphorylation.

Catalysis. The process whereby a substance (catalyst) facilitates a chemical reaction without itself being consumed in the process. See also **Enzyme, Ribozyme.**

Cell division. The process whereby a cell, after duplication of its genetic material, divides into two daughter cells.

Cell sap. See **Cytosol.**

Cellular defecation. The discharge, by exocytosis, of the contents of old lysosomes or digestive vacuoles. See also **Exocytosis, Lysosome.**

Cell wall. A solid shell, made of murein in eubacteria and of a different material in archaebacteria, that surrounds most prokaryotic cells. See also **Archaebacteria, Eubacteria, Murein.**

Central Dogma. The rule, so named by the physicist and molecular biologist Francis Crick, according to which molecular information can be transferred from nucleic acids to nucleic acids or to proteins, but not from proteins to proteins or to nucleic acids.

Centriole. A cell structure related to the root, or basal body, of eukaryotic flagella and cilia. See also **Cilium, Flagellum, Undulipodia.**

Cerebral cortex. The part of folded neural tissue that envelops the brain and is particularly developed in higher primates and humans.

Chemical transmitter. A substance whereby a cell influences another cell. See also **Hormone, Neurotransmitter.**

Chemoautotrophy. The property of autotrophs able to support their biosynthetic processes with the help of energy provided by the transfer of electrons between a mineral donor and oxygen or some other mineral acceptor. See also **Autotrophy.**

Chemotaxis. The movement of a cell in the direction of a chemical source (positive chemotaxis) or away from it (negative chemotaxis).

Chirality. The property of molecules that exist in two forms whose spatial configurations are mirror images of each other. Examples are L- and D-amino acids. See also **Amino acid, Protein.**

Chlorophyll. A porphyrin derivative related to heme but with magnesium replacing iron; the green, light-absorbing substance that serves as the principal catalyst of phototrophy. See also **Heme, Phototrophy.**

Chloroplast. The central, light-utilizing organelles of all eukaryotic phototrophs. Evolutionarily derived from endosymbiotic cyanobacteria, chloroplasts are dense, oblong particles, about one ten-thousandth of an inch in size, surrounded by two membranes and filled with stacks (grana) of flat, membranous pouches (thylakoids) bearing membrane-embedded photosystems I and II, together with associated electron-transfer chains. See also **Chlorophyll, Cyanobacteria, Electron-transfer chain, Endosymbiont, Photosystem I, Photosystem II.**

Chordates. Animals possessing a notochord at least at some stage of development. They include all vertebrates and a few lower forms. See also **Notochord.**

Chromosome. A cell structure containing genetic DNA. In eukaryotes, chromosomes are situated in the nucleus and are detectable as compact rods during mitosis. Prokaryotic chromosomes are circular and have a simpler structure. See also **Mitosis.**

Cilium. Plural: cilia. A short, rythmically beating motor appendage of eukaryotic cells, constructed from microtubules and other components according to the same architecture as the eukaryotic flagellum. See also **Centriole, Flagellum, Microtubule, Undulipodia.**

***Cis* splicing.** The joining together, after removal of the intron, of two segments of the same RNA molecule separated by an intron. See also **Intron, RNA splicing, *Trans* splicing.**

Clathrin. A structural protein that polymerizes into trellis or basket arrangements typically involved in certain phenomena of membrane invagination and vesiculation. See also **Cytoskeleton, Endocytosis, Vesicular transport.**

Clone. A population of cells derived by successive divisions from a single ancestral cell.

Codon. A triplet of bases in messenger RNA coding for an amino acid or (3 out of the total of 64 triplet combinations) for a chain termination signal. See also **Amino acid, Anticodon, Base pairing, Genetic code, Messenger RNA, Translation.**

Coelom. An internal body cavity, lined by mesodermal cells, found, at least at some developmental stage, in all animals except diploblasts and primitive worms. See also **Diploblasts, Mesoderm, Triploblasts.**

Coenzyme. An organic substance, often containing a vitamin as the main component, that acts as a cofactor in enzyme-catalyzed reactions. Most coenzymes are either electron or group carriers. See also **Carrier, Enzyme, Vitamin.**

Coenzyme A. A pantetheine-containing thiol combination serving as group carrier in a number of synthetic reactions involving organic acids. See also **Carrier, Group transfer, Pantetheine, Thiol.**

Condensation reaction. A chemical process in which two molecules become linked to each other with the loss of a water molecule.

Condensing agent. A water-removing substance that helps a condensation reaction.

Conjugation. The phenomenon whereby two bacterial cells join temporarily and exchange genetic material. See also **Pilus, Plasmid.**

Cotranslational transfer. The translocation of a protein across a membrane that occurs while the protein is being synthesized. See also **Posttranslational transfer, Targeting sequence.**

Coulomb force. The electrostatic force responsible for the attraction between oppositely charged or polarized groups and for the mutual repulsion of similarly charged or polarized groups. See also **Hydrophilic, Van der Waals force.**

Crossing-over. A phenomenon whereby homologous stretches of DNA are exchanged between chromosomes in the course of meiosis. See also **Meiosis, Recombination.**

Cyanobacteria. Phototrophic bacteria containing photosystems I and II, and capable, therefore, of using light energy to extract hydrogen from water, with the production of molecular oxygen. Ancestral to chloroplasts. See also **Chlorophyll, Chloroplast, Photosystem I, Photosystem II, Phototrophy.**

Cytochrome. A heme-containing protein acting as electron carrier in electron-transfer reactions. Cytochrome *c* is the best-known cytochrome. See also **Carrier, Electron transfer, Heme.**

Cytomembrane system. The system of internal membranes in eukaryotic cells. It comprises a large number of saclike structures, which communicate with each other by permanent connections or, more frequently, by vesicular transport. See also **Endoplasmic reticulum, Endosome, Golgi, Lysosome, Vesicular transport.**

Cytoplasm. The total content of a eukaryotic cell, with the exception of the nucleus.

Cytosine. A pyrimidine base found in both DNA and RNA. See also **Base pairing, Deoxyribonucleic acid, Pyrimidines, Ribonucleic acid, Thymine, Uracil.**

Cytoskeleton. The set of internal structures, made of proteins, that support eukaryotic cells. See also **Actin, Clathrin, Microtubule, Tubulin.**

Cytosol. Or cell sap, the unstructured part of the cytoplasm.

Darwinian selection. See **Natural selection.**

Dendrite. The bushy extension whereby a neuron receives signals. See also **Axon, Neuron, Synapse.**

Deoxyribonucleic acid. Or DNA, a nucleic acid made of deoxyribonucleotides, the depository of genetic information in all living cells and a number of viruses. See also **Deoxyribonucleotide, Ribonucleic acid.**

Deoxyribonucleotide. A nucleotide containing deoxyribose as 5-carbon sugar. See also **Deoxyribose, Nucleotide.**

Deoxyribose. The 5-carbon sugar characteristically found in DNA. Corresponds to ribose without an oxygen atom in position 2. See also **Deoxyribonucleic acid, Deoxyribonucleotide, Nucleotide, Ribose.**

Determinism. The doctrine according to which like causes always produce like effects and, conversely, events are entirely explainable by their antecedent causes.

Deuterostomes. Animals in which the blastopore gives rise to the anus in the course of embryological development. They comprise chordates and a few lower forms (protochordates), as well as echinoderms. See also **Blastopore, Chordates, Protostomes.**

Differentiation. A process whereby a stem cell acquires the characteristic features of a given cell type.

Digestion. The process whereby complex nutrient molecules are broken down by hydrolysis. See also **Hydrolysis.**

Diploblasts. Animals arising from an embryonic structure made of two cell layers, an ectoderm and an endoderm. They include the placozoa, sponges, and coelenterates and related organisms. See also **Ectoderm, Endoderm, Triploblasts.**

Diploid. The character of a cell having two sets of chromosomes. See also **Chromosome, Haploid.**

DNA. See **Deoxyribonucleic acid.**

DNA polymerase. The DNA-synthesizing enzyme involved in DNA replication.

Double fertilization. A characteristic phenomenon, involving the fusion of two haploid cells and that of a haploid and a diploid cell, in the formation of a fruit from a flower. See also **Diploid, Fertilization, Flower, Fruit, Haploid.**

Double helix. The characteristic structure of DNA, made of two intertwined, complementary strands stabilized by base pairing. See also **Base pairing, Deoxyribonucleic acid.**

Dualism. The doctrine, first advocated by Descartes, that views the human body and mind as two distinct entities connected in the brain. See **Monism.**

Dynein. An ATP-splitting protein that serves as the motor element in connection with microtubules, in cilia and flagella. See also **Cilium, Flagellum, Microtubule, Undulipodia.**

Ectoderm. The embryonic cell layer that gives rise to the skin and derived structures and to neural tissue. See also **Endoderm, Mesoderm.**

Egg cell. The female gamete. See also **Fertilization, Gamete, Sexual reproduction, Sperm cell.**

Electron. A negatively charged elementary particle, about one two-thousandth the mass of a proton. With protons and neutrons, which form the atomic nucleus, electrons are one of the three constituents of atoms, in which they occupy a peripheral position. See also **Neutron, Nucleus, Proton.**

Electron acceptor. See **Electron transfer.**

Electron carrier. See **Carrier, Coenzyme, Electron transfer.**

Electron donor. See **Electron transfer.**

Electron transfer. A key metabolic process in which electrons are transferred from a donor to an acceptor. See also **Oxidation-reduction, Phosphorylation.**

Electron-transfer chain. A chain of several electron carriers organized within the fabric of a membrane in a way that allows electrons to move from one carrier to another. A major system in biological energy transfers. See also **Carrier-level phosphorylation.**

Endocytosis. Engulfment of extracellular materials by a membrane invagination that closes, sometimes with the help of clathrin, into an intracellular vesicle containing the captured materials. See also **Clathrin, Exocytosis, Phagocytosis, Pinocytosis.**

Endoderm. The embryonic cell layer that gives rise to the lining of the alimentary canal and derived tissues. See also **Ectoderm, Mesoderm.**

Endoplasmic reticulum. Or ER, a part of the cytomembrane system of eukaryotic cells, consisting of flat, membranous sacs or vesicles serving in the temporary storage and early processing of newly synthesized secretory and lysosomal proteins, which are further transferred to the Golgi. See also **Cytomembrane system, Golgi, Lysosome, Protein, Rough ER, Secretion, Smooth ER, Vesicular transport.**

Endosome. An intracellular, membrane-bounded vesicle arising from the cell membrane by endocytosis. See also **Endocytosis.**

Endosymbiont. A cell, most often of prokaryotic type, adopted as a stable component of a eukaryotic cell. Several cell organelles, including chloroplasts, mitochondria, and, perhaps, peroxisomes and hydrogenosomes, are derived from endosymbiotic bacteria. See also **Chloroplast, Hydrogenosome, Mitochondrion, Peroxisome.**

Enzyme. A catalytic protein involved in metabolism. See also **Catalysis, Protein.**

Epigenesis. The process whereby traits not strictly determined by genetic factors are acquired in the course of development. The exact wiring of neurons in a brain is epigenetic.

ER. See **Endoplasmic reticulum.**

Ester. The substance arising by combination, with loss of water, of the carboxyl group of an organic acid with the hydroxyl group of an alcohol, with formation of an ester bond, -O-CO-. See also **Carboxyl group, Hydroxyl group, Thioester.**

Ester lipid. A lipid in which fatty acids are linked to hydroxyl groups of glycerol by ester bonds. See also **Glycerol, Lipid.**

Ether. The combination, with loss of water, of the hydroxyl group of an alcohol with the hydroxyl group of another alcohol, with formation of an ether bond, -O-. See also **Hydroxyl group.**

Ether lipid. A lipid in which alcohols derived from fatty acids are linked to hydroxyl groups of glycerol by ether bonds. See also **Glycerol, Lipid.**

Eubacteria. A major group of prokaryotes, recognized as a separate kingdom or domain. See also **Archaebacteria, Bacteria.**

Eukaryotes. Living organisms—comprising protists, plants, fungi, and animals, including humans—made of large cells having a fenced-off nucleus and a cytoplasm containing cytomembranes, cytoskeletal elements, and, most often, mitochondria, peroxisomes, and, in algae and plants, chloroplasts. See also **Chloroplast, Cytomembrane system, Cytoskeleton, Mitochondrion, Nucleus, Peroxisome, Prokaryotes.**

Evolutionary convergence. The independent evolutionary development of the same trait in two or more separate lines.

Exocytosis. Or reverse endocytosis, the extracellular discharge of the contents of intracellular vesicles by fusion of the vesicle membrane with the cell membrane. See also **Endocytosis.**

Exon. An expressed part of a split gene. See also **Intron, RNA splicing, Split genes.**

Fermentation. A metabolic process in which one intermediary serves as the acceptor for electrons donated by another intermediary, thus not needing an outside electron acceptor. See also **Electron transfer, Metabolism.**

Ferric iron. An iron ion bearing three positive charges, Fe^{3+}.

Ferrous iron. An iron ion bearing two positive charges, Fe^{2+}.

Fertilization. The combination of a male gamete, or sperm cell, with a female gamete, or egg cell. See also **Egg cell, Gamete, Sexual reproduction, Sperm cell.**

Flagellum. Plural: flagella. In prokaryotes, a rigid, helical rod, made of protein, that propels the cell by axial rotation. In eukaryotes, a long, waving motor appendage constructed from microtubules and other components according to the same architecture as cilia. See also **Cilium, Undulipodia.**

Flower. The sex organ of angiosperms. See also **Angiosperms, Fruit.**

Fork organism. An ancestral organism that has, by mutation, initiated a new evolutionary branch.

Fruit. The seed-containing structure that develops from a flower after fertilization in angiosperms. See also **Angiosperms, Double fertilization, Flower.**

Gamete. A haploid germ cell. See also **Egg cell, Fertilization, Haploid, Sexual reproduction, Sperm cell.**

Gastrula. An early stage in embryonic development, derived from the blastula, consisting of a double-walled pouch opening to the outside by the blastopore. See also **Blastopore, Blastula.**

Gene. A unit of hereditary material.

Genetic code. The set of equivalences between proteinogenic amino acids and the nucleotide triplets, or codons, that code for the amino acids in messenger RNAs. See also **Amino acid, Codon, Messenger RNA, Translation.**

Genetic drift. The evolutionary spread of mutations without selective screening. See also **Mutation, Natural selection.**

Genome. The sum total of the genes of an organism. See also **Genotype.**

Genotype. Equivalent to genome, this term is used in opposition to phenotype to designate the hidden genetic information responsible for the expressed properties of an organism (phenotype). See also **Phenotype.**

Germination. The initiation of development in a dormant plant embryo.

Glycerol. A 3-carbon substance bearing three hydroxyl groups; a major constituent of lipids. See also **Ester lipid, Ether lipid, Hydroxyl group, Lipid, Phospholipid.**

Golgi. From the name of the Italian scientist who discovered it, a part of the cytomembrane system, consisting of stacked, flat, membranous sacs, involved in the processing of newly synthesized lysosomal and secretory proteins. See also **Cytomembrane system, Endoplasmic reticulum, Lysosome, Protein, Secretion.**

Gram-negative bacteria. Bacteria with two outer membranes that react negatively to a test devised by the Danish bacteriologist Hans Christian Joachim Gram.

Gram-positive bacteria. Bacteria with a single outer membrane that react positively to the Gram test. See also **Gram-negative bacteria.**

Greenhouse effect. The trapping of light energy as heat by an atmospheric screen (for example, carbon dioxide or methane) that, like the glass cover of a greenhouse, lets in visible light of lower wavelength but partly blocks the exit of infrared light of higher wavelength.

Group. A characteristic part of a molecule.

Group carrier. See **Carrier, Group, Group transfer.**

Group transfer. A key biosynthetic process whereby a chemical group is transferred from a donor to an acceptor. See also **Carrier, Group.**

Guanine. A purine base found in both DNA and RNA. See also **Adenine, Base pairing, Deoxyribonucleic acid, Purines, Ribonucleic acid.**

Gymnosperms. Seed-producing plants whose seeds are not contained within a fruit. See also **Angiosperms.**

Haploid. The character of a cell having a single set of chromosomes. See also **Diploid.**

Heme. A porphyrin derivative with a central iron atom, involved, when combined with a protein (hemoprotein), in electron transfer, oxygen transport, and related reactions. See also **Chlorophyll, Cytochrome, Electron transfer.**

Hemoprotein. See **Heme.**

Hermaphrodite. An organism with both male and female sex organs.

Heterotrophy. The characteristic of organisms that require organic nutrients produced by living organisms to survive and develop. See also **Autotrophy.**

Homeobox. A highly conserved sequence of 180 nucleotides common to many regulatory genes and coding for the DNA-binding part of the corresponding regulatory proteins. See also **Homeotic gene, Transcription factor.**

Homeostasis. The ability of living organisms to keep constant certain of their physical or chemical properties by self-regulation.

Homeotic gene. A regulatory gene containing a homeobox sequence. See also **Homeobox.**

Horizontal gene transfer. The transfer of genes from one organism to another, as opposed to vertical gene transfer, from parent to offspring.

Hormone. A chemical transmitter transported by the blood stream or some other humoral connection from the cells that secrete it to the cells on which it acts. See also **Chemical transmitter.**

Hydrocarbon. A substance, typically found in petroleum, made only of carbon and hydrogen.

Hydrogen bond. A special kind of electrostatic bond between a hydrogen-bearing group and a negatively charged or polarized group.

Hydrogen ion. See **Proton.**

Hydrogenosome. A membrane-bounded cytoplasmic organelle found in some protists and fungi, and characterized by the ability to produce molecular hydrogen. Possibly of endosymbiont origin. See also **Endosymbiont.**

Hydrolase. An enzyme catalyzing a hydrolytic reaction. See also **Hydrolysis.**

Hydrolysis. The splitting of a chemical bond with the help of a water molecule.

Hydrophilic. The water-binding property common to all electrically charged or polarized substances (including water itself). See also **Amphiphilic, Hydrophobic.**

Hydrophobic. The water-excluding property common to all nonpolar substances. See also **Amphiphilic, Hydrophilic, Van der Waals force.**

Hydrothermal vents. Crevices in the ocean bottom through which pressurized hot water is ejected from subterranean regions.

Hydroxyl group. -OH, characteristic of alcohols.

Hydroxyl ion. The negatively charged OH^- ion, arising, together with a hydrogen ion, or proton, from the dissociation of water. See **Proton.**

Intron. Or intervening sequence, a gene segment that is transcribed but is subsequently excised from the resulting RNA and, therefore, is not expressed. See also **Exon, RNA splicing, Split genes.**

Ion. An atom or molecule that is electrically charged due to the loss or gain of one or more electrons.

Isotopes. Atoms of the same element with different atomic masses. Isotopes have the same number of protons and electrons, but differ by the number of neutrons in the atomic nucleus. See also **Electron, Neutron, Proton.**

Kerogen. Carbon-containing geological material of organic origin.

Lipid. A fatty substance.

Lipid bilayer. A characteristic fabric of all biological membranes, consisting of two monomolecular layers of amphiphilic lipids joined by their hydrophobic faces. See also **Amphiphilic.**

Liposome. An artificial vesicle made of one or more lipid bilayers. See also **Lipid bilayer.**

Lysosome. A membrane-limited vesicle containing a variety of hydrolytic enzymes acting best in an acidic medium; present in all eukaryotic cells where it serves in intracellular digestion. See also **Cytomembrane system, Digestion, Endocytosis, Endoplasmic reticulum, Golgi.**

Lysozyme. An enzyme that hydrolyzes murein. See also **Murein.**

Meiosis. A special kind of mitotic division, involved in gamete maturation, by which a single diploid cell gives rise, after duplication of its genome, to four haploid cells. See also **Crossing-over, Diploid, Egg cell, Fertilization, Gamete, Haploid, Mitosis, Sperm cell.**

Membrane. A biological, sheetlike structure made of a lipid bilayer and a number of inserted proteins. See also **Lipid bilayer.**

Membrane potential. The electrochemical imbalance created by a disparity of electric charges on the two faces of a membrane. See also **Proton potential.**

Mesoderm. The embryonic cell layer, present in triploblasts but not in diploblasts, that gives rise to many internal tissues, including muscle, connective tissue, blood, and bone. See also **Diploblasts, Ectoderm, Endoderm, Triploblasts.**

Messenger RNA. An RNA molecule coding for a protein. See also **Codon, Genetic code, Translation.**

Metabolism. The set of enzyme-catalyzed chemical reactions that support a living organism. See also **Enzyme.**

Microbody. A morphological term designating an intracellular membrane-bounded organelle most often possessing the functional properties of a peroxisome. See also **Peroxisome.**

Microtubule. A hollow, tubular structure, made by the lateral association of thirteen linear arrangements of tubulin, and serving, among others, in the construction of the eukaryotic cytoskeleton, mitotic apparatus, cilia, and flagella. See also **Centriole, Cilium, Cytoskeleton, Flagellum, Mitosis, Neurotubule, Tubulin, Undulipodia.**

Mitochondrion. Plural: mitochondria. An endosymbiont-derived organelle present in most eukaryotic cells and playing a major role in respiratory metabolism and energy retrieval. Mitochondria are relatively dense, oblong particles, about one twenty-thousandth of an inch in size, surrounded by two membranes, of which the inner one, a site of actively phosphorylating respiratory chains, forms many infoldings, or cristae. See also **Endosymbiont, Phosphorylation, Respiration.**

Mitosis. The mode of division of eukaryotic nuclei by means of a complex, spindle-shaped apparatus built with microtubules. See also **Microtubule.**

Mitotic apparatus. See **Mitosis.**

Mitotic spindle. See **Mitosis.**

Monism. The doctrine, opposed to dualism, that views the human body and mind as consisting of a single entity. See also **Dualism.**

Monoclonal antibody. An antibody belonging to a population of identical molecules made by a clone of lymphocytes derived from a single cell. See also **Antibody, Clone.**

Multimer. A term used in this book to designate a chainlike molecule of moderate length made predominantly of amino acids but possibly less regular than a peptide and containing other components besides representatives of the twenty proteinogenic amino acids. See also **Oligomer, Peptide, Polymer.**

Murein. The main component of the cell wall of eubacteria. See also **Cell wall, Eubacteria, Lysozyme.**

Mutation. An alteration of a gene, transmissible by replication. See also **Gene, Genetic drift, Natural selection.**

Myosin. An ATP-splitting protein that serves as the motor part in connection with actin fibers, in muscle fibrils and other cellular motor elements. See also **Actin.**

Natural selection. Or Darwinian selection, the natural screening process whereby organisms that produce more progeny progressively crowd out genetically different organisms that have a lower reproductive success. See also **Mutation.**

Neoteny. Retention of juvenile characteristics produced by retardation of body development.

Neural tube. An ectoderm-derived embryonic structure that develops into the central nervous system in higher animals. See also **Ectoderm.**

Neuron. A nerve cell, characteristically made of a cell body, an emitter extension, or axon, and a ramified receiving extension made of dendrites. See also **Axon, Dendrite, Synapse.**

Neurotransmitter. A chemical, typically secreted at synapses, involved in the transmission of a nerve impulse to a neighboring cell. See also **Chemical transmitter, Synapse.**

Neurotubule. A special kind of microtubule supporting neuronal extensions. See also **Microtubule.**

Neutron. A component of the atomic nucleus, of mass equal to that of a proton but devoid of electric charge. See also **Electron, Isotopes, Nucleus, Proton.**

Notochord. A rodlike dorsal structure that serves as backbone in lower chordates and as temporary backbone in the embryological development of vertebrates. See also **Chordates.**

Nucleic acid. A long chainlike molecule made by the association of a large number of nucleotides. See also **Deoxyribonucleic acid, Nucleotide, Ribonucleic acid.**

Nucleolus. An inner structure of eukaryotic nuclei in which ribosomal RNAs are transcribed and processed. See also **Nucleus, Ribosomal RNA, Transcription.**

Nucleotide. A molecule made of a purine or pyrimidine base, a 5-carbon sugar consisting of ribose or deoxyribose, and a phosphate molecule; a building block of nucleic acids. See also **Deoxyribonucleotide, Deoxyribose, Nucleic acid, Purines, Pyrimidines, Ribonucleotide, Ribose.**

Nucleus. In physics, the core of an atom, made of protons and neutrons. See also **Electron, Isotopes, Neutron, Proton.** In biology, the central structure of eukaryotic cells housing the chromosomes and the enzyme systems involved in DNA replication and transcription and in RNA processing. See also **Chromosome, Replication, Transcription.**

Oligomer. A molecule made by the association of a small number of identical or similar molecular units. See also **Multimer, Polymer.**

Ontogeny. The life history of an individual, as opposed to phylogeny, the evolutionary history of a group of organisms. See also **Phylogeny, Recapitulation.**

Osmotic pressure. The force developed by the tendency of a solvent—for example, water—to diffuse from a region where dissolved substances are less concentrated to a region where they are more concentrated.

Oxidation. The removal of one or more electrons or hydrogen atoms (electrons + protons) from an atom or molecule. See also **Electron, Proton, Reduction.**

Oxidation-reduction. The electron-transfer process necessarily linking the oxidation of an electron donor to the reduction of an electron acceptor. See also **Electron transfer, Oxidation, Reduction.**

Oxidative phosphorylation. The process coupling the assembly of ATP from ADP and inorganic phosphate to an electron-transfer reaction, usually, but not obligatorily, with oxygen as the final electron acceptor. See also **ATP, Carrier-level phosphorylation, Electron transfer, Substrate-level phosphorylation.**

Ozone. O_3, a gaseous molecule, made of three oxygen atoms, which forms a layer in the upper atmosphere that shields the Earth against excessive ultraviolet radiation.

Pantetheine. A complex thiol combination found in the form of pantetheine phosphate either in combination with a protein or as part of coenzyme A, and serving as the group carrier in a number of synthetic reactions involving organic acids. See also **Carrier, Coenzyme A, Group transfer, Thiol.**

Peptide. The substance arising from the combination of two or more amino acids joined together by peptide bonds (-CO-NH-), which are bonds formed, with loss of water, between the carboxyl group of one amino acid with the amino group of another. See also **Amino acid, Amino group, Carboxyl group, Polypeptide, Protein.**

Periplasmic space. The space between inner and outer membrane in gram-negative bacteria. See also **Gram-negative bacteria.**

Peroxisome. A membrane-bounded, intracellular organelle of eukaryotic cells, possibly derived from an endosymbiont and involved in hydrogen peroxide metabolism and related reactions. See also **Endosymbiont, Microbody.**

Phagocyte. A cell depending largely for its supply of nutrients on phagocytosis and lysosomal digestion of the material taken up. See also **Digestion, Lysosome, Phagocytosis.**

Phagocytosis. The ability of a cell to engulf large objects. See also **Endocytosis, Pinocytosis.**

Phenotype. Term used in opposition to genotype, the hidden genetic information, to designate the expressed properties of an organism. See also **Genotype.**

Phospholipid. A phosphate-containing fatty substance with amphiphilic properties, the typical component of lipid bilayers in biological membranes. See also **Amphiphilic, Lipid bilayer, Membrane.**

Phosphorylation. In general, the attachment of a phosphate group to a molecule. In particular, the assembly of ATP from ADP and inorganic phosphate coupled to electron transfer. See also **ATP, Carrier-level phosphorylation, Electron transfer, Oxidative phosphorylation, Substrate-level phosphorylation.**

Photophosphorylation. Light-supported phosphorylation. See also **Phosphorylation.**

Photosynthesis. Self-construction with the help of light energy. See also **Phototrophy.**

Photosystem I. The more primitive catalytic machinery for the utilization of light energy, common to all chlorophyll-dependent phototrophs. See also **Phototrophy.**

Photosystem II. The more advanced catalytic machinery, which uses light energy to extract hydrogen from water with the release of molecular oxygen, absent from a number of phototrophic bacteria, but present in cyanobacteria, eukaryotic algae, and green plants. See also **Chloroplast, Cyanobacteria, Phototrophy.**

Phototrophy. Or photoautotrophy, light-supported autotrophy. See also **Autotrophy, Chemoautotrophy.**

Phylogeny. The evolutionary history of a group of organisms, as opposed to ontogeny, the life history of an individual. See also **Ontogeny, Recapitulation.**

Pilus. Plural: pili. A long, slender, hairlike projection on the surface of certain bacterial cells and involved in cellular attachment and conjugation. See also **Conjugation.**

Pinocytosis. The ability of a cell to engulf fluid droplets. See also **Endocytosis, Phagocytosis.**

Placozoa. A small phylum comprising the simplest diploblasts and believed to include the most primitive animal forms. See also **Diploblasts.**

Plasma membrane. Or cell membrane, the peripheral membrane surrounding all cells.

Plasmid. A small, circular piece of genetic DNA, distinct from the chromosome, that is present in certain bacteria and is transmissible by conjugation. See also **Conjugation.**

Polymer. A molecule made by the association of a large number of identical or similar molecular units. See also **Multimer, Oligomer.**

Polynucleotide. See **Nucleic acid.**

Polypeptide. A macromolecule made by the association of a large number of amino acids joined by peptide bonds. See also **Amino acid, Peptide, Protein.**

Polyphosphate. A molecule consisting of a large number of phosphate molecules joined by pyrophosphate bonds. See also **Pyrophosphate bond.**

Polysaccharide. A macromolecule made by the association of a large number of sugars or similar molecular units. See also **Carbohydrates.**

Posttranslational transfer. The translocation of a protein across a membrane that occurs after the protein has been synthesized. See also **Cotranslational transfer, Targeting sequence.**

Prokaryotes. As opposed to eukaryotes, microorganisms of bacterial type. See also **Eukaryotes.**

Protein. A macromolecule made by the association of a large number of amino acids—selected from a set of twenty defined L-amino acids—joined by peptide bonds. See also **Amino acid, Chirality, Peptide, Polypeptide.**

Protists. A term encompassing all unicellular eukaryotic microorganisms. See also **Eukaryotes.**

Protometabolism. The set of chemical reactions that supported emerging life until the development of enzyme-catalyzed metabolism. See also **Metabolism.**

Proton. A positively charged elementary particle of mass equal to that of a neutron and almost two thousand times that of an electron; a component, with neutrons, of all atomic nuclei. The nucleus of the hydrogen atom. A hydrogen ion, H^+, arising either through the removal of the peripheral electron from a hydrogen atom or, together with a hydroxyl ion, from the dissociation of a water molecule. See also **Electron, Hydroxyl ion, Isotopes, Neutron, Nucleus.**

Protonmotive force. The force generated by protons moving down a proton potential. See also **Carrier-level phosphorylation, Proton potential.**

Proton potential. The energy difference created by a combined imbalance of protons and positive electric charges (membrane potential) across a membrane. See also **Membrane potential, Proton, Protonmotive force.**

Proton pump. A membrane-embedded machinery that forcibly drives protons across the membrane with the help of energy derived from the splitting of ATP or from electron transfer. See also **Pump.**

Protostomes. Animals in which the blastopore gives rise to the mouth in the course of embryological development. They comprise all invertebrates with the exception of echinoderms. See also **Blastopore, Deuterostomes.**

Pump. An active transport system. See also **Active transport.**

Purines. A group of bases comprising two main constituents of nucleic acids, adenine and guanine. See also **Base pairing, Nucleic acid, Nucleotide.**

Pyrimidines. A group of bases comprising three main constituents of nucleic acids, cytosine, uracil (present only in RNA), and thymine (present only in DNA). See also **Base pairing, Nucleic acid, Nucleotide.**

Pyrophosphate. A molecule formed by the association, with loss of a water molecule, of two phosphate molecules.

Pyrophosphate bond. The bond joining phosphates in pyrophosphate and polyphosphates, also joining the three terminal phosphates of ATP, where it serves as the main conveyer of biological energy. See also **ATP, Polyphosphate, Pyrophosphate.**

Recapitulation. The phenomenon by which, according to the German nineteenth-century biologist and philosopher Ernst Haeckel, the course of ontogeny follows some of the steps of phylogeny. See also **Ontogeny, Phylogeny.**

Receptor. A substance, most often situated on the cell surface, through which an agonist exerts its biological effect. See also **Agonist.**

Recombination. The rearrangement of DNA molecules by the exchange of homologous segments, as occurs in crossing-over. See also **Crossing-over.**

Reduction. The gain of one or more electrons or hydrogen atoms (electrons + protons) by an atom or molecule. See also **Electron, Oxidation, Proton.** In cell biology, chromosome reduction refers to the halving of the chromosome number that occurs at meiosis. See also **Diploid, Haploid, Meiosis.**

Replicase. The viral enzyme that replicates RNA molecules. See also **Replication, Virus.**

Replication. The copying of nucleic acids by the synthesis of a molecule complementary to a template strand of the same nature (DNA on DNA, RNA on RNA). See also **Base pairing.**

Repressor. A molecule, usually of protein nature, that blocks the transcription of certain genes. See also **Transcription.**

Respiration. The utilization of oxygen as an electron acceptor, or the breathing of air that provides oxygen to fulfill this function. See also **Electron transfer.**

Respiratory chain. See **Electron-transfer chain.**

Retrovirus. A virus with an RNA genome replicated and expressed by way of the complementary DNA synthesized by a viral reverse transcriptase. See also **Reverse transcriptase, Virus.**

Reverse electron transfer. The uphill transfer of electrons, supported by the splitting of ATP or by protonmotive force, from a donor occupying a lower energy level to an acceptor occupying a higher energy level. See also **ATP, Electron transfer, Protonmotive force.**

Reverse transcriptase. The enzyme, characteristic of retroviruses, that catalyzes reverse transcription. See also **Enzyme, Retrovirus, Reverse transcription.**

Reverse transcription. The synthesis of DNA on an RNA template. See also **Base pairing, Transcription.**

Ribonucleic acid. Or RNA, a nucleic acid made of ribonucleotides, most often transcribed from DNA, except in certain viruses, and serving mostly in protein synthesis and RNA splicing. See also **Messenger RNA, Ribonucleotide, Ribosomal RNA, Ribozyme, RNA splicing, Transcription, Transfer RNA, Virus.**

Ribonucleotide. A nucleotide containing ribose as 5-carbon sugar. See also **Nucleotide, Ribose.**

Ribose. The 5-carbon sugar constituent of ribonucleotides and RNAs.

Ribosomal RNA. An RNA constituent of ribosomes. See also **Nucleolus, Ribonucleic acid, Ribosome.**

Ribosome. A compact particle, one-millionth of an inch in size, that serves as the site of protein assembly in all living cells. Consists of two subunits of unequal size, each made of protein and RNA molecules. See also **Protein, Ribosomal RNA.**

Ribozyme. A catalytic RNA. See also **Catalysis, Enzyme, Ribonucleic acid.**

RNA. See **Ribonucleic acid.**

RNA replicase. See **Replicase.**

RNA splicing. The end-to-end joining of two RNA stretches belonging either to distinct molecules (*trans* splicing) or to the same molecule in which they were separated by an intron, which is removed in the process (*cis* splicing). See ***Cis* splicing, Exon, Intron, Split genes, *Trans* splicing.**

Rough ER. A part of the endoplasmic reticulum whose membranes are studded with attached ribosomes and show a rough appearance in cross section for this reason. See also **Endoplasmic reticulum, Ribosome, Smooth ER.**

Secretion. The process whereby materials synthesized inside cells are discharged extracellularly. Protein secretion, in particular, depends, in prokaryotes, on cotranslational or posttranslational transfer across the cell membrane, and, in eukaryotes, on cotranslational transfer across rough ER membranes, subsequent processing and transfer through smooth ER, Golgi, and secretory vesicles, ending in discharge by exocytosis. See also **Cotranslational transfer, Exocytosis, Golgi, Posttranslational transfer, Rough ER, Smooth ER.**

Seed. A structure containing a plant embryo, together with nutritive material, within a protective hull. Seeds are naked in gymnosperms and embedded in a fruit in angiosperms. See also **Angiosperms, Fruit, Gymnosperms.**

Sequencing. The determination of the sequence of nucleotides in nucleic acids and of amino acids in proteins. See also **Nucleic acid, Protein.**

Sexual reproduction. The mode of reproduction dependent on the fusion of haploid gametes—sperm cell and egg cell—into a diploid fertilized egg cell. See also **Diploid, Egg cell, Fertilization, Gamete, Haploid, Meiosis, Sperm cell.**

Smooth ER. A part of the endoplasmic reticulum whose membranes are not studded with attached ribosomes and show a smooth appearance in cross section for this reason. See also **Endoplasmic reticulum, Ribosome, Rough ER.**

Sperm cell. The male gamete. See also **Egg cell, Fertilization, Gamete, Sexual reproduction.**

Splicing. See **RNA splicing.**

Split genes. Genes made of two or more exons—parts that are expressed—separated by introns—parts that are not expressed. See also **Exon, Intron, RNA splicing.**

Spore. A protected, dormant form of a prokaryotic or eukaryotic microorganism. In plants, a protected, haploid germ cell that develops, when conditions are favorable, into a gamete-generating structure or organism. See also **Alternation of generations, Diploid, Gamete, Haploid.**

Stomata. Openings in leaves that allow absorption of carbon dioxide and release of oxygen.

Stromatolite. A stratified rock formation derived, by fossilization, from superimposed bacterial mats topped by phototrophic bacteria.

Substrate-level phosphorylation. A coupled process whereby ATP is assembled from ADP and inorganic phosphate, often by way of thioesters, with the help of energy supplied by a downhill electron-transfer process between a metabolic substrate and an electron carrier. See also **ATP, Carrier-level phosphorylation, Electron transfer, Phosphorylation, Thioester.**

Synapse. A connection between the axon of one neuron and a dendrite of another neuron. As a rule, communication is established by means of a neurotransmitter diffusing from axon to dendrite across a small gap, the synaptic cleft, that separates the two cellular extensions. See also **Axon, Dendrite, Neuron, Neurotransmitter.**

Targeting sequence. A sequence of amino acids, in a protein, that directs the protein, during or after its synthesis, toward its intracellular or extracellular location. See also **Cotranslational transfer, Posttranslational transfer.**

Thioester. The substance arising by combination, with loss of water, of the carboxyl group of an organic acid with the thiol group of a thiol, with formation of a thioester bond, -S-CO-. See also **Carboxyl group, Ester, Substrate-level phosphorylation, Thiol.**

Thiol. A substance carrying a thiol group, -SH. See also **Thioester.**

Thymine. Or 5-methyl-uracil, a pyrimidine base found only in DNA. See also **Base pairing, Cytosine, Pyrimidines, Ribonucleic acid, Uracil.**

Transcriptase. The enzyme that catalyzes transcription. See also **Enzyme, Transcription.**

Transcription. The synthesis of RNA on a DNA template. See also **Base pairing, Reverse transcription.**

Transcription factor. A substance, usually of protein nature, that regulates gene expression by controlling DNA transcription. See also **Homeobox, Homeotic gene, Transcription.**

Transfection. The artificial introduction of foreign DNA into a cell.

Transfer RNA. An RNA molecule serving as transporter of a defined kind of amino acid in protein synthesis and bearing a triplet of bases, or anticodon, that, when the molecule is strategically situated on the ribosome surface, is specifically recognized by a messenger-RNA codon. See also **Anticodon, Base pairing, Codon, Messenger RNA, Protein, Ribosome, Wobble.**

Translation. The process whereby the information written in the nucleotide sequence of a gene is expressed in the amino-acid sequence of the corresponding polypeptide or protein. See also **Genetic code, Polypeptide, Protein.**

***Trans* splicing.** The end-to-end joining of two distinct RNA molecules. See also ***Cis* splicing, RNA splicing.**

Triploblasts. Animals arising from an embryonic structure made of three cell layers, an ectoderm, an endoderm, and a mesoderm. They include all animals except diploblasts. See also **Diploblasts, Ectoderm, Endoderm, Mesoderm.**

Tubulin. The protein building block of microtubules. See also **Microtubule.**

Undulipodia. A term grouping eukaryotic cilia and flagella. See also **Cilium, Flagellum, Microtubule.**

Uracil. A pyrimidine base found only in RNA. See **Cytosine, Pyrimidines, Ribonucleic acid, Thymine.**

Van der Waals force. The short-range force responsible for the attraction between hydrophobic groups. See also **Coulomb force, Hydrophobic.**

Vesicular transport. The transport of material between two intracellular, membrane-bounded enclosures by means of vesicles that detach from one enclosure and fuse with the other. See also **Cytomembrane system.**

Virus. An infectious particle consisting essentially of a protein-coated or membrane-enveloped DNA or RNA genome that codes for the constituents of the virus. Some viruses contain additional components, and their genomes additional genes. RNA viruses, in particular, have to provide the replicase or reverse transcriptase needed for their replication. Viruses enter certain prokaryotic or eukaryotic cells and multiply within these cells with the help of the local metabolism and enzyme machinery, often causing injuries or death of the infected cells. See also **Replicase, Retrovirus, Reverse transcriptase.**

Vitamin. An essential substance that an organism is unable to synthesize and must find in its food.

Wall. See **Cell wall.**

Wobble. In codon-anticodon association, the imperfect pairing of the third base of a codon with a noncomplementary anticodon base. This phenomenon allows some anticodons to be recognized by more than one codon and explains why the number of transfer RNAs (anticodons) is lower than the number of codons. See also **Anticodon, Base pairing, Codon, Transfer RNA.**

Yolk. The nutrient material in an egg cell. See also **Egg cell.**

Zygote. The diploid product of fertilization of an egg cell by a sperm cell. See also **Diploid, Egg cell, Fertilization, Gamete, Haploid, Sexual reproduction, Sperm cell.**

Zymogen. The inactive precursor of a hydrolytic enzyme, or hydrolase. See also **Hydrolase.**

Additional Reading

The books listed here are a selected few, chosen for their informational content and presentation. Additional sources are mentioned in the notes.

COSMOLOGY AND GEOCHEMISTRY

1. *Earth's Earliest Biosphere,* edited by J. W. Schopf (Princeton, N.J.: Princeton University Press, 1983).
This massive opus is a gold mine of valuable data. It may be too detailed for the general reader.

2. *The Solar System,* by R. Smoluchowski (New York: Scientific American Books, 1983).
This lavishly illustrated book covers the birth and main properties of the sun and the planets.

3. *Sun and Earth,* by H. Friedman (New York: Scientific American Books, 1986).
Complementing the preceding entry, this book offers an attractive description of the sun and its influences on the Earth.

4. *Biogeochemistry,* by W. H. Schlesinger (San Diego: Academic Press, 1991).
Organized as a textbook for college and graduate students, this is a dry but instructive and well-documented compendium of how the chemistry of the Earth affects life, and vice versa.

THE ORIGIN OF LIFE

5. *Origins: A Skeptic's Guide to the Creation of Life on Earth,* by R. Shapiro (New York: Summit Books, 1986).

Although now slightly dated, this entertaining and critical overview of competing theories on the origin of life remains an excellent introduction to the topic. It includes vignettes of the main researchers in the field, together with accounts of a number of personal interviews.

6. *Blueprint for a Cell,* by C. de Duve (Burlington, N.C.: Neil Patterson Publishers, Carolina Biological Supply Company, 1991).
A condensed description of the main properties common to all living cells forms the subject of the first half of this book. The second half is devoted to a critical examination of the origin of life. It documents in greater detail the theories proposed in the present book.

7. *Major Events in the History of Life,* edited by J. W. Schopf (Boston: Jones & Bartlett, 1992).
Six chapters, written by experts at a level appropriate for use in first- and second-year college coursework, address some important events in the evolution of life.

8. *Frontiers of Life,* edited by J. Trân Thanh Vân, J. C. Mounolou, J. Schneider, and C. McKay (Gif-sur-Yvette, France: Editions Frontières, 1992).
Proceedings of a symposium that brought together in Blois, France, in October 1991, a number of cosmologists, physicists, chemists, biologists, and theoreticians with a common interest in the origin and early evolution of life.

BIOCHEMISTRY

9. *Principles of Biochemistry,* by H. R. Horton, L. A. Moran, R. S. Ochs, J. D. Rawn, and K. G. Scrimgeour (Englewood Cliffs, N.J.: Neil Patterson Publishers, Prentice-Hall, 1993).
Exactly what the title promises, in clear, up-to-date, and nicely illustrated form.

10. *Biochemistry,* 3d ed., by L. Stryer (New York: W. H. Freeman, 1988).
It initiated a new style in textbooks when it first appeared in 1975. Clearly written and regularly updated, it remains a favorite.

CELLULAR AND MOLECULAR BIOLOGY

11. *The Eighth Day of Creation: The Makers of the Revolution in Biology,* by H. F. Judson (New York: Simon & Schuster, 1979).
An excellent history of molecular biology, based largely on interviews with some of the main protagonists and written in a vivid, journalistic style.

12. *A Guided Tour of the Living Cell,* by C. de Duve (New York: Scientific American Books, 1984).

Organized as a visit by "cytonauts," this abundantly illustrated account attempts to combine structure and function in a coherent view of cellular organization, with special emphasis on bioenergetics.

13. *Molecular Cell Biology,* 2d ed., by J. Darnell, H. Lodish, and D. Baltimore (New York: Scientific American Books, 1986).
When it came out in 1983, this excellent textbook was the first to extend molecular biology to eukaryotic cells. It includes interesting excursions into immunology, developmental biology, and neurobiology.

14. *Molecular Biology of the Cell,* 3d ed., by B. Alberts, D. Bray, J. Lewis, M. Raff, K. Roberts, and J. D. Watson (New York: Garland, 1994).
Comparable to the preceding entry in the scope, level, and quality of its coverage, this textbook differs from it in the relative emphasis given to different topics. It is often profitable to consult both.

15. *Molecular Biology of the Gene,* 2 vols., 4th ed., by J. D. Watson, N. H. Hopkins, J. W. Roberts, J. A. Steitz, and A. M. Weiner (Menlo Park, Calif.: Benjamin/Cummings, 1987).
A classic first published by Watson in 1965, enlarged to include the molecular biology of both prokaryotes and eukaryotes. RNA biochemistry and molecular biology are treated in particular detail.

EVOLUTION

16. *An Evolutionary Survey of the Plant Kingdom,* by R. F. Scagel, R. J. Bandoni, G. E. Rouse, W. B. Schofield, J. R. Stein, and T. M. C. Taylor (Belmont, Calif.: Wadsworth, 1966).
Except for a few outdated notions, this compendium, illustrated with beautiful line drawings, provides an invaluable overview of the evolution of plants, and also of unicellular algae and fungi.

17. *Ontogeny and Phylogeny,* by S. J. Gould (Cambridge, Mass.: Harvard University Press, 1977).
A scholarly disquisition on the relationship between development and evolution, inspired by the "recapitulation law" of the German biologist and philosopher Ernst Haeckel.

18. *Biological Science,* 2 vols., 5th ed., by W. T. Keeton (New York: Norton, 1993).
A good, comprehensive college-level textbook covering the whole of biology. Its emphasis is rightly put on general aspects, rather than on taxonomic minutiae.

19. *The Extended Phenotype,* by R. Dawkins (San Francisco: W. H. Freeman, 1982).
A companion and complement to the previously published *Selfish Gene* by the same author (reference 22).

20. *The Blind Watchmaker,* by R. Dawkins (New York: Norton, 1986).
A clear exposition of modern Darwinian theory, written by an enthusiast convinced "that our own existence once presented the greatest of all mysteries, but that it is a mystery no longer because it is solved."

21. *Extinction,* by S. M. Stanley (New York: Scientific American Books, 1987).
Periodically in the history of life on Earth, mass extinctions of plant and animal forms of life have occurred. This book tries to relate the paleontological evidence of such events with their possible causes, mostly climatic changes associated with tectonic movements of the Earth's crust.

22. *The Selfish Gene,* by R. Dawkins (New York: Oxford University Press, 1976; new expanded edition, 1989).
The work of a young, uncompromisingly dedicated Darwinian, this book has put the "tyranny of the genes" on a firm conceptual basis. At the same time, it has raised many a hackle by extending that notion to human evolution and behavior.

23. *Symbiosis in Cell Evolution,* by L. Margulis (San Francisco: W. H. Freeman, 1981; 2d ed., 1992).
Publication of the first edition of this work was a landmark in the development of the now generally accepted theory of the endosymbiotic origin of mitochondria and chloroplasts.

24. *On Methuselah's Trail,* by P. D. Ward (New York: W. H. Freeman, 1992).
A pleasantly written, eminently readable introduction to "living fossils and the great extinctions."

25. *From Egg to Adult,* report no. 3 (Bethesda, Md.: Howard Hughes Medical Institute, 1992).
This slim collection of six short, beautifully illustrated articles brings into sharp focus the mysteries of animal development, as well as the promising new approaches of molecular biology toward unraveling developmental processes.

BIODIVERSITY AND ECOLOGY

26. *The Ages of Gaia,* by J. Lovelock (New York: Norton, 1988).
The father of Gaia defends his theory in appealingly eloquent terms in which the poetic is not always clearly distinguishable from the scientific.

27. *Biodiversity,* edited by E. O. Wilson (Washington, D.C.: National Academy Press, 1988).
The proceedings of the National Forum on BioDiversity convened in 1986 by the National Academy of Sciences and the Smithsonian Institution and including participants from many different fields, ranging from laboratory scientists to field workers, economists, policymakers, philosophers, and even poets.

28. *Discordant Harmonies,* by D. B. Botkin (New York: Oxford University Press, 1990).
An ecologist confronts fact and theory in the light of his own unsettling experiences. He stresses the fact that life and environment are related by a complexity of mutual relationships.

29. *The Diversity of Life,* by E. O. Wilson (Cambridge, Mass.: Harvard University Press, 1992).
A leading biologist shows how human onslaughts on the diversity of life are inflicting immensely greater damage than the natural catastrophes that have signaled major extinctions in the past. His solidly documented plea deserves to be heard.

30. *Green Delusions,* by M. W. Lewis (Durham, N.C.: Duke University Press, 1992).
A former radical environmentalist dissects "eco-radicalism," of which he distinguishes five variants: antihumanist anarchism, primitivism, humanist eco-anarchism, eco-marxism, and eco-feminism. The excursion is entertaining, but more anecdotal than timely, as it deals mostly with unimportant fringe movements.

HUMAN ORIGINS

31. *Origins,* by R. E. Leakey and R. Lewin (New York: Dutton, 1977).
In this book, Richard Leakey, the son of the famous Kenyan couple Louis and Mary Leakey and a distinguished paleoanthropologist in his own right, has teamed up with the science writer Roger Lewin to give a vivid, personal account of the search for human origins.

32. *Lucy: The Beginnings of Humankind,* by D. C. Johanson and M. Edey (London: Granada, 1981).
The story of Lucy, a 3.5-million-year-old young female, as told by her discoverer, the American Donald Johanson, Leakey's main rival in the hunt for human remains.

33. *The Rise and Fall of the Third Chimpanzee,* by J. Diamond (London: Radius, 1991).
A zoologist with an interest in New Guinea birds takes a hard look at the human species, its peculiar characteristics, and the reasons for its extraordinary evolutionary success, which could also be the causes of its future downfall. The author steers a course between bleak pessimism and cautious optimism.

34. *Origins Reconsidered,* by R. E. Leakey and R. Lewin (New York: Doubleday, 1992).
An updated version of reference 31 with a more philosophical bent. This book, like that by Johanson and Edey (reference 32), also conveys some of the drama and rivalries that surround the search for human origins.

BRAIN AND MIND

35. *The Question of Animal Awareness,* by D. Griffin (New York: Rockefeller University Press, 1976).
A leading ethologist looks at animal behavior to uncover the birth and development of consciousness. See also references 37 and 44.

36. *The Self and Its Brain,* by K. R. Popper and J. C. Eccles (New York: Springer International, 1977).
A philosopher and a neurobiologist, both leaders in their fields, explain in detail their "Argument for Interactionism." After two parts written separately by each author, the book concludes with twelve dialogues between the two. It makes impressive reading and continues to deserve close attention, even though the main thesis of the book—which favors a dualistic view of the mind-brain relationship—is now widely discredited among specialists. See also reference 42.

37. *Animal Thinking,* by D. Griffin (Cambridge, Mass.: Harvard University Press, 1984).
More on the author's frequent topic (see also references 35 and 44).

38. *Neuronal Man: The Biology of Mind,* by J.-P. Changeux, translated by L. Garey (New York: Pantheon, 1985). Original edition: *L'Homme Neuronal* (Paris: Librairie Arthème Fayard, 1983).
An excellent exposition of the functioning of the brain, addressed to the cultured layperson. It advocates a strictly materialistic view of the mind. See also reference 41.

39. *The Mind's New Science,* by H. Gardner (New York: Basic Books, 1985).
In this "history of the cognitive revolution," an expert psychologist gives a penetrating analysis of human intelligence and of attempts to replace it by computers.

40. *Mind from Matter?* by M. Delbrück (Palo Alto, Calif.: Blackwell Scientific Publications, 1986).
Assembled by a group of his disciples from a series of lectures given by the author a few years before his death in 1981, these notes offer a number of interesting thoughts on the mind-brain problem by a physicist who became the acknowledged father of modern molecular biology.

41. *Matière à Pensée,* by J.-P. Changeux and A. Connes (Paris: Odile Jacob, 1989).
An interesting dialogue, not yet translated into English, between an uncompromisingly materialistic neurobiologist (see reference 38) and a mathematician with a more Platonist view of mathematical truth. See also reference 43 in connection with this topic.

42. *Evolution of the Brain: Creation of the Self,* by J. C. Eccles (London and New York: Routledge, 1989).

This clear and well-documented introduction to the structure and development of the human brain by a Nobel Prize–winning neurobiologist gives the author an opportunity to reiterate his firm adherence to Cartesian dualism (see also reference 36).

43. *The Emperor's New Mind,* by R. Penrose (New York: Oxford University Press, 1989).
This ambitious book by a leading mathematician discusses the mind-brain problem within the vast context of mathematics, quantum mechanics, and theoretical physics. The author believes he has found in an unsolved problem of quantum theory a possible loophole for consciousness to slip into the deterministic functioning of the neurons. His philosophical views are decidedly Platonist.

44. *Animal Minds,* by D. Griffin (Chicago: University of Chicago Press, 1992).
A continuation of references 35 and 37. Note the progression in the titles, from awareness to thinking to mind.

45. *The Rediscovery of the Mind,* by J. R. Searle (Cambridge, Mass.: MIT Press, 1992).
This book, written by a philosopher, offers a clear, informative, and critical survey of the modern literature on mind and brain. Few leaders in the field escape unscathed from the analysis. The author defines useful guidelines for future research, but with no end yet in sight.

46. *Mind and Brain,* a special issue of *Scientific American* 267, no. 3 (1992): 48–159.
A collection of very readable articles on different aspects of the mind-brain problem, with an excellent introduction by G. D. Fischbach.

47. *Bright Air, Brilliant Fire,* by G. M. Edelman (New York: Basic Books, 1992).
The author, who first gained a Nobel Prize for deciphering the structure of an antibody molecule and later turned his attention to the human brain, summarizes for a wider readership the results—described in three previous, highly technical books—of his investigations on the wiring and functioning of the brain. Even thus simplified, it does not make easy reading but offers a rich spectrum of ideas inspired by a Darwinian perspective.

48. *The Astonishing Hypothesis: The Scientific Search for the Soul,* by F. Crick (New York: Scribner's, 1994).
World-renowned for his discovery, with James D. Watson, of the double-helical structure of DNA, the author summarizes the results of his more-recent exploration of the brain. Largely devoted to the mechanisms of vision, the book marshals an impressive number of facts in defense of the "astonishing hypothesis" that mind and consciousness are no more than the products of neuronal functioning. His conclusion: The "scientific search for the soul"—a topic hardly mentioned—leads to nothing.

SOCIAL AND POLITICAL ISSUES

49. *The Population Bomb,* by P. R. Ehrlich (New York: Ballantine, 1968).
One of the first books to sound a truly alarming warning on the impending consequences of demographic expansion. See also reference 56.

50. *Sociobiology: The New Synthesis,* by E. O. Wilson (Cambridge, Mass.: Harvard University Press, 1975).
This monumental survey of the social behavior of animals attempts to explain the origin of behavioral traits in terms of Darwinian selection. The author's extension of this approach to human behavior has caused an outcry (see reference 51).

51. *The Sociobiology Debate,* edited by A. L. Caplan (New York: Harper & Row, 1978).
A fascinating collection of essays selected from the past literature or written in response to Wilson's *Sociobiology* (reference 50).

52. *On Human Nature,* by E. O. Wilson (Cambridge, Mass.: Harvard University Press, 1978).
Undismayed by a volley of attacks, the author develops in greater and more forceful detail his view of human behavior as the product of natural selection.

53. *Promethean Fire,* by C. J. Lumsden and E. O. Wilson (Cambridge, Mass.: Harvard University Press, 1983).
A condensed version, for a wider public, of a specialized monograph in which the physicist Lumsden has joined with the biologist Wilson to analyze the evolutionary development of the human mind and the nature-nurture dichotomy, within the somewhat edulcorated framework of sociobiological doctrine.

54. *Not in Our Genes,* by R. C. Lewontin, S. Rose, and L. J. Kamin (New York: Pantheon, 1984).
An empassioned attack on sociobiology, based on solid scientific evidence but with strong political overtones, by an American geneticist, a British neurobiologist, and an American psychologist.

55. *The Dialectical Biologist,* by R. Levins and R. Lewontin (Cambridge, Mass.: Harvard University Press, 1985).
Even more radical than the preceding entry, this book tries to reveal the hidden social and political biases that lie behind the conventional, "bourgeois" scientific approach. The authors do not hesitate to publicize their own bias, which is openly Marxist. A dialectic masterpiece, it may not convince the unbeliever, but it forces reflection.

56. *The Population Explosion,* by P. R. Ehrlich and A. H. Ehrlich (New York: Simon & Schuster, 1990).
Disheartened by the lack of effective public response to his earlier warning (reference 49), Paul Ehrlich, now joined by his wife, Anne, sounds an even more urgent alarm: "In 1968, the fuse was burning; now the population bomb has detonated."

57. *Only One World, Our Own to Make and to Keep,* by G. Piel (New York: W. H. Freeman, 1992).
A sober and fact-crammed account of the problems of our planet by a man who was for four decades the publisher of *Scientific American.*

58. *Earth in the Balance,* by A. Gore (Boston: Houghton Mifflin, 1992).
Then a senator, now vice president, Al Gore covers much the same ground as Piel (reference 57), but with more evangelistic fervor and fewer hard facts. He ends with a passionate political plea for an environmentally inspired global Marshall Plan.

59. *Preparing for the Twenty-first Century,* by P. Kennedy (New York: Random House, 1993).
Different from the previous two entries, this book considers our predicament from the point of view of an economist, with an emphasis not only on global issues but also on problems specific to each major part of the world.

COMPLEXITY

60. *Complexity: The Emerging Science at the Edge of Order and Chaos,* by M. M. Waldrop (New York: Simon & Schuster, 1992)

61. *Complexity: Life at the Edge of Chaos,* by R. Lewin (New York: Macmillan, 1992).
Two science writers simultaneously offered books on the same topic—a think tank known as the Santa Fe Institute—under virtually the same title and with the same emphasis on the people involved. Inevitably, there is plenty of overlap, as well as some differences. Reference 60 is primarily concerned with the foundation and history of the institute, whereas reference 61 deals more with the work that goes on within its walls. Both communicate a feeling of fervor and excitement typical of new converts of some esoteric creed.

62. *The Origins of Order: Self-Organization and Selection in Evolution,* by S. A. Kauffman (New York: Oxford University Press, 1993).
In this scholarly work, one of the stars of the Santa Fe Institute gives a dense account of his attempts at modeling the inherent tendency of complex systems to self-organize under certain critical conditions—when evolving "at the edge of chaos." He shows with many examples the biological relevance of such behavior, which, in his opinion, forms an essential complement to Darwinian selection.

63. *AI: The Tumultuous History of the Search for Artificial Intelligence,* by D. Crevier (New York: Basic Books, 1993).
An enthusiastic account of the history of the new field, written by an insider. See also reference 39.

PHILOSOPHY

64. *Chance and Necessity,* by J. Monod, translated by A. Wainhouse (New York: Knopf, 1971). Original edition: *Le Hasard et la Nécessité* (Paris: Editions du Seuil, 1970).
In this book, which sparked major controversies when it first came out, a master of modern biology (who died in 1976) defends a stoically and romantically despairing existentialist view of the human condition. Somewhat outdated but still beautiful reading.

65. *The Anthropic Cosmological Principle,* by J. D. Barrow and F. J. Tipler (New York: Oxford University Press, 1986).
In this huge work, the authors marshal evidence from history, philosophy, religion, biology, physics, astrophysics, cosmology, quantum mechanics, and biochemistry in support of the general concept, of which several distinct formulations exist, that the universe is built the way it is because, if it were not, there would be no intelligent life whereby the universe could be known.

66. *Death of the Soul: From Descartes to the Computer,* by W. Barrett (Garden City, N.Y.: Anchor Press/Doubleday, 1986).
A philosopher shows how human thought about itself has become progressively divorced from fact, to the point that "you entertain and support in argument an intellectual position that you could not possibly live." The book also provides a useful introduction to three centuries of philosophy for the layperson. It profitably complements reference 45.

67. *Science as Salvation,* by M. Midgley (London and New York: Routledge, 1992).
Halfway between treatise and pamphlet, this clever book by a distinguished philosopher seeks to put science back where it belongs. It expresses much of the distrust science has evoked in recent years and, as such, should be read by every scientist.

68. *Disturbing the Universe,* by F. Dyson (New York: Harper & Row, 1979).
In this entertainingly written, largely autobiographical account, a physicist uses his rich experience to reflect on a wide variety of problems, including the function of science in society and the place of life and mind in the universe.

Index